Molecular Cloning

Molecular Cloning

Edited by **Erik Pierre**

R CALLISTO
REFERENCE

New York

Published by Callisto Reference,
106 Park Avenue, Suite 200,
New York, NY 10016, USA
www.callistoreference.com

Molecular Cloning
Edited by Erik Pierre

International Standard Book Number: 978-1-63239-468-2 (Hardback)

Printed in the United States of America.

Contents

Preface

This book has been an outcome of determined endeavour from a group of educationists in the field. The primary objective was to involve a broad spectrum of professionals from diverse cultural background involved in the field for developing new researches. The book not only targets students but also scholars pursuing higher research for further enhancement of the theoretical and practical applications of the subject.

Technology is the boon of modern times. The expansion of molecular cloning technology in the late 20th century created an uprising in the biological and biomedical sciences that extends till today. This book supplies the reader with a viewpoint on how extensive the functions of molecular cloning have become. The data in this book is arranged in sections based on functions, and range from cancer biology and immunology to plant and evolutionary biology. This book even covers a broad variety of technological techniques such as situational cloning and exceptional tools for recombinant protein appearance. This book intends to provide some useful information for students and experts.

It was an honour to edit such a profound book and also a challenging task to compile and examine all the relevant data for accuracy and originality. I wish to acknowledge the efforts of the contributors for submitting such brilliant and diverse chapters in the field and for endlessly working for the completion of the book. Last, but not the least; I thank my family for being a constant source of support in all my research endeavours.

Editor

Part 1

Technological Advances

Non-Viral Vehicles: Principles, Applications, and Challenges in Gene Delivery

Abbas Padeganeh[1], Mohammad Khalaj-Kondori[2],
Babak Bakhshinejad[1] and Majid Sadeghizadeh[1]
[1]Department of Genetics, Faculty of Biological Sciences,
Tarbiat Modares University, Tehran
[2]Department of Biology, Faculty of Natural Sciences, University of Tabriz, Tabriz,
Iran

1. Introduction

Gene therapy is often referred to as transfer of transgenes into the somatic cells of a patient to obtain a therapeutic effect .One of the goals for all such therapies is efficient and safe delivery of the desired extrinsic genes into target cells, thereby increasing the therapeutic efficiency (Robson & Hirst, 2003). This has been a major obstacle in gene therapy experiments (Sarbolouki et al., 2000; Sadeghizadeh et al., 2008).
To address this problem, there has been an increasing number of reports in the development of efficient gene delivery vehicles in recent years (Sadeghizadeh et al., 2008). Clinical trials have also focused on the delivery of genes directly to the target area e.g. tumor sites by intratumoral administration of both viral and non-viral delivery agents. But the problem remains to be overcome yet, as majority of tumors are not accessible for direct injection. A number of strategies are now being developed to target viral and non-viral delivery agents to tumor sites. These include genetically modifying viral carriers and incorporating a novel tumor-specific ligand into the viral coat proteins to direct the viral system to a tissue receptor and also incorporation of tissue specific ligands and monoclonal antibodies onto the surface of non-viral carriers (Robson & Hirst, 2003). The delivery of the carrier system to the target site is however not the end of the goal. Efficient entrance of the gene or drug into the cells and expression of therapeutic gene are also the next hurdles to be overcome. There are several techniques for delivery of genes as well as drugs into eukaryotic cells using similar carriers practiced in-vitro and in-vivo. The in-vivo efficacy of a gene or drug delivery system depends on its capability to pass the main extracellular and intracellular barriers encountered from the site of administration to entry into the nucleus of desired cells (Sarbolouki et al., 2000; Sadeghizadeh et al., 2008).
The therapeutic effect of a gene therapy experiment would be expected once the introduced transgene in target cells is considered as part of the genetic component of host cell and leads to the production of a new functional protein. To date, this type of gene transfer known as transfection   (Lewin, 2007; Singleton & Sainsbury, 1995) has been studied widely and various techniques have been developed for it, each possessing its own advantages and shortcomings. Generally speaking, gene delivery techniques are classified into viral and

non-viral categories. Non-viral systems include physical, chemical and biological methods. Conventional physical techniques include electroporation and microinjection. Among chemical methods are calcium-phosphate precipitation, use of diethylaminoethyl (DEAE), polyethylenimines (PEIs), polybrene/dimethyl sulfoxide, liposomes, cationic amino polymers, polyamidoamin (PAMAM), dendrimers and dendrosomes (Guyden, 1993; Hammarskjold, 1991). Biological methods embrace viral transfer systems which utilize SV40-based vectors, adenoviral vectors, vaccina virus vectors, BPV vectors and retroviral carriers (Hammarskjold, 1991; Wong & Neuman, 1982) and non-viral systems which bacteriophages are among the most important. A detailed description all of these delivery systems lies beyond the scope of this paper (for review see references Hammarskjold, 1991; Wong & Neuman, 1982).

Nanotechnology referred to as the creation of useful materials, devices and systems used to manipulate matter at an extremely small size -between 1 and 100 nm- (Alivisatos, 1996; Suntherland, 2002) offers great opportunity in the field of drug and gene delivery. The problems and shortcomings of current anti-cancer treatment strategies such as systemic administration of drugs or genes which do not greatly differentiate between cancerous and normal cells leading to systemic toxicity and adverse effects, have caused limitations in allowable dose of drugs (Sinha et al., 2006) and led to a resurgence of interest in developing safe and efficient nano-scale gene and drug porters capable of detecting target sites and delivering proper genes and/or drugs to diseased cells (Sadeghizadeh et al., 2008).

In recent years, a number of nanoparticle-based therapeutic agents have been developed for treatment of cancer, diabetes, asthma, allergy, infection, etc. (Brannon-Peppas & Balanchette, 2004; Kawasaki & Player, 2005) . These nano-scale agents may provide more efficient and/or more convenient routes of administration, have lower toxicity, extend the product life-cycle and ultimately allow targeted and controlled release of therapeutic gene or drug (Zhang et al., 2007).

It has previously been reported that dendrosomes are capable of easily delivering genes into human cells (e.g. human hepatocytes, kidney cells and several cancer cell lines) and animal models in-vivo. They are easily synthesized, highly stable (nearly 4 years at ambient conditions) and extremely convenient to handle and use (Sarbolouki et al., 2000; Sadeghizadeh et al., 2008).

In this chapter, we discuss some of the most commonly used non-viral gene delivery systems (some also used as drug-carriers), with a focus on features of the recently introduced dendrosomes as novel gene porters shedding more light onto future perspectives of these promising nanocarriers.

Non-viral biological gene delivery methods include bacteria, bacteriophages, virus-like particles (VLPs), erythrocyte ghosts and exosomes. Elaboration on all of these approaches is beyond the scope of this chapter. Unavoidably, in this regard we will narrow our debate to bacteriophages being one of our research interests. Hence, at the end of the chapter, bacteriophages as one of the most significant non-viral biological systems or strategies for gene delivery developed over the recent years will be discussed (Seow & Wood, 2009).

2. Chemical strategies

2.1 Calcium-phosphate precipitation

This technique is the most common way to transfect foreign genes into eukaryotic cells mainly due to its simplicity and inexpensiveness. In this technique, foreign genes are precipitated on the surface of cell monolayer. Briefly, calcium chloride, DNA and phosphate buffer are mixed at a neutral pH, the calcium-phosphate-DNA complex then precipitates on the cells and would enter the cells through endocytosis. These complexes are then transported to different organelles including the nucleus (Guyden, 1993). In order to increase the efficiency of transfection, addition of glycol/dimethyl sulfoxide to the monolayer following precipitation of the complex and removal of the old medium has been effective (Okayama et al., 1991; Li & Thacker, 1997). However, this method suffers from frailties including transient and unstable expression of tranfected genes following their degradation, low percentage of stably-transformed cells (only 0.001- 1%) and the need to determine the optimum conditions for transfectoin of each cell type.

Nevertheless, calcium phosphate nanoparticles introduced by Roy et al. (2003) are ultra-low size entities around 80 nm in diameter and it seems to be able to protect encapsulated DNA from environmental DNases with capability of surface modifications. This new class of nanoparticles have been used for gene delivery to liver.

2.2 Diethylaminoethyl (DEAE)-dextran

This method is based upon the negative charge of DNA and the positive charge of diethyl aminoethyl-dextran leading to the formation of a complex and adherence of the complex to the cell surface followed by endocytosis. This technique benefits from advantages such as simple and rapid preparation steps, low cost and reproducibility. However, cells show different sensitivities to the toxicity of this compound, therefore the proper ratio of diethylaminoethyl-dextran-DNA must be determined for each cell line (Holter et al., 1989). Moreover, it is preferred for transfection of adherent cells over cells in suspension (Gulick, 2003).

2.3 Polycations

Many libraries perform DNA trasfection experiments using artificial polycations such as polybren/dimethyl sulfoxide which have been shown to enhance retroviral infection in cell cultures by making an electrostatic bridge between the negatively charged viral particles and neutral components of the recipient cell membrane. It also binds DNA and attaches it to the cell surface. Finally, exposing the cells to dimethyl sulfoxide increases the speed of DNA uptake. It is a proper method when dealing with very limited amounts of DNA (ng DNA). The efficiency of this technique is sometimes 0.01-0.1% (Roy et al., 2003). Other polycation-based systems may also utilize poly-lysine compounds (Sarbolouki et al., 2000). Of the most widely used gene carriers of this category are polyethylenimines (PEIs). Linear or branched PEIs have been efficiently used for in vitro transfection of genes. However, in vivo application of PEIs, leads to non-specific interactions of the PEI/DNA complex with components of the host blood and results in failure of proper gene delivery. Thus, various surface modified derivatives of PEIs (polyethylene glycol-conjugated PEIs) have been emerging to overcome these issues (Kichler, 2004).

2.4 Polymeric L-lysin vehicles

Poly L-lysin, also referred to as PLL, has been shown to form complexes with DNA as a result of interaction of negatively charged DNA and positively charged amino groups of lysine (Tae et al., 2006). These polyelectrolyte complexes have also been shown to suffer from drawbacks e.g. high degree of cytotoxicity, rapid clearance and self aggregation (Liu et al., 2001). Similar to that of PEIs, surface conjugation of PLL with polyethylene glycol, has been shown to improve properties of this class of gene delivery vehicles. In addition, various strategies have been employed to target PLL polymers to specific cell types or tissues. These include conjugating sugar moieties e.g. lactose or galactose to target PPL/DNA complexes to hepatocytes (Nishikawa et al., 1998; Hashida et al., 1998). There have also been efforts to conjugate antibodies (e.g. leukemia-specific- antigen antibody, anti JL-1 antibody) to PLL complexes showing higher leukemia-specific cell uptake and specificity (Suh et al., 2001).

2.5 Polysaccharides as gene carriers
There are known examples of using polymers composed of sugar molecules, e.g. chitosan and cyclodextrin, for gene delivery purposes. Cyclodextrin is in fact a cyclic entity made up of oligomeric glucose units forming a hydrophobic internal cavity and a hydrophilic extremity. Another example is chitosan, a polysaccharide made up of repetitive units of D-glucosamine linked to N-acetyl-D-glucosamine. Both above mentioned polysaccharide-based structures can interact with DNA to form stable complexes and have been reported to have comparable or higher transfection efficiencies in regard to PEI or PLL (Gonzalez et al., 1999).

3. Liposome-based gene/drug delivery systems

Liposomes are spherical lipid vesicles with bilayer membrane structure composed of natural or synthetic amphiphilic lipid molecules (Zhang & Granick, 2006). Liposomes have been widely used as pharmaceutical carriers in the past decade because of their unique abilities in encapsulating both hydrophilic and hydrophobic agents with a high efficiency, protecting the encapsulated drugs from undesirable side effects of external conditions, being functionalized with specific ligands that can target specific cells, and being coated with inert biocompatible polymers (Roy et al., 2003; Moghimi & Szebeni, 2003). Liposomes are also used as gene carriers. An efficient strategy to encapsulate DNA within liposome is the reverse phase evaporation method (REV) in which phospholipids are resolved in ether making up an organic phase and DNA is added to PBS making up an aqueous phase. Then the aqueous and the organic phases are emulsified in a sonicator followed by mixing the lipids with DNA which leads to the formation of lipid vesicles containing the DNA molecules inside (Guyden, 1993). Since liposomes are usually not fused to the cell surface but rather phagocytosized by cells, the carried nucleotides would be exposed to the lysosomic enzymes and therefore digested, reducing the efficiency of successful transfection/expression process. Other problems of liposomes include the possibility of formation of small-sized liposomes uncapable of encapsulating large macromolecules such as DNA and the multistep difficult processes of their production. They also have a low gene transfer efficiency and usually exhibit cytotoxicity in lymphoma cells (Buttgereit et al., 2000).

3.1 Cationic liposomes
Cationic liposomes such as lipofectins have also been developed to overcome some of mentioned shortcomings of liposomes. Lipofectins contain positively charged lipids like dioleoylphosphatidylethanolamine   (DOPE)   and   N-(1-2,3-dioleyloxypropyl)-N,N,N-trimethylammonium (DOTMA). DOTMA for example, is designed as stable cationic bilayer

vesicles spontaneously interacting with polyanionic DNA and RNA molecules and therefore forming liposome/polynucleotide complexes. These complexes are taken in by the anionic surface of host cells with an efficiency of about 10-100 fold higher than that of negatively or neutrally charged liposomes. Large DNA molecules such as baculoviral DNA (130 kb) and genomes of RNA-viruses have also been introduced into cell-cultures using DOTMA confirmed by the formation of viral particles (Felgner, 1991; Strauss et al., 1994).

3.2 Modified and targeted liposomes

One drawback of the use of liposomes is the fast clearance of liposomes from blood by phagocytic cells of the reticuloendothelial system, resulting in unfavorable therapeutic index (Roy et al., 2003). One of the widely used strategies is to formulate long-circulating liposomes by coating the liposome surface with inert biodegradable polymers such as polyethylene glycol. The polymer layer provides a protective shell over the liposome surface and suppresses liposome recognition by opsonins and therefore subsequent clearance by the reticuloendothelial system (Dutta et al., 2006). Another strategy is to increase the accumulation of liposomes in the target site by attaching targeting ligands such as antibodies and small moiety molecules (e.g. folate and transferrin) to the liposome surface. Targeted liposomes have been developed for differential drug and gene delivery (Saunak et al., 2004).

4. Nanopolymer-based carriers

4.1 Dendrimers

Dendrimers are a class of artificial, highly branched and reactive three dimensional polymers, with all bonds originating from a central core. The term dendrimer comes from the Greek word "dendron" which means tree and the suffix "mer" from meros referring to smallest repeating units. In recent years, there has been much interest in dendrimers; since due to the large number of terminal functional groups (e.g. amino groups) on their surface, they are easily linkable to antibodies and reactive therapeutic agents making them proper for use in biomedical research (Bousif et al., 1995; Buttgereit et al., 2001; Massumi et al., 2005). Other attractive features such as nanoscale size, highly controllable molecular weight and possibility of encapsulating a guest molecule (e.g. a gene or drug) in their internal cavities (Tomalia, 2005) give dendrimers a distinctive advantage over other polymers for delivery of drugs and genes (Strauss, 1994). To use DNA therapeutically, it must pass some barriers in the body of host, including capillary endothelial cells, phagocytes, reticuloendothelial system and eventually the membrane of the target cell (Dutta et al., 2006). The nanoscopic size of dendrimers not only helps them evade the reticuloendothelial system, but also generates benefit for them in intracellular delivery [34]. Amino-dendrimers have been specifically attractive due to their defined structures and the large number of surface amino groups ((Sarbolouki et al., 2000; Sadeghizadeh et al., 2008; Bielinska et al., 1997). Ployamidoamin (PAMAMs) dendrimers are a member of this family of dendrimers known as water soluble constructs covered with a large number of amino groups on their surface due to which they are positively charged at physiologic pH and therefore thought to interact with DNA (Sadeghizadeh et al., 2008; Kukowska et al., 1996) . Another member of this family "poly (propyleneimine) dendrimers" (PPI) are also highly branched and globular with primary amino groups on the periphery (Saunak, 2004). As a result, these dendrimers readily form complexes with DNA and are capable of transfecting cell cultures with low toxicity and higher efficiency. Dendrimers have different generations and modification of the large number of surface functional groups by conjugation to guest molecules has led to the production of dendrimer conjugates (Buttgereit et al., 2001; Massumi et al., 2005). For example, in a recent study, labeled biotin-conjugated PAMAM dendrimers were constructed and used to target tumor cells. As biotin specific receptors are overexpressed on the surface of cancer cells, results have shown an increased carrier uptake by these cells (Pourasgari et al., 2007).

There are few studies on the antigenic properties of nanoparticles such as dendrimers. An early study on PAMAM dendrimers did not reveal overt antigenicity of generations 3, 5 and 7 amino-terminated dendrimers (Dutta et al., 2006). In a study on immunosuppressive properties of dendrimers, generations 3 and 5 PAMAM dendrimers conjugated to glucosamine strongly inhibited induction of inflammatory cytokines and chemokines in human macrophages and dendritic cells exposed to bacterial endotoxins. Also, dendrimers have been reported to possess hemolytic toxicity and cytotoxicity owing to their cationic nature. Moreover, interaction with oppositely charged macromolecules in plasma may result in premature release of their cargo (e.g. plasmid DNA or carried drug) within the blood (Buttgereit et al., 2001). In addition, degradation of the plasmid DNA by plasma DNases leads to poor gene expression in-vivo (Massumi et al., 2005; Pourasgari et al., 2007).

4.2 Dendrosomes: New generation of nanoscale gene porters

Dendrosomes, are a novel family of non-viral vehicles and gene porters that form hyperbranched spherical entities hence the term derndrosome is applied to them (Sarbolouki et al., 2000; Sadeghizadeh et al., 2008).

Dendrimers could be presumably considered as the primary ancestors of these novel gene delivery systems. Dendrosomes possess valuable advantages over other carriers which include ease of synthesis, stability (nearly 4 years at ambient conditions), nontoxicity, inexpensiveness, biodegradability, neutrality, spherical structure, capability of easily delivering genes and being extremely convenient to handle and use (Sarbolouki et al., 2000; Sadeghizadeh et al., 2008; Dobrovolskaia & McNeil., 2007).

According to atomic force microscopy (AFM) observations, dendrosomes are expectedly nanoscopic particles 10-100 nm in size . A unique feature of dendrosomes is the ease with which they provide suitable inert gene porters for various DNA sizes and target cells. There have been several reports by our group and other researchers all showing that dendrosomes may serve as promising high efficient candidates for transfection and therapy (Sarbolouki et al., 2000; Sadeghizadeh et al., 2008). In early studies, three generations of dendrosomes named Den450, 700 and 123 were synthesized and used and their applicability and efficiency were assessed by studies on transfection of human cell cultures as well as vaccination of mice against hepatitis B. To assess their cytotoxicity, cells were treated with void dendrosomes. These experiments showed that bare dendrosomes Den450 and Den700 when exposed to A7r5 cells (rat aortic somatic muscle cells) not only showed no deleterious effects, but even seemed to help their propagation. This surprising effect probably implies the fact that these agents can act as adjuvants and improve uptake of nutrients by cells. Southern blot analysis also clearly demonstrated the episomal presence of the carried gene in the cytoplasm of transfected cells and therefore the capability of dendrosmes in delivery of genes into cells (Sarbolouki et al., 2000). Several advantages of dendrosomes confer them other potentials for use in DNA vaccination. These include protection of plasmid DNA from nuclease degradation, efficient delivery of their contents to antigen presenting cells (APCs) and extended release of cargo.

Mixing the HBsAg gene-harboring plasmid with a small amount of dendrosomes followed by intramuscular or intradermal administration of the mixture into BALB/c mice, resulted in a much higher production of anti-HBs antibodies compared to the administration of recombinant antigen itself (Sarbolouki et al., 2000). More recent studies by another group on the protective efficiency of dendrosomes as novel nano-sized adjuvants, have also approved their capability for DNA vaccination against allergy (Dutta et al., 2006). Conventional immunotherapies suffer from the drawbacks of use of an active antigen, such as sever IgE-mediated side effects like anaphylactic reactions induced by cross-linking of pre-existing IgE antibodies on the surface of mast cells (Buttgereit et al., 2001). However, use of Den123 for delivery of allergen-encoding plasmid for DNA vaccination, yielded promising results as these nanoparticles showed IgE inhibition while maintaining Th1/Th2 balance after DNA vaccination, sustained release of DNA plasmids and augmentation of the IgG2a level gradually by prolongation of the intracellular presence of the plasmid (Dutta et al., 2006; Balenga et al., 2006). In another study, the dendrosome Den123 was used to deliver and enhance transfection of DNA vaccine plasmid encoding gB gene of Herpes Simplex Virus type-1 along with Bax-encoding plasmid in order to evaluate the apoptosis induction effect on DNA vaccination efficiency (Pourasgari et al., 2007). Another group has recently synthesized and used dendrosomes containing entrapped PPI dendrimer-DNA complexes in genetic immunization against hepatitis B as well. The dendrosome formulation chosen for this study was DF3 as it possessed optimum size and entrapment efficiency. Animals immunized with PPI dendrimer-DNA complex entrapped within DF3 dendrosomes underwent maximum immune response in terms of total IgG compared to those immunized with plasmid DNA alone and/or PPI dendrimer-DNA complex. Higher level of IFN-γ in DF3-immunized animals also suggested that the immune response was strictly Th1-mediated (Dutta et al., 2006). These results are in accordance with our observations about the superiority of dendrosomes in genetic immunization and DNA vaccination compared to other strategies. The dendrosomes DF3 prepared by the reverse phase evaporation method have also been used in transfection of HEK-293 cells with PGL2 showing that they possess a superior transfection against other non-viral delivery systems under study .

In a comparative study, apoptosis induction in human lymphoma and leukemia cell lines was assessed using dendrosomes carrying wild type p53(Dend+p53) along with other very commonly used non-viral carrier lipofectin (Lipo+p53). The rate of apoptosis in Dend+p53 transfected K562 cells (human erythroleukemia cell line) which do not produce the p53 protein (Buttgereit et al., 2001) was twice that of the Lipo+p53 transfected cells, which was indicative of higher transfection efficiency of dendrosomes. In toxicity assessment studies, lipofectin showed a higher cytotoxicity on CCRF and MOIT-4 cells (belonging to T-lymphocyte cell lines) compared to the dendrosomes used (Massumi et al., 2005). Another study reported successful and efficient transfection of A549 (a human lung cell line) by dendrosomes containing the recombinant rotavirus VP2 gene equal to that of lipofectin results where dendrosomes showed a lower cytotoxicity (Pourasgari et al., 2007). In our recently published work, dendrosomes prepared at the IBB center, Iran, were studied and assessed in several aspects including interaction of dendrosomes with plasmid and genomic DNA, their ability in delivery and expression of genes into Huh7, VERO, Bowes, Raw, U-937, CCRF-CEM, MOLT-4 and K562 cells, comparison of their performance with a commercial gene porter lipofectin and bacterial ghosts, their non-toxicity against human cells and animal models and their performance as adjuvant in immunization of BALB/c mice against hepatitis C (Sadeghizadeh et al., 2008).

CD spectra of the studied dendrosomes entrapping DNA molecules indicated that the commonly used Den123 (made of amphipatic monomers) would moderately cause a transition from B- to A-form in linear DNA. Also, sensitivity of the interaction with the GC-content of DNA was assessed according to the CD spectra information. Results showed that dendrosomes generally and mildly transform high and low GC-content DNA from B-form into another B-form DNA. Transfection and expression studies also demonstrated that all the dendrosomes used could perform best at very low levels and at weight ratios of Den/DNA ranging from 1/1 to 1/10 therefore minimizing the chance of undesired side effects on the host (Holter & Fordis, 1989). Superiority of Den55 and Den10 compared to lipofectin and bacterial ghosts was also seen in these studies. Results from non-toxicity and immunization experiments also demonstrated that cells exposed to dendrosomes did not show any signs of toxicity whereas those exposed to lipofectin revealed sever toxicity and that animal immunization with dendrosomes caused long term immunization of mice treated with Den123 carrying the HBsAg or the HCV core pcDNA3 without developing signs of toxicity or discomfort (Sadeghizadeh et al., 2008).

Recently, based on the mentioned desired properties of dendrosomes, our group has employed this system for the delivery of a hydrophobic anticancer agent, curcumin into tumor cell lines (manuscript in preparation). This new formulation of curcumin, hereafter referred to as dendrosomal curcumin, is prepared in a very simple mixing-sonication step, and has been shown to significantly improve curcumin water solubility, an important limiting factor for the use of free curcumin as a chemotherapeutic. Using the intrinsic fluorescence property of curcumin (Bisht et al., 2007) and by fluorescence microscopy, cellular uptake levels of dendrosomal curcumin were shown to increase significantly compared to that of free curcumin, More importantly, using FACS analysis and MTT assays, it was demonstrated that as a result of the treatment of human gastric adenocarcinoma cell line, with dendrosomal curcumin, free curcumin or void dendrosomes in vitro, the rate of apoptosis and cell cycle arrest induced by dendrsomal curcumin was significantly higher than that of free curcumin and that treatment of void dendrosomes induced no sign of toxicity on the cells. Similarly, administration of dendrosomal curcumin into tumor-bearing mice in vivo, abolished tumor progression and toxicological analysis indicated that this novel formulation of curcumin did not cause any severe side effects or cytotoxicity in mice.

Following in vitro assays, our group performed experiments to confirm apoptosis induction and tissue uptake of dendrosomal curcumin in vivo. To this end, we injected cancerous cell line WEHI-146 (fibrosarcoma) intraperitoneally into BALB-c mice which led to generation of tumors in mice. Administration of dendrosomal curcumin into the mice gave rise to decrease of size or elimination of tumors. The results of FACS analysis, performed to determine the type of cell death, exhibited the occurrence of apoptosis. In comparison with negative control samples (void dendrosome and curcumin), the results of real-time PCR on genes underlying apoptosis (both apoptosis stimulatory and inhibitory genes) confirmed the induction of apoptosis (manuscript in preparation).

All together, our data suggest that dendrosomes are not only promising gene carriers, but also could be used as efficient drug delivery vehicles for hydrophobic agents.

4.3 Bacteriophages

Bactreiophages, also abbreviated as phages, are amongst non-viral biological agents employed for gene delivery. Bacteriophages are the most abundant life forms in the biosphere and exist in various environments as part of a complex microbial ecosystem

(Clokie et al., 2011). These particles are consisting of a DNA or RNA genome contained within a protein coat. They infect bacteria and either incorporate viral DNA into the host genome, replicating as part of the host (lysogenic cycle), or multiply inside the host cell before releasing phage particles by budding from the membrane or actively lysing the cell (lytic cycle).

They could be found almost in all environments even in the human or animal body. The normal host of routinely used phages as gene delivery vectors (M13 and lambda) is the well characterized bacterium E. coli which is a vital component of the intestinal flora. Therefore it could be suggested that at least some bacteriophage species are stable in the gastro-intestinal tract and should be transferred intact via lymph into blood circulation. If so, this would raise the possibility that phages containing therapeutic genes could be administered orally.

Although still in the early research stages, bacteriophage vectors offer an attractive alternative to various viral and non-viral vectors, because they can potentially overcome the drawbacks of either approach (Table 1). For example, a major advantage of bacteriophages over animal viral vectors is their lack of natural tropism for animal cells, a major concern for current animal viral-vector targeting. Even if non-specific internalization of phage were to occur, the production of phage proteins or replicative phages is unlikely in the foreign milieu of an animal cell (Larocca & Baird, 2001).

	Animal Virus	Synthetic Vector	Phage
Potential toxicity	High	Low	Low
Generation of competent virus	Yes	No	No
Viral protein expressed	Yes	No	No
Complexity	High	Low	Low
Cost	High	Low	Low
Gene transfer efficiency	High	Low	Low
Reproducibility	High	Low	High
Genetic targeting	Limited	Limited	Yes
Directed evolution	No	Yes	Yes

Table 1. A comparison of features of phage vectors with animal and synthetic vectors (Larocca & Baird, 2001).

Also they possess some other advantages. These include the presence of a capsidic structure surrounding the DNA. This is absent in non-viral systems, whose DNA is therefore sensitive to degradation, which impairs gene transfer efficacy, especially in gene therapy experiments performed in vivo (Schmidt-Wolf & Schmidt-Wolf, 2003). The genetic or structural modification of capsid proteins can be easily achieved in phages, allowing selective targeting of eukaroytic receptors and consequent tissue-specific transduction. Furthermore, bacteriophage DNA does not contain eukaryotic sequences and it can be completely replaced by exogenous sequences. These characteristics avoid recombination phenomena, which have contributed to the doubts raised regarding the clinical safety of eukaryotic viruses-derived vectors (Thomas et al., 2003). Phage vector production is expected to be simpler and more cost effective than many existing vectors, because phage can be produced to high titer in the supernatant of bacterial cultures and easily purified at a large scale. The baceteriophage has generous packaging capacity beyond what most viral vectors like adenoviruses (8 kb), adeno-associated viruses (4.7 kb), or lentiviruses (8–10 kb) offer (Chauthaiwale et al., 1992). In addition, phages have been experimentally administered to animals and safely used in humans for applications that include the treatment of bacterial infections (Stone, 2002).

The greatest limitation of phage-mediated gene delivery which has restricted its use is low efficiency compared with typical viral vectors. To partially address this problem, one great technological achievement called phage display is utilized. Phage display is based on expressing recombinant proteins or peptides fused to phage coat proteins. The genetic information encoding for the displayed molecule is physically linked to its product via the displaying particle.

Since a major issue in gene therapy is the development of effective vehicles to deliver a therapeutic gene to specific target cells to achieve adequate and sustained selective expression in the diseased tissue, with minimal toxicity in other tissues, many attempts have been prompted in which, by using phage display, natural ligands for cell-surface receptors have been linked to phages to enhance the efficiency and selectivity of gene transduction. In 1999, Larocca et al first described the transduction of a mammalian cell by a genetically targeted filamentous bacteriophage. They modified the phage genome to display fibroblast growth factor on its surface coat as an N-terminal fusion to the minor coat protein pIII, and showed that such phages undergo receptor-mediated endocytosis, resulting in the expression of phage-encoded genes in mammalian cells.

One of appealing characteristics of phage display is to identify novel cell-targeting ligands, which increase the rate and specificity for the transport of macromolecules. Phage-display technology offers large collections of potential ligands including short peptides, antibody fragments and randomly modified physiological ligands able to bind to cell receptors. Thus, novel ligands can be selected from phage libraries by their ability to deliver a reporter gene to targeted cells. With the incorporation of targeting ligands, specificity, delivery and efficiency of cell transduction can be dramatically improved. For example, the specificity of EGF (epidermal growth factor) -targeted phages is expected to be limited to cells expressing EGF receptor, which is elevated in many tumors.

Previous studies have demonstrated targeted gene-delivery to mammalian cells using modified phage-display vectors. Specificity is determined by the choice of the genetically displayed targeting ligand. Without targeting, phage particles have virtually no tropism for mammalian cells. This raises the possibility of developing phage vectors for targeted gene therapy.

5. Concluding remarks and future perspectives

It is well established that the key to gene therapy research is development of proper gene carriers. Calcium-phosphate precipitation, use of DEAE, PEIs, PLLs, polybrene/dimethyl sulfoxide, liposomes, cationic amino polymers and dendrimers are major chemical methods used for transfection of genes. However, only liposomes and dendrimers are widely used for efficient in vivo gene delivery and the other methods usually have a lower efficiency and can be applied only in vitro. Lipsomes and dendrimers also have limitations such as fast clearance of liposomes from blood, degradation of their content by lysosomic enzymes as they are phagocytosized, hemolytic toxicity and cytotoxicity of dendrimers owing to their cationic nature and premature release of the dendrimer's cargo in plasma.

Dendrosomes, however, might be an exception in this regard. Dendrosomes are non-viral nanovehicles that form stable, biodegradable, neutral, readily synthesized hyperbranched spherical nanostructurs. They not only can deliver genes of various sizes and drugs into target cells but also have been proven to be efficient porters for DNA vaccination in vivo. A unique feature of dendrosomes is the ease with which they provide suitable inert gene porters for various DNA sizes and target cells. Dendrosomes may trap DNA molecules within their arms or/and form clusters entrapping DNA inside with no strong electrostatic bonds, but rather with weak interactions with DNA. This is important as most DNA carriers make bonds with DNA and condense it to a proper size for carriage and this reduces the expression of the condensed DNA even when successfully transfected into cells. But, dendrosomes do not condense DNA while entrapping it efficiently enough for transfection both in vitro and in vivo with no toxicity on target cells. However, optimizing the DNA/dendrosome ratio and safe amounts of the carrier is an essential step to be considered.

Bacteriopahges present tantalizing opportunities for therapeutic applications given further research and development. Exploiting the natural properties of these biological entities for specific gene delivery applications will also expand the repertoire of gene therapy vectors available for clinical uses. The discovery that targeted displayed phage can deliver genes to cells has created new opportunities for expanding the use of phage-display in gene discovery and now gene therapy.

6. Acknowledgements

This work is dedicated to the lovable and unforgettable memory of the late Dr. Mohammad Nabi Sarbolouki. The authors are deeply grateful to Research Council of Tarbiat Modares University and Center of Excellence of Biotechnology for their help. Also, we kindly acknowledge Dr. Farhood Najafi for his gift giving various kinds of dendrosomes. The current address of A.P is Laboratory of Mitotic Mechanisms and Chromosome Dynamics, Institute for Research in Immunology and Cancer (IRIC), University of Montreal, H3T 1J4, Quebec, Canada.

7. References

Alivisatos A (1996). Perspectives on the physical chemistry of semiconductor nanocrystals. J Phys Chem. 100:13226-39.

Balenga NAB, et al (2006). Protective efficiency of dendrosomes as novel nano-sized adjuvants for DNA vaccination against birch pollen allergy. J Biotechnol. 124: 602-614.

Bielinska AV, et al (1997). Regulation of invitro gene expression using antisense oligonucleotides for antisense expression plasmids transfected using starburst PAMAM dendrimers. Nucleic Acid Res. 24: 2176-2184.

Bisht S, Feldmann G, Soni S, Ravi R, Karikar C, Maitra A, Maitra (2007). A polymeric nanoparticle-encapsulated curcumin ("nanocurcumin"): a novel strategy for human cancer therapy. J Nanobiotechnol. Apr 17,5:3.

Bousif O, et al (1995). A versatile vector for gene and oligonucleotide transfer into cells in culture and in vivo. Proc Natl Acad Sci USA. 92: 7294-7301.

Brannon-Peppas L, Balanchette JO (2004). Nanoparticles and targeted systems for cancer therapy. Adv Drug Deliv Rev. 56: 1649-1659.

Buttgereit P, Weineck S, Ropke G, Marten A, Brand K (2000). Efficient gene transfer into lymphoma cells using adenoviral vectors combined with lipofectin. Cancer Gene Ther. 7:1145-1155.

Buttgereit P, et al (2001). Effects of adeniviral wild-type p53 gene transfer in p53 mutated lymphoma cells. Cancer Gene Ther.8:430-439.

Chauthaiwale VM, Therwath A, Deshpande VV (1992). Bacteriophage lambda as a cloning vector. Microbiol Rev. 56: 577–591.

Clokie M, Millard A, Letarov A, Heaphy S (2011). Phages in nature. Bacteriophage. 1: 31-45.

Dobrovolskaia MA, McNeil AE (2007). Immunological properties of engineered nanomaterials. Nat Nanotechnol. 2: 469-478.

Dutta T, et al (2006). Dendrosome-based gene delivery. J Experimental Nanosci. 1: 235

Felgner PL (1991). Cationic liposome-mediated transfection with lipofectin reagent. Inc: Methods in molecular biology, Vol 7: Gene transfer and expression, edited by Murray EJ. Humana Press Inc. Clifton, New Jersey, United Stated.

Gonzalez H, Hwang SJ, Davis ME (1999). New class of polymers for the delivery of macromolecular therapeutics, Bioconjug Chem. 10:1068– 1074.

Gulick T (2003). Curr Protoc Cell Biol. Transfection using DEAE-dextran. Aug;Chapter 20:Unit 20.4.

Guyden J (1993). Techniques for gene cloning and expression. In: Recombinant DNA technology edited by Steinberg M, PTR Printic Hall, New Jersey, United States, pp: 81-97.

Hammarskjold ML (1991). Manipulation of SV40 vectors. In: Methods in molecular biology, Vol7: Gene transfer and expression protocols edited by Murray EJ Clifton, Humana Press Inc, pp: 169-173.

Hashida M, Takemura S, Nishikawa M, Takakura Y (1998). Targeted delivery of plasmid DNA complexed with galactosylated poly(l-lysine). J Control Release 53:301– 310.

Holter W, Fordis CM , et al (1989). Efficient gene transfer by sequential treatment of mammalian cell with DEAE-Dextran and deoxyribonucleic acid. Exp Cell Res. 184:546-551.

Kawasaki ES, Player A (2005). Nanotechnology, nanomedicine and the development of new therapies for cancer. Nanomedicine. 1: 101-109.

Kichler A (2004). Gene transfer with modified polyethylenimines. J Gene Med. Feb;6 Suppl 1:S3-10.

Kukowska J, et al (1996). Efficient transfer of genetic material into mammalian cells using starburst polyamidoamine dendrimers. Proc Natl Acad Scie USA. 93:4897-4902.

Larocca D, Baird A (2001). Receptor-mediated gene transfer by phage-display vectors: applications in functional genomics and gene therapy. Drug Discov Today. 6: 793-801.

Lewin B, Genes IX (2007). Johnes & Bartlett Publishers, Inc, Oxford University Press, United States, New York.

Li S,Thacker LN (1997). High efficiency stable DNA transfection using cationic detergent and glycerol. Biohem Biophys Res Commun. 231: 531-534.

Liu G, Molas M, Grossmann GA, Pasumarthy M, Perales JC, Cooper MJ,.Hanson RW (2001). Biological properties of poly-l-lysine–DNA complexes generated by cooperative binding of the polycation, J Biol Chem. 276: 34379–34387.

Massumi M, et al (2005). Apoptosis induction in human lymphoma and leukemia cell lines by transfection via dendrosomes carrying wild type p53 cDNA. Biotechnol Lett. 28: 61-66.

Moghimi SM, Szebeni (2003). Stealth liposomes and long circulating nanoparticles: critical issues in pharmacokinetics, opsonization and protein-binding properties. Prog Lipid Res. 42:463-478.

Nishikawa M, Takemura S, Takakura Y, Hashida M (1998) . Targeted delivery of plasmid DNA to hepatocytes in vivo: optimization of the pharmacokinetics of plasmid DNA/galactosylated poly(l-lysine) complexes by controlling their physicochemical properties. J Pharmacol Exp Ther. 287:408– 415.

Okayama, et al (1991). Calcium phosphate mediated gene transfer into established cell lines. Inc: Methods in molecular biology, Vol 7: Gene transfer and expression, edited by Murray EJ. Humana Press. New Jersey, Clifton, United States.

Pourasgari F, et al (2007). Low cytotoxicity effect of dendrosome as an efficient carrier for rotavirus VP2 gene transferring into a human lung cell line: Dendrosome, as a novel intranasally gene porter. Mol Biol Rep. 2007 Oct 7. [Epub ahead of print].

Robson T, Hirst DG (2003). Transcriptional targeting in cancer gene therapy. J Biomed Biotechnol. 2:110-137.

Roy I, Mitra S, Maitra A, Mozumdar S (2003). Calcium phosphate nanoparticles as novel non-viral vectors for targeted gene delivery. Int J Pharm. Jan 2:25-33.

Sadeghizadeh M, et al (2008). Dendrosomes as novel gene porters-III. J Chem Technol Biotechnol. 83:912-920.

Sarbolouki MN, Sadeghizadeh M, Yaghoubi MM, Karami A, Lohrasbi T (2000). Dendrosomes: a novel family of vehicles for transfection and therapy. J Chem Technol Biotechnol. 75:919-922.

Saunak S, et al (2004). Polyvalent dendrimer glucosamine conjugated prevent scar tissue formation. Nat Biotechnol. 22: 977-984.

Schmidt-Wolf G, Schmidt-Wolf I (2003). Non-viral and hybrid vectors in human gene therapy: an update. Trends Mol Med. 9:67–72.

Seow Y, J.Wood M (2009). Biological gene delivery vehicles: beyond viral vectors. Mol Ther. 17: 767-777.

Singelton P, Sainsbury D (1995). Dictionary of microbiology and molecular biology, John Wiley & Sons, pp: 902-903, UK.

Sinha R, et al (2006). Nanotechnology in cancer therapeutics: bioconjugated nanoparticles for drug delivery. Mol Cancer Ther. 5: 1909-1917.

Stone R (2002). Bacteriophage therapy. Stalin's forgotten cure. Science. 298: 728–731.

Strauss PL, et al (1994). Transfection of mammalian cells via lipofectin. Inc: Methods in molecular biology, Vol 54: YAC protocols, edited by Markie D Totowa. Humana Press Inc.

Suh W, Chung JK, Park SH, Kim SW (2001). Anti-JL1 antibody-conjugated poly(l-lysine) for targeted gene delivery to leukemia T cells. J Control Release 72: 171– 178.

Suntherland A (2002). Qunatum dots as luminescent probes in biological systems. Curr Opin Solid State Mater Sci. 6:360-370.

Tae B, Park G, Jeong JH, Kim SW (2006). Current status of polymeric gene delivery systems. Advanced Drug Delivery Reviews. 58:467– 486.

Thomas C, Ehrhardt A, Kay M (2003). Progress and problems with the use of viral vectors for gene therapy. Nat Rev Genet. 4:346–358.

Tomalia DA (2005). Birth of a new macromolecular architecture: dendrimers as quantized blocks for nanoscale synthetic polymer chemistry. Prog Polym Sci. 30: 294-324.

Wong TK, Neuman E (1982). Electric field mediated gene transfer. Biochem Biophys Res Commun. 107: 584-587.

Zhang L, Granick S (2006). How to stabilize phospholipids liposome (using nanoparticle) Nano Lett. 6:694-698.

Zhang L, et al (2007). Nanoparticles in medicine: therapeutic applications and developments. Clinic Pharmacol Therapeutics. Doi:10.1038/sj.clpt.6100400.

2

Screening of Bacterial Recombinants: Strategies and Preventing False Positives

Sriram Padmanabhan, Sampali Banerjee and Naganath Mandi

Lupin Limited, Biotechnology, R & D, Ghotawade Village, Mulshi Taluka,
India

1. Introduction

Complete decoding of complex eukaryotic genomes is a prerequisite for understanding varied gene functions. Gene silencing (point mutations, gene deletions, etc), sub cellular localization of proteins, gene expression pattern analysis by promoter activity assay, structure-function analysis, and *in vitro* or *in vivo* biochemical assays (Hartley et al., 2000; Curtis & Grossniklaus, 2003; Earley et al., 2006) are some of the approaches followed for understanding gene functions.

Typically, all the above approaches require the cloning of target genes with or without selective mutations, or cloning their promoter fragments into specialized vectors for further characterization. While the traditional approach for constructing expression cassettes that is based on the restriction enzyme/ligase cloning method is laborious and time consuming, the process is often hampered by length of the gene of interest, GC content of the gene, toxicity of the gene product to the expressing host and lack of relevant restriction sites for cloning purposes. All these factors render the production of expression constructs a significant technical obstacle for large-scale functional gene analysis.

After generating successful cloning/expression constructs, several steps followed are screening high number of colonies, avoiding false positive recombinants and requirement of dephosphorylation of vectors in case of single site cloning to ensure the generation of recombinants with rightly oriented gene of interest and to minimize vector background (non-recombinants).

Screening for recombinants is one of the most crucial and time-consuming steps in molecular cloning and several approaches available for this purpose include colony PCR screening, blue white screening, screening of recombinants, which have the gene of interest in the MCS region of the cloning vehicle, in such a way that the toxic gene reading frame is interrupted making the toxic gene inactivated upon insertion of any foreign gene; GFP fluorescence vectors wherein upon cloning, the GFP fluorescence disappears, etc. The method for screening of bacterial transformants that carry recombinant plasmid with the gene of interest, has become more rapid and simple by the use of vectors with visually detectable reporter genes.

2. Molecular cloning

A recombinant DNA comprises of two entities namely a vector and the gene of interest (GOI). The process of joining vector and any GOI is by making a phosphodiester bond by a process called ligation. The ligation reaction is facilitated with the help of T4 DNA ligase in the presence of ATP. If a vector and any target DNA fragments are generated by the action of the same restriction endonuclease, they will join by base-pairing due to the compatibility of their respective ends. Such a construct is then transformed into a prokaryotic cell, where unlimited copies of the construct, an essentially the target DNA sequence is made inside the cell.

2.1 Steps in molecular cloning
The conventional restriction and ligation cloning protocol involves four major steps namely fragmentation of DNA with restriction endonucleases, ligation of DNA fragments to a plasmid vector, introduction into bacterial cells by transformation and screening and selection of recombinants.

2.1.1 Selection and preparation of vector and insert
A cloning vehicle, also termed as a vector, can be classified as a carrier carrying a gene to be transferred from one organism to another. Other cloning vectors include plasmids, cosmids, bacteriophage, phagemids and artificial chromosomes. In the early days of producing proteins in E. coli, limitations to transcription initiation were believed to lead to lower protein expression levels (Gralla, 1990). This event resulted in efforts put into construction of expression vectors, which carried strong promoters to enhance mRNA yield and a stable mRNA eventually. The promoters used included phage promoters like T7 and T5, the synthetic promoters tac and trc, and the arabinose inducible araBAD (Trepe, 2006). Additional vectors that were made available included Lambda promoters, PR and PL, (Elvin et al., 1990), rhamnose promoter (Cardona & Valvano, 2005), Trp-lac promoter (Chernajovskyi et al., 1983) etc. Certain promoter variants as seen in the expression vector pAES25 yield the maximum level of soluble active target protein (Broedel & Papciak , 2007).

Downstream of each specific promoter, there is a multiple cloning site (MCS) for cloning the gene to be expressed. While the inducible promoters are used to drive the foreign gene expression, the constitutive promoters (Liang et al,., 1999) are used mainly to express the antibiotic expression marker genes for plasmid maintenance.

TA cloning vectors (Zhou & Gomez-Sanchez, 2000; Chen et al., 2009) that takes advantage of the well-known propensity of non-proofreading DNA polymerases (e.g., Taq, Tfl, Tth) to add a single 3′-A to PCR products are also employed for cloning large PCR fragments. The proof-reading polymerases lack 5'-3' proofreading activity and are capable of adding adenosine triphosphate residues to the 3' ends of the double stranded PCR product. Such a PCR amplified product can then be cloned in any linearized vector with complementary 3' T overhangs.

The GC cloning technology is based on the recent discovery that the above proof-reading enzymes similarly add a single 3′-G to DNA molecules, either during PCR or as a separate G-tailing reaction to any blunt DNA. GC cloning vectors pSMART® GC and pGC™ Blue (commercialized by Lucigen, USA) contain a single 3′-C overhang, which is compatible with the single 3′-G overhang on the inserts.

Mead and coworkers (Mead et al., 1991) report cloning of PCR products without any restriction digestion taking advantage of the single 3' deoxyadenylate extension that

Thermus aquaticus, Thermus flavus, and Thermococcus litoralis DNA polymerases add to the termini of amplified nucleic acid.

Gateway cloning system is a relatively new trend in the field of molecular cloning, where in the site specific recombination system of lambda phage is used (Katzen 2007). This system enables the researchers to efficiently transfer DNA fragments between different vector and expression systems, without changing the orientation of the gene or its reading frame. The specific sequences are called "Gateway att sites" and recombination is facilitated by two enzymes "LR clonase" and "BP clonase". This easy Ligase-free cloning system is very beneficial for cloning, combining and transferring of DNA segments between different expression platforms in a high-throughput manner, but making the gateway entry clone usually involves conventional restriction enzyme based cloning, and this is a major drawback of this system.

DNA vectors that are used in many molecular biology gene cloning experiments need not necessarily result in protein expression. Expression vectors are often specifically designed to contain regulatory sequences that act as enhancer and promoter regions, and lead to efficient transcription of the gene that is carried on the expression vector. The regularly used cloning cum expression vectors include pET vectors, pBAD vectors, pTrc vectors etc wherein the GOI is cloned with a suitable promoter of the vector using the start codon of the vector or using a gene of interest with its own start codon into an apopropriate restriction site in the MCS.

RNA polymerases are enzyme complexes that synthesize RNA molecules using DNA as a template. The transcription begins when RNA polymerase binds to the DNA double helix which is at a promoter site just upstream of the gene to be transcribed. While in prokaryotes, one DNA-dependent RNA polymerase transcribes all classes of DNA molecules and the core *Escherichia coli* enzyme called *E. coli* RNA polymerase consists of three types of subunit, α, β, and β', and has the composition $\alpha_2\beta\beta'$; the holoenzyme contains an additional σ subunit or sigma factor (Aaron, 2001). The phage RNA polymerase like T7 RNA polymerase found in pET based expression vectors are much smaller and simpler than bacterial ones: the polymerases from phage T3 and T7 RNA, e.g., are single polypeptide chains of <100 kDa.

The DNA fragment to be cloned is first isolated by a number of ways like cDNA preparation, nuclease fragments of genomic DNA, synthetic DNA's, amplified DNA fragments by means of polymerase chain reaction. After appropriate restriction enzyme digestion and purification, the purified inserts are ligated to the vector of choice.

2.1.2 Ligation of vector and insert

The ligation step is carried out with bacteriophage T4 DNA ligase using ATP required for the reaction and a suitable buffer condition. This process involves the joining of two DNA molecule ends with a phosphodiester bond between the 3' hydroxyl of one nucleotide and the 5' phosphate of the other. The ligation event is of two types sticky or blunt based on the types of restriction enzyme used for digestion of the vector and insert.

2.1.3 Transformation

Following ligation, the ligation product (recombinant plasmid) is transformed into bacteria for propagation. The transformed bacteria are then plated on selective agar to select for bacteria that have the plasmid of interest. Individual colonies are picked up and tested for the desired insert. The transformation is achieved by chemical method (Hanahan, 1983; Inoyue, 1997; Bergmans, 1981) or electroporation (Morrison, 2001).

2.1.3.1 Chemical transformation

For transformation of bacterial cells by chemical means, the cells are grown to mid-log phase, harvested and treated with divalent cations such as $CaCl_2$ to make them competent. After mixing DNA with such competent cells, on ice, followed by a brief heat shock at 42 ^{0}C, the cells are incubated with rich medium for 30-60 minutes prior to plating on suitable antibiotic containing LB agar plates. The biggest advantage of this method includes no special equipment for transformation with no requirement to remove salts from the DNA used for transformation.

2.1.3.2 Electroporation

For electroporation, cells are also grown to mid-log phase but are then washed extensively with water to eliminate all salts from the growth medium, and glycerol added to the water to a final concentration of 10% so that the cells can be stored frozen and saved for future experiments. To electroporate DNA into cells, washed *E. coli* are mixed with the DNA to be transformed and then pipetted into a plastic cuvette containing electrodes. A short electric pulse, about 2400 volts/cm, is applied to the cells causing small pores in the membrane through which the DNA enters. The cells are then incubated with broth as above before plating. For electroporation, the DNA must be free of all salts so the ligations are first precipitated with alcohol before they are used.

2.2 Types of E. coli host cells used for transformation

For most cloning applications, *E. coli* k12 hosts like DH5α which are OmpT protease expressing cells (Salunkhe etal., 2010) are used. These cells are compatible with *lacZ* blue/white selection procedures, are easily transformed, and good quality plasmid DNA can be recovered from transformants. DH5α is one of the most preferred strains for plasmid propagation, because it is an *EndA1* knockout which inactivates the endonucleases ,and a recA knock out which prevents rapid homologous recombination, hence ensuring that the plasmids are stable inside the cells. One notable exception is when transforming with plasmid constructs containing recombinant genes under control of the T7 polymerase, these constructs are typically transformed into DH5α cells during the cloning stage and later introduced into a bacterial strain expressing T7 RNA polymerase for expression of the recombinant protein. The derivatives available for this purpose include BL21(DE3), BL21A1 which are all lon and OmpT protease negative strains (Banerjee et al., 2009; Mandi et al, 2008) and ER2566 (Yu et al., 2004) strains.

3. Selection of recombinants

The need to identify the cells that contain the desired insert at the appropriate and right orientation and isolate these from those not successfully transformed is of utmost importance to researchers. Modern cloning vectors include selectable markers (most frequently antibiotic resistance markers) that allow only cells in which the vector, but not necessarily the insert, has been transformed to grow. Additionally, the cloning vectors may contain color selection markers which provide blue/white screening (via α-factor complementation) on X-gal medium. Nevertheless, these selection steps do not absolutely guarantee that the DNA insert is present in the cells. Further investigation of the resulting colonies is required to confirm that cloning was successful. This may be accomplished by means of PCR, restriction fragment analysis and/or DNA sequencing.

3.1 Colony immunoassay

Telford et al., (1977) reported identification of bacterial plasmids carrying DNA upon lysis and mixing with molten agar with ethidium bromide stain while in as early as 1978, a colony hybridization was developed for screening recombinants by Cami and Kourilsky, (1978) where upon blotting onto nitrocellulose filters and hybridization with a highly radioactive probe, screening of many thousands of colonies per plate for the presence of a DNA sequence carried by a plasmid and complementary to the probe was achieved.

An immunological approach to screen recombinant clones is possible if the gene of interest encodes a polypeptide for which specific antibodies can be prepared. In one approach, DNA complementary to mRNA is inserted in frame with the coding regions of genes present in E. coli plasmids. These results in "fused polypeptides" consisting of the N-terminal region of an E. coli polypeptide covalently linked to a sequence encoded by the cloned cDNA segment. The identification of cloned genes by colony immunoassays has not been common and one limitation of all previous colony immunoassays is that each fused polypeptide molecule must simultaneously bind to two different antibody molecules. Typically, the first antibody, immobilized on a solid support such as chemically activated paper, is used to entrap the fused polypeptide at the site of the lysed colony, and a second labelled antibody is then bound to the fused polypeptide and detected by autoradiography. A potential disadvantage of all immunological methods is that only one in six sequences inserted at random into the vector would have the orientation and frame consistent with translation into a recognizable fused polypeptide. Kemp & Cowman (1981) have described a method by which fused polypeptides can be detected by a colony immunoassay that demands binding of only one antibody molecule. E. coli colonies containing recombinant plasmids are lysed in situ, and proteins in the lysate are immobilized by binding directly to CNBr-activated paper. Antigens attached to the paper are then allowed to react with antiserum, and the antibodies that bind to them are in turn detected by reaction with ^{125}I-labeled protein A (^{125}I-protein A) from Staphylococcus aureus, followed by autoradiography. The limitations of this method are mRNA instability, inefficient translation, or rapid proteolytic degradation of the fused polypeptides that restrict their accumulation within the cells.

A simple immunoassay has been developed by Reggie and Comeron (1986) for isolation of particular gene(s) from a clone bank of recombinant plasmids. A clone bank of the DNA is constructed with a plasmid vector in Escherichia coli and filtered onto a hydrophobic grid membrane and grown up into individual colonies, and a replica was made onto nitrocellulose paper. The bacterial cells upon lysis are immobilized onto the nitrocellulose paper which is reacted with a rabbit antibody preparation made against the particular antigenic product to detect the recombinant clone which carries the corresponding gene. The bound antibodies can be detected easily by a colorimetric assay using goat anti-rabbit antibodies conjugated to horseradish peroxidase.

3.2 Visual screening
3.2.1 The blue-white screening

The blue white screening is one of the most common molecular techniques that allow detecting the successful ligation of gene of interest in vector (Langley et al., 1975; Zamenhof & Villarejo 1972; Ausubel et al., 1988). The α-Complementation plasmids are among the most commonly used vectors for cloning and sequencing the DNA fragments, as they generally have a good multiple cloning site and an efficient blue-white screening system for identification of recombinants in presence of a histochemical dye, 5-bromo-4-chloro-3-indolyl-β-d-galactoside (X-gal), and binding sites for commercially available primers for direct sequencing of cloned fragments (Manjula 2004).

The molecular mechanism for blue/white screening is based on a genetic engineering of the *lac* operon in the *E. coli* as a host cell combined with a subunit complementation achieved with the cloning vector. The *lacZ* product, a polypeptide of 1029 amino acids, gives rise to the functional enzyme after tetramerization (Jacobson et al., 1994) and is easily detected by chromogenic substrates either in cell lysates or directly on fixed cells *in situ* (Ko et al., 1990). The tetramerization is dependent on the presence of the *N*-terminal region spanning the first 50 residues. Deletions in the *N*-terminal sequence generate a so-called omega peptide that is unable to tetramerize and does not display enzymatic activity. The activity of the omega peptide can be fully restored either in bacteria or *in vitro,* if a small fragment (called alpha peptide) corresponding to the intact N-terminal portion is added *in trans* (Gallagher et al., 1994). The phenomenon is called α-complementation and the small *N*-terminal peptide is called alpha peptide. This effect has been widely exploited for studies in prokaryotes, where special strains that constitutively express omega peptide exist and allow the detection of expression of the small alpha peptide.

The vector (e.g. pBluescript) encodes the α subunit of LacZ protein with an internal multiple cloning site (MCS), while the chromosome of the host strain encodes the remaining omega subunit to form a functional β-galactosidase enzyme upon complementation. The foreign DNA can be inserted within the MCS of *lacZα* gene, thus disrupting the formation of functional β-galactosidase. The chemical required for this screen is X-gal, a colorless modified galactose sugar that is metabolized by β-galactosidase to form 5-bromo-4-chloro-indoxyl which is spontaneously oxidized to the bright blue insoluble pigment 5,5'-dibromo-4,4'-dichloro-indigo and thus functions as an indicator. Isopropyl β-D-1-thiogalactopyranoside (IPTG) which functions as the inducer of the Lac operon, can be used to enhance the phenotype. The hydrolysis of colorless X-gal by the β-galactosidase causes the characteristic blue colour in the colonies indicating that the colonies contain vector without insert. White colonies indicate insertion of foreign DNA and loss of the cells ability to hydrolyze the marker. Bacterial colonies in general, however, are white, and so a bacterial colony with no vector will also appear white. These are usually suppressed by the presence of an antibiotic in the growth medium. Blue white screening is thus a quick and easy technique that allows for the screening of successful cloning reactions through the color of the bacterial colony. However, the correct type of vector and competent cells are important considerations when planning a blue white screen.

Although the *lacZ* and many other systems have been extensively used for gram negative bacteria like *E. coli*, there are limited options available for screening recombinants transformed in Gram positive bacteria. Chaffin & Rubens (1998) have developed a gram positive cloning vector pJS3, that utilizes the interruption of an alkaline phosphatase gene, *phoZ*, to identify recombinant plasmids. A multiple cloning site (MCS) was inserted distal to the region coding for the putative signal peptide of phoZ where the alkaline phosphatase protein expressed from the *phoZ* gene (*phoZMCS*) retained activity similar to that of the native protein and cells displayed a blue colonial phenotype on agar containing 5-bromo-4-chloro-3-indolyl phosphate (X-p). Introduction of any foreign DNA into the MCS of phoZ produced a white colonial phenotype on agar containing X-p and allowed discrimination between transformants containing recombinant plasmids versus those maintaining self-annealed or uncut vector. This cloning vector has improved the efficiency of recombinant DNA experiments in gram-positive bacteria.

Cloning inserts into the multiple cloning region of the pGEMÂ®-Z Vectors disrupts the alpha-peptide coding sequences, and thus inactivates the beta-galactosidase enzyme resulting in white colonies. Recombinant plasmids are transformed into the appropriate strain of bacteria (i.e. JM109, DH5α), and subsequently plated on indicator plates containing 0.5 mM IPTG and 40 μg/ml X-gal.

A new version of TA cloning vector with directional enrichment and blue-white color screening has been reported by Horn (2005).

3.2.2 Limitations of blue-white screening

The "blue screen" technique described above suffers from the disadvantage of using a screening procedure (discrimination) rather than a procedure for selecting the clones. Discrimination is based on visually identifying the recombinant within the population of clones on the basis of a color. The *LacZ* gene, in the vector used for generating recombinants, may be non-functional and may not produce β-galactosidase. As a result, these cells cannot convert X-gal to the blue substance so the white colonies seen on the plate may not be recombinants but just the background vector.

A few white colonies might not contain the desired recombinant but a small piece of DNA to be ligated into the vector's MCS might change the reading frame for *LacZα*, and thus prevent its expression giving rise to false positive clones. Furthermore, a few linearized vectors may get transformed into the bacteria, the ends "repaired" and ligated together such that no LacZα is produced as a result, these cells cannot convert X-gal to the blue substance. On the other hand, in some cases, blue colonies may contain the insert, when the insert is "in frame" with the *LacZα* gene and is devoid of stop codon. This could sometimes lead to the expression of a fusion protein that is still functional as LacZα. Small inserts which happen to be in frame with the alpha-peptide coding region may produce light blue colonies, as beta-galactosidase activity is only partially inactivated.

Last but not the least, this complex procedure requires the use of the substrate X-gal which is very expensive, unstable and is cumbersome to use.

3.3 Reporter gene based screening

Another method for screening and identification of recombinant clones is by using the green fluorescent protein (GFP) obtained from jellyfish *Aequorea victoria*. It is a reporter molecule for monitoring gene expression, protein localization, protein-protein interaction etc. GFP has been expressed in bacteria, yeast, slime mold, plants, drosophila, zebrafish and in mammalian cells. Inouye et al., (1997) have described a bacterial cloning vector with mutated *Aequorea* GFP protein as an indicator for screening recombinant plasmids. The pGREENscript A when expressed in *E. coli* produced colonies showing yellow color in day light and strong green fluorescence under long-UV. Inserted foreign genes are selected on the basis of loss of the fluorescence caused by inactivation of the GFP production. The vector used in the study is a derivative of pBluescript SK(+) (Short et al., 1998) and encodes for the same MCS flanked by T3 and T7 promoters, but lacks the *lacZ* gene and the f1 origin region (for single strand DNA production). Instead, the GFP-S65A cDNA is substituted in its place and is under the control of the *lac* promoter/operator. The insertion of foreign DNAs into MCS of pGreenscript A interferes with the production of GFPS65A and causes a loss in the green fluorescence and yellow color of *E. coli* colonies. While GFP solubility appears to be one of the limiting factors in whole cell fluorescence, Davis & Vierstra (1998) have reported about soluble derivatives of GFP for use in *Arabidopsis thaliana*.

A system for direct screening of recombinant clones in *Lactococcus lactis*, based on secretion of the staphylococcal nuclease (SNase) in the organism, was developed by Loir and co-workers (Loir et al, 1994). *L. lactis* strains containing the nuc' plasmids secrete SNase and are readily detectable by a simple plate test. An MCS was introduced just after the cleavage site between leader peptide and the mature SNase, without affecting the nuclease activity. Cloning foreign DNA fragments into any site of the MCS interrupts nuc gene and thus results in nuc mutant clones which are easily distinguished from nuc' clones on plates. The biggest advantage of this vector is the possibility of assessing activity of the fusion protein since the nuclease activity is not diminished by its N-terminal tail and is also reported to be unaffected by denaturing agents such as sodium dodecyl sulfate (SDS) or trichloroacetic acid.

3.3.1 Limitations of the reporter gene based screening

All the above described plasmids could also result in false positive clones, which is a major concern for researchers. Loss of GFP fluorescence due to medium composition is also known to lead to false positive results. Although the SNase based screening would give absolute 100% recombinants, the active nuc fusion protein expression might render the cell fragile and enhance its susceptible to the lethal action of the fusion protein upon hyper expression. Also, for all the above cases, there is a requirement of transfer the genes of interest from the cloning vector to the expression vectors which calls for fresh cloning followed by screening for recombinants.

Hence, it is evident that the commonly used method for screening and identification of recombinant clones are associated with problems of false positive results forcing researchers to look for alternate methods of screening bacterial recombinants and availability of vectors that would act both as cloning vectors and expression vector are user friendly and advantageous.

3.4 Reporter gene based screening- new concepts

The recent approach of screening recombinants is the use of vector for one-step screening and expression of foreign genes (Banerjee et al., 2010) (Fig. 1). The strategy uses the cloning of GFP gene into any expression vector with a stop codon other than the amber stop codon upstream of the ORF of GFP. Upon induction, the GFP would not express and hence would not fluoresce due to presence of the stop codon. Upon in frame cloning of any gene of interest upstream of GFP in such a vector, would then excise the initial stop codon and the resultant fusion protein would fluoresce. The gene of interest contains an amber stop codon and the recombinant screening is carried out in an amber suppressor *E. coli* strain. For expression studies, the same clone can be used for expression in a non amber suppressor *E. coli* host like LE392 cells.

This report describes a vector where screening the transformants *in situ* for the presence of recombinants is possible without any false positive results. All other commercially available vectors show loss of color or loss of fluorescence that may not be unfailing while the major advantage of the vector described in this report, takes the property of color or fluorescence obtained after cloning. This unique vector would also be applicable with any other reporter genes like beta galactosidase gene, luciferase gene, DsRed protein instead of the described gene for GFP in the same vector constructed similarly. It also provides researchers to skip setting up the control ligation mix (without insert) and the dephosphorylation step (CIP or SAP step) since the religated vector would never glow and all the fluorescing colonies are indicative of only the recombinants and also indicative of correct reading frame of the inserted target gene. Since GFP fluorescence is brightest when it is expressed in soluble form (Davis & Vierstra 1998), the intensity of the fluorescence after cloning any foreign gene would also indicate the extent of solubility of the fusion protein. The major advantage of the vector described in this report, takes the property of color or fluorescence obtained after cloning, which is hitherto unreported. Such a vector used for identification, selection and expression of recombinants has been patented by the authors (Deshpande etal., 2010) and is published under PCT (Patent no. WO2010/0226601A2)

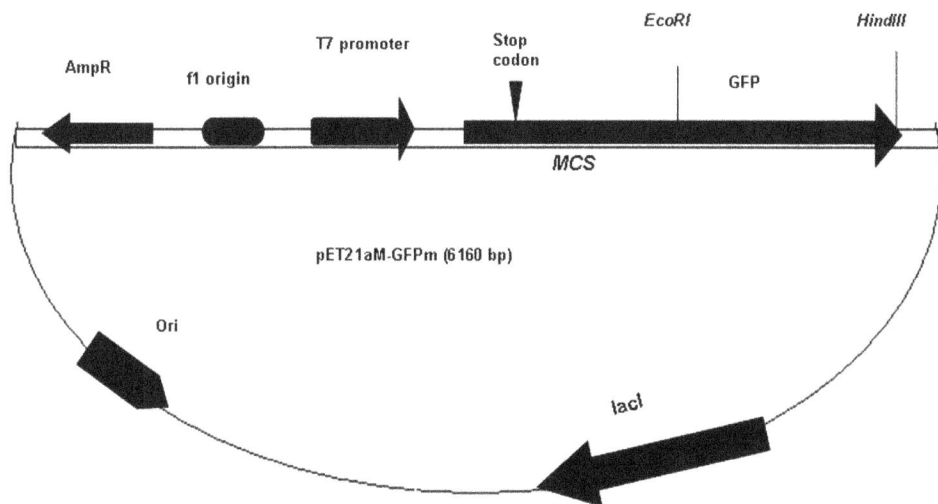

Fig. 1. Plasmid map of the vector pET21aM-GFPm that carries the reporter gene GFP at the 3' end of the MCS. Amp^R refers to ampicillin resistant marker gene.

3.5 Screening clones by positive selection

A variety of positive selection cloning vectors has been developed that allow growth of only those bacterial colonies that carry recombinant plasmids. Typically, these plasmids express a gene product that is lethal for certain bacterial hosts and insertion of any DNA fragment that insertionally inactivates the expression of the toxic gene product resulting in growth of colonies.

Positive selection has been a powerful method of screening insert containing transformants. Here the toxic property of the molecule to the host cells is utilized for recombinant selection. The DNA sequence coding for the toxic product is directly cloned under the promoter elements recognized by the host cells. Positive selection in these vectors is achieved by either inactivation or replacement of toxic gene by the target gene. In general former is much more convenient than the latter (For a detailed review on positive selection vectors ; see reference Choi et al., 2002). The advantage of these systems is that no background colonies (non recombinants) appear on vector alone plates since the religated vectors carrying toxic intact genes are lethal to the host cells.

The genes of toxin-antitoxin system (Finbarr, 2003) of *E. coli* are being utilized for the development of positive selection vectors. A cloning vector carrying cloned *ccdB* gene which encodes poisonous topoisomerase II (DNA gyrase) causing unrecoverable DNA damage has been developed and is widely used as zero background cloning kit from Invitrogen (Bernard 1995). The other toxins from this system which are used successfully are *parD* and *parE* toxins (Gabant et al., 2000; Kim et al., 2004). Apart from this system, the other toxic gene used for the development of positive selection vectors are Colicin E3 and mutated form (E181Q) of catabolite activator protein (Ohashi-Kunihoro et al., 2006; Schlieper et al., 1998). These powerful selection strategies, however, are often only suitable for cloning and require a special host strain for propagation which carry a gene encoding antidote to the product of toxic gene.

The authors of this chapter have developed a positive selection vector which would be used for cloning and expression of heterologous genes simultaneously (Mandi et al., 2009). Here the toxic gene is derived from the antisense strand of *ccdB* gene and cloned under tightly regulated araBAD promoter downstream to the multiple cloning sites (Fig. 2). Multiple cloning sites facilitate cloning of foreign genes and doesn't affect the lethality of toxic gene. A simple method is used to screen the recombinants in that the transformed cells are plated on LB agar medium containing 13 mM arabinose. Moreover, this vector is used also for the expression of genes with authentic *N*-terminus and does not require special host strain for its propagation. Recently, Haag & Ostermeier (2009) have also reported the development of a novel positive selection vector, RHP-Amp[s], that is suitable for cloning and high level protein expression in *E. coli*. Although some limitations exist, positive selection vectors are useful in recombinant DNA experiments thereby reducing the time, effort and cost spent on identifying the correct clones.

Some of other well documented lethal genes include bacteriophage λ repressor, EcoRI methylase, EcoRI endonuclease, galactokinase, colicin E3, transcription factor GATA-1, lysis protein of φX174, barnase (Sambrook et al, 1989), SacB protein of *Bacillus subtilis* (Gay et al., 1985), RpsL protein of *E. coli* (Hashimoto-Gotoh et al., 1993), and also the ParD system of the R1 plasmid (Gabant et al., 2000).

A recently described strategy by Manjula, (2004) involves the use of galactose sensitivity exhibited by galactose epimerase (galE) mutants of *Escherichia coli*. Here, the *E. coli* cells that are lacZ+ galE, but not lacZ− galE are killed upon addition of lactose due to the accumulation of a toxic intermediate, UDP-galactose, by hydrolysis of lactose. Such a method has been suggested to be useful during primary cloning experiments such as construction of genomic or cDNA libraries and also in instances involving selection for rare recombinants.

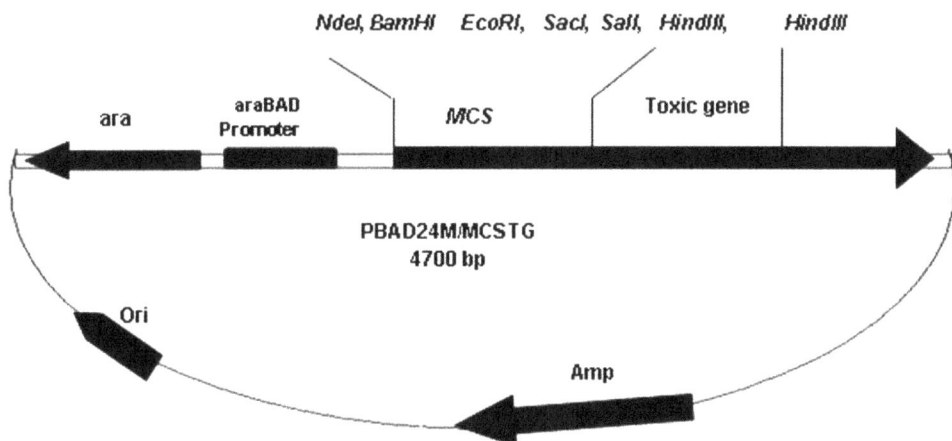

Fig. 2. Plasmid map of pBAD24MMCSTG which has a toxic gene under arabinose promoter as a HindIII/HindIII fragment and having a MCS for cloning any foreign gene to interrupt the expression of the TG gene to select recombinants based on cell survival in presence of the added inducer.

3.5.1 Potential limitations of screening recombinants based on positive selection
Requirements of tightly controlled promoter for expression of the anti-dote of the lethal gene. Requirements of special host cells for the lethal gene inserted or integrated in the bacterial genome is also one of the potential drawbacks and renders this system with limited applications. Moreover, since different genes respond to different promoters, requiring different kinds of host RNA polymerases, the modification of the host with the required lethal gene becomes a prerequisite with various genes which involves cost and is time-consuming. However, for efficient positive selection, the lethality of the marker gene must be strong enough to completely kill the clones harboring vector without insert.

4. Prediction of solubility of recombinant clones during screening

The structural and functional genomics require large supply of soluble, pure and functional proteins for high throughput analysis and as far as screening of soluble or insoluble recombinants is concerned, soluble protein production in E. coli is still a major bottleneck for investigators and a couple of efforts have been reported to improve the solubility or folding of recombinant protein produced in E. coli (Smith 2007). These include strategies like co-expression of chaperone proteins such as GroES, GroEL, DnaK and DnaJ lowering incubation temperature, use of weak promoters, addition of sucrose and betaine in a growth media, use of a richer media with phosphate buffer such as terrific broth (TB), use of signal sequence to export the protein to the periplasmic fraction, fermentation at extreme pH's and use of fusion tags to aid in expression and protein purification (De Marco et al., 2004).

A colony filtration (CoFi) blot method for rapid identification of soluble protein expression in E. coli, based on a separation of soluble protein from inclusion bodies by a filtration step at the colony level is described to screen a deletion mutagenesis library by Cornvik et al.,

(2005) while Coleman et al (2004) report a fluorescent based screening of soluble protein expression where specifically labeled proteins in cellular lysates are detected in one of three formats: a microplate using a fluorescence plate reader, a dot-blot using a fluorescence scanner or a microarray using a laser scanner.

Fusion tags have become indispensable tools for structural and functional proteomics initiatives (Banerjee et al., 2010). Fusion tags that are available for the ease of expression and purification of recombinant proteins include His tag (6-10 aa) (Smith et al., 1988), thioredoxin (109 aa) (Lavallie et al., 1993), Glutathione S-transferase (236 aa) (Smith & Johnson 1988), maltose binding protein (363 aa) (di Guan et al., 1998), NusA (435 aa) (Davis et al., 1999) etc. Although these tags are mainly used to promote the solubility of the target proteins and there by prevent the formation of inclusion bodies in *E. coli*, its use in screening recombinants for solubility of protein of interest is also demonstrated. Maxwell et al., (1999) report cloning of a gene of interest as chloramphenicol acyl transferase (CAT) fusion. Based on the resistance to higher concentration of chloramphenicol, it would be easier to predict the solubility of target protein. The authors have developed an *in vivo* system for assessing protein or protein domain of interest with CAT, the enzyme responsible for conferring bacterial resistance to chloramphenicol. CAT is highly soluble homodimeric protein that has shown to maintain activity when fused to various other proteins. It has been observed that CAT fusions to insoluble proteins confer lower chloramphenicol resistance than that of a fusion with highly soluble partner.

Similarly, Banerjee et al., (2010) have shown that the solubility of the target protein could be predicted *in situ* at the time of recombinant screening based on the intensity of the GFP fusion proteins. This work demonstrates that higher the solubility of the target protein , the intensity of the GFP fluorescence on the agar plate is higher rendering the screening of the recombinants a dual objective of identification and also predicting the solubility of the gene of interest attached to the reporter gene. The article as described by Banerjee et al., (2010) demonstrates this clearly. While GCSF, a human granulocyte colony stimulating factor gene which is known to get expressed as insoluble aggregates in *E. coli* shows lesser fluorescence as GFP fusion (Figure 3, panel B) the *E. coli* methionine amino peptidase, the well soluble *E. coli* protein exhibits higher intensity of GFP fusion (Figure 3, panel A).

A	B
MAP-GFP fusion clones	**GCSF-GFP fusion clones**

Fig. 3. Photograph showing two examples of GFP fusions in the pBAD24mGFP vector to demarcate the solubility difference by means of fluorescence. Panel A shows the *E. coli* methionine amino peptidase GFP fusion *E. coli* clone under long UV while Panel B shows GCSF-GFP fusion clone fluorescence under long UV. Higher glow indicates higher solubility. Clones #3 refers to a non-recombinant in case of MAP-GFP fusion while clone #2 and 5 are non-recombinants of GCSF-GFP fusions.

A similar strategy has been reported by Colman and Homes (2005) for integration confirmation in Pseudomonas strains. Here a pUC-based reporter plasmid (pUS23) was developed containing a recombination site [*aadB* 59 base element (59-be)] upstream of promoterless *aadB* [gentamicin (Gm) resistance] and *gfp* (green fluorescence) genes, and this construct was used to investigate the recombination and expression activities of the CI in *Pseudomonas stutzeri* strain Q capture of pUS23 at *attI* by an integron results in Pc-mediated expression of the aadB and gfp genes, which are silent in the initial construct. The final end result is gentamicin resistant and green fluorescent recombinants for positive integrants.

5. Conclusions

In this article, we have summarized some updated information about (1) new vectors that are commercially available to make the screening system of bacterial recombinants antibiotic free (2) new concepts about easy screening of recombinants utilizing the solubility property of the protein of interest and (3) specialized host strains and using the same clone for expression studies.

Developments in recombinant DNA technology have allowed rapid progress in the analysis of gene structure and function, and the production of potentially useful polypeptides in *Escherichia coli*. Often the experimentally limiting step has been the lack of a suitable screening methodology for expressed cDNAs. The extensive variety of screening bacterial colonies for recombinant plasmids arises from the fact that there is no single method for achieving fool-proof recombinant clone. Conventionally the screening methods employed routinely in academia and industry, for bacterial recombinants include colony hybridization, PCR and plasmid preparations. While all the methods involve cloning the gene of interest in a cloning vehicle and then reintroduction of the recombinant clone into another host cell for expression of the interest making the entire process time-consuming, laborious and expensive. While colony hybridizations require several days to a week and may involve the use of radioactivity, PCR based methods are expensive and have lengthy set-up and reaction times, plasmid preps require considerable hours for cell growth and preparation of mini-prep plasmids. An additional step of screening such recombinants for the solubility of the protein of interest makes the entire process labor-intensive and challenging.

6. Acknowledgements

The authors thank Dr. Kamal Sharma, Managing Director, Lupin Limited for being a constant source of encouragement. Thanks are also due to KrishnaMohan Padmanabha Das for critical reading of the manuscript, formatting the main text and the references.

7. References

Aaron K A Marvellous Machine for Making Messages.(2001) *Science* ,Vol. 292, No. 5523, pp. 1844-1846.

Ausubel, F. M., R. Brent, R. E. Kingston, D. D. Moore, J. G. Seidman, & K. Struhl. 1988. Current protocols in molecular biology. Wiley Interscience, John Wiley and Sons, NY. Cambridge, MA.

Banerjee S, Deshpande AA, Mandi N & Padmanabhan S (2009) A novel cytokine derived fusion tag for over-expression of heterologous proteins in E. coli. International Journal of Bioogical Life Sciencesl Vol. 5, pp.129-133.

Banerjee S, Kumar J, Deshpande AA, & Padmanabhan S (2010) A novel prokaryotic vector for identification and selection of recombinants: Direct use of the vector for expression studies in E. coli. Microbial Cell Factories, Vol. 9, No. 30

Bergmans HE, Van Die IM, & Hoekstra WP (1981) Transformation in Escherichia coli: stages in the process. Journal of Bacteriology. Vol.146, No.2, pp. 564–570.

Bernard, P (1995). New ccdB positive-selection cloning vectors with kanamycin or chloramphenicol selectable markers. Gene, 162, pp. 159-160.

Broedel, Jr SE & Papciak,SM (2007) ACES™ Promoter Selection Vector Technical Brief

Cami B & Kourilsky P (1978) Screening of cloned recombinant DNA in bacteria by in situ colony hybridization. Nucleic Acids Research, Vol. 5, No. 7, pp. 2381–2390.

Cardona ST & Valvano MA (2005) An expression vector containing a rhamnose-inducible promoter provides tightly regulated gene expression in Burkholderia cenocepacia. Plasmid, Vol. 54, No. 3, pp. 219-228.

Chaffin DO & Rubens CE (1998) Blue/white screening of recombinant plasmids in Gram-positive bacteria by interruption of alkaline phosphatase gene (phoZ) expression. Gene, Vol. 28, pp.91-99.

Chen S, Songkumarn P, Liu J & Wang Guo-Liang (2009) A Versatile Zero Background T-Vector System for Gene Cloning and Functional Genomics. Plant Physiologyl. Vol. 150, pp. 1111-1121.

Chernajovsky Y, Mory Y, Vaks B, Feinstein SI, Segev D & Revel M (1983) Production of human interferon in E. coli under lac and tryplac promoter control. Annals in New York Academic Sciences., Vol. 413, pp. 88-96.

Choi YJ, Wang TT & Lee BH (2002) Positive selection vectors. Critical Reviews in Biotechnology, Vol. 22, pp. 225-244.

Coleman MA, Lao VH, Segelke BW, & Beernink PT (2004) High-Throughput, Fluorescence-Based Screening for Soluble Protein Expression. Journal of Proteome Research, Vol 3, No. 5, pp. 1024–1032.

Coleman, NV & Holmes AJ (2005) The native Pseudomonas stutzeri strain Q chromosomal integron can capture and express cassette-associated genes. Microbiology. Vol 151, pp. 1853-1864.

Cornvik T, Dahlroth Sue-Li, Magnusdottir A, Herman MD, Knaust R, Ekberg M & Nordlund P (2005) Colony filtration blot: a new screening method for soluble protein expression in Escherichia coli. Nature Methods, Vol 2, 507 – 509.

Curtis MD, Grossniklaus U (2003) A Gateway cloning vector set for high-throughput functional analysis of genes in planta. Plant Physiology ,Vol 133, pp. 462–469.

Davis GD, Elisee C, Newham DM & Harrison RG (1999) New fusion protein systems designed to give soluble expression in *Escherichia coli*. *Biotechnology and Bioengineering*, Vol 65, No. 4, pp 382-388.

Davis SJ & Vierstra RD (1998) Soluble, highly fluorescent variants of green fluorescent protein (GFP) for use in higher plants. *Plant Molecular Biology*, Vol. 36, pp. 581–588

De Marco, V., G. Stier, S. Blandin, & A. de Marco (2004) The solubility and stability of recombinant proteins are increased by their fusion to NusA. *Biochemical and Biophysical Research Communications*, Vol. 322, pp. 766-771.

Deshpande AA, Sampali Banerjee, Jitendra Kumar & Padmanabhan S (2010) Vector for identification, selection and expression of recombinants. International Publication number WO2010/026601 A2.

di Guan C, Li P, Riggs PD & Inouye H. (1998) Vectors that facilitate the expression and purification of foreign peptides in *Escherichia coli* by fusion to maltose-binding protein. *Gene*, Vol. 67, No. 1, pp. 21-30.

Earley KW, Haag JR, Pontes O, Opper K, Juehne T, Song K & Pikaard CS (2006) Gateway-compatible vectors for plant functional genomics and proteomics. *The Plant Journal*, Vol. 45, pp. 616–629.

Elvin CM, Thompson PR, Argall ME, Hendry P, Stamford NP, Lilley PE, Dixon NE (1990) Modified bacteriophage lambda promoter vectors for overproduction of proteins in *Escherichia coli*. *Gene*, Vol.87, No.1, pp. 123-126.

Finbarr H (2003) Toxins-antitoxins: plasmid maintenance, programmed cell death and cell cycle arrest. *Science*, Vol. 310, pp. 1496–1499.

Gabant PT, Van Reeth PL, Dreze M F, Szpirer C & Szpirer J (2000). New positive selection system based on the parD(kis/kid) system of the R1 plasmid. *BioTechniques*, Vol. 28 , pp. 784-788.

Gallagher C N, Roth N J & Huber R E (1994) A rapid method for the purification of large amounts of an alpha-complementing peptide derived from beta-galactosidase (*E. coli*). *Preparative Biochemistry*, Vol. 24, No. 34, pp 297-304.

Gay P, Le Coq D, Steinmetz M, Berkelman T & Kado CI (1985) Positive selection procedure for entrapment of insertion sequence elements in gram-negative bacteria. *Journal of Bacteriology*, Vol 164, pp. 918-921.

Gralla, J.D. (1990) Promoter Recognition and mRNA Initiation by *Escherichia coli* Es70. *Meth Enzymol.*, Vol. 185, pp.37-54.

Haag AF & Ostermeier C (2009) Positive selection vector for direct protein expression. *Biotechniques* Vol. 46, pp 453-457.

Hartley JL, Temple GF & Brasch MA (2000) DNA cloning using in vitro site-specific recombination. *Genome Reearch.*, Vol.10, 1788–1795.

Hanahan, D. (1983) Studies on transformation of *Escherichia coli* with plasmids. *Journal of Molecular Biology*. Vol.166, 557-580.

Hashimoto-Gotoh, T, Tsujimura A, Kuriyama K & Matsuda S (1993) Construction and characterization of new host-vector systems for the enforcement-cloning method. *Gene* , Vol.137, pp. 211-216.

Horn D (2005) Directional enrichment of directly cloned PCR products. BioTechniques, Vol. 39, No. 1, pp. 40–46.

Inouye S, Ogawa H, Yasuda K, Umesono K & Tsuji, F (1997) A bacterial cloning vector using mutated *Aequorea* green fluorescent protein as an indicator. *Gene* Vol. 189, pp 159-162.

Jacobson RH, Zhang XJ, Dubose RF & Matthews BW (1994) Three-dimensional structure of β-galactosidase from *E. coli. Nature*, Vol. 369, No. 6483, pp 761-766.

Katzen F (2007) Gateway recombination cloning: a biological operating system. *Expert Opinion in Drug Discovery*, Vol . 2, No. 4, pp 571-581

Kemp DJ & Cowman AF (1981) Direct immunoassay for detecting *E. coli* colonies that contain polypeptides encoded by cloned DNA segments. *Proceedings of National Academy of Sciences USA*, Vol. 78, No. 7, pp. 4520-4524.

Kim HG, Hwang HJ, Kim MS, Lee DY, Chung SK, Lee JM, Park JH & Chung DK (2004) pTOC-KR: a positive selection cloning vector based on the ParE toxin. *BioTechniques*, Vol. 36, pp. 60-64.

Ko MS, Nakauchi H & Takahashi N (1990) The Dose dependence of glucocorticoid-inducible gene expression results from changes in the number of transcriptionally active templates. EMBO Journal, Vol 9, No. 9, pp. 2835-2842

Langley KE, Villarejo MR, Fowler AV, Zamenhof PJ & Zabin I (1975) Molecular basis of beta-galacosidase α-complentation. *Proceedings of Naional Academy of Sciences USA*, Vol. 72, No. 4, pp. 1254-1257.

LaVallie ER, DiBlasio EA, Kovacic S, Grant KL, Schendel PF & McCoy JM (1993) A Thioredoxin gene fusion expression system that circumvents inclusion body formation in the *E. coli* cytoplasm. *Nature Biotechnology* Vol. 11, pp. 187 – 193.

Liang S, Bipatnath M, Xu Y, Chen S, Dennis P, Ehrenberg M & Bremer H. (1999) Activities of constitutive promoters in *Escherichia coli. Journal of Molecular Biology*, Vol. 292, No.1, pp.19-37.

Loir YLe, Gruss A, Ehrilch SD & Langella P (1994) Direct Screening of Recombinants in Gram-Positive Bacteria Using the Secreted Staphylococcal Nuclease as a Reporter gene. *Journal of Bacteriology*, Vol.176, No. 16, pp. 5135-5139.

Mandi N, Kotwal P & Padmanabhan S (2009) Construction of a novel zero background prokaryotic expression vector: potential advantages. *Biotechnology Letters*, Vol. 31, No. 12, pp.1905-1910.

Mandi N, Soorapaneni S, Rewanwar S, Kotwal P, Prasad B, Mandal G & Padmanabhan S (2008) High yielding recombinant staphylokinase in bacterial expression system-cloning, expression, purification and activity studies. *Protein Express Purification*, Vol. 64, pp. 69-75.

Manjula R (2004) Positive selection system for identification of recombinants using α-complementation plasmids *BioTechniques*, Vol. 37, No. 6, pp. 948-952.

Maxwell KL, Mittermaier AK, Forman-Kay JD & Davidson AR (1999) A simple in vivo assay for increased protein solubility. *Protein Science*, Vol. 8, pp. 1908–1911.

Mead DA, Pey NK, Herrnstadt C, Marcil RA & Smith LM (1991) A universal method for the direct cloning of PCR amplified nucleic acid. *Biotechnology* (N Y), Vol. 9, No.7, pp. 657-663.

Morrison SL. Transformation of *E. coli* by electroporation. *Current Protocols in Immunology,* 2001 May; Appendix 3: Appendix 3N.

Ohashi-Kunihiro S, Hagiwara H, Yohda M, Masaki H & Machida M. (2006) Construction of a positive selection marker by a lethal gene with the amber stop codon(s) regulator. *Bioscience Biotechnology Biochemistry,* Vol. 70, No. 1, pp. 119-125.

Reggie YCLo & Cameron LA (1986) A simple immunological detection method for the direct screening of genes from clone banks. *Biochemistry and Cell Biology,* Vol 64, pp 73-76.

Salunkhe S, Raiker VA,Rewanwar S, Kotwal P, Kumar A & Padmanabhan S (2010) Enhanced fluorescent properties of an OmpT site deleted mutant of green fluorescent protein. *Microbial Cell Factories,Vol.* 9, pp. 26

Sambrook J, Fritsch EF, & Maniatis. T (1989). Molecular Cloning: A Laboratory Manual, 2nd ed. CSH Laboratory Press, Cold Spring Harbor, NY.

Schlieper D, Schmidt M, Sobek H & Brigitte von Wilcken-Bergmann (1998) Blunt end cloning of PCR fragments using a new positive-selection vector. *Biochemica,* No. 2, pp. 34-35.

Short JM, Fernandez JM, Sorage JA & Huse WD (1998) λ ZAP: a bacteriophage λ expression vector with in vivo excision properties. *Nucleic Acid Research,* Vol. 16, No. 15, pp. 7583-7600.

Smith DB & Johnson KS (1988) Modified glutathione S-transferase fusion proteins for simplified analysis of protein - protein interactions. *Gene,* Vol. 67, pp. 31-40.

Smith HE (2007) The transcriptional response of *Escherichia coli* to recombinant protein insolubility. *Journal of Structural and Functional Genomics,* Vol. 8, No. 1, pp. 27-35.

Smith MC, Furman TC, Ingoliaq TD, & Idgeony CP (1988) Chelating peptide-immobilized metal Ion affinity chromatography: A new concept in affinity chromatography for recombinant proteins. *The Journal of Biological Chemistry,* Vol. 263, No. 15, pp. 7211-7215.

Telford J, Boseley P, Schaffner W & Birnstiel M (1977) Novel screening procedure for recombinant plasmids. Science, Vol. 195 no. 4276, pp. 391-393.

Trepe, K, (2006) Overview of bacterial expression systems for heterologous protein production: from molecular and bio- chemical fundamental to commercial systems. *Applied Microbiology and Biotechnology.* ,Vol.72, No.2, pp. 211-222.

Yu RJ, Hong A, Dai Y & Gao Y (2004) Intein-mediated rapid purification of recombinant human pituitary adenylate cyclase activating polypeptide. B*iochimica et Biophysica Acta,* Vol. 36, No.11, pp. 759-66.

Zamenhof P & Villarejo M (1972) Construction and properties of *E. coli* strains exhibiting α-complementation of beta-galactosidase fragments in-vivo. *Journal of Bacteriology,* Vol. 110, No. 1, pp. 171-178

Zhou MY & Gomez-Sanchez (2000) Universal TA Cloning. *Current   Issues in Molecular Biology*, Vol. 2, No.1, pp. 1-7.

Part 2

Cancer and Cell Biology

3

Molecular Cloning and Overexpression of WAP Domain of Anosmin-1 (a-WAP) in *Escherichia coli*

Srinivas Jayanthi[1], Beatrice Kachel[1], Jacqueline Morris[1],
Igor Prudovsky[2] and Thallapuranam K. Suresh Kumar[1]
[1]*Department of Chemistry & Biochemistry, University of Arkansas, Fayetteville*
[2]*Maine Medical Centre Research Institue, Scarborough,
USA*

1. Introduction

Fibroblast Growth Factor (FGF) signaling plays a vital role in a wide range of cellular responses (Detillieux et al., 2003; Freeman et al., 2003). The activation of FGF Receptor (FGFR) is a crucial step in the diverse FGF signaling pathway (Ayari & Soussi-Yanicostas, 2007). Various intra- and extracellular modulators are involved in the FGF signaling pathway. Anosmin-1, an extracellular matrix associated glycosylated protein, is a newly identified modulator of the FGF-mediated signaling process (Bribian et al., 2006; Gonzalez-Martinez et al., 2004). Anosmin-1 is the product of KAL1 gene. The Kallman syndrome (KS) is a manifestation of the different loss-of-function mutations in the KAL1 gene. The most characteristic features of the Kallman syndrome are anosmia (lack of smell) and hypogonadotrophic hypogonadism (Maestre, 1856; Kallmann et al., 1944). The biological role(s) of Anosmin-1 has been studied both at the cellular and at the biochemical level. Anosmin-1 is suggested to be involved in cell adhesion, cell migration, cell proliferation as well as cell differentiation (Andrenacci et al., 2004; Andrenacci et al., 2006; Ardouin et al., 2000; Bribian et al., 2006; Cariboni et al., 2004; Ernest et al., 2007; Hardelin et al., 1999; Hu et al., 2004; Okubo et al., 2006; Robertson et al., 2001; Rugrali et al., 1996; Soussi-Yanicostas, 1996; Soussi-Yanicostas et al., 1998; Soussi-Yanicostas et al., 2002; Yanicostas et al., 2008;). Anosmin-1 is shown to be a heparin binding protein and formation of the heparin-Anosmin binary complex is believed to be crucial for the function of the protein (Bulow et al., 2002). Interestingly, recent reports suggest a direct interaction between the N- terminus of Anosmin-1 and the FGF-2/FGFR-1/HS ternary complex (Hu et al., 2009; Hu & Bouloux, 2010).

The structure of Anosmin-1(Figure 1) comprises of an N-terminal cysteine-rich domain (CR), whey acidic like-protein domain (a-WAP), four fibronectin type III (FnIII) repeats and C-terminal histidine rich region. Recent reports suggest direct interactions of some of the structural domains of Anosmin-1, including a-WAP, with the fibroblast growth factor receptor(s). Although the Anosmin-FGFR binding interface is still not mapped, it is strongly believed that the interactions of Anosmin-1 with FGFR(s) regulates the activation of the receptor.

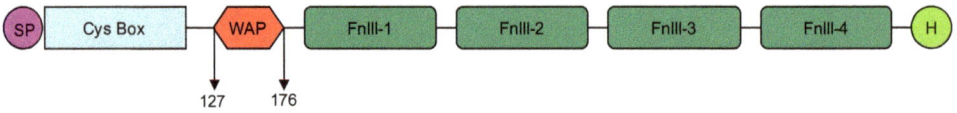

Fig. 1. Schematic representation of human full length Anosmin-1 protein with all the domains, the a-WAP domain labeled in red color spans from 127 to 176 amino acid.

Evolutionarily, Anosmin-1 sequences show the highest conservation within the N-terminal a-WAP and the FnIII-1 domain. Therefore, it is contemplated that these structural domains are functionally significant. The WAP domain of Anosmin-1 (a-WAP) contains 8 cysteine residues which are disulphide bonded to form the four-disulphide core (FDSC) (Figure 2). Most of the proteins containing WAP domain exhibit protease inhibitory activity and consequently regulate the cell proliferation, cell differentiation and tissue remodeling processes (Whitlock et al., 2005). In marked contrast, Anosmin-1 containing the a-WAP domain has been shown to promote the uPA (uro-plasminogen activator) proteolytic activity (Hu et al., 2004).

A.

```
MVPGVPGAVLTLCLWLAASSGCLAAGPGAAAAARRLDESLSAGSVQRARCASRCLSLQITRISAFFQHFQNNGSL
VWCQNHKQCSKCLEPCKESGDLRKHQCQSFCEPLFPKKSYECLTSCEFLKYILLVKQGDCPAPEKASGFAAACV
ESCEVDNECSGVKKCCSNGCGHTCQVPKTLYKGVPLKPRKELRFTELQSGQLEVKWSSKFNISIEPVIYVVQRR
WNYGIHPSEDDATHWQTVAQTTDERVQLTDIRPSRWYQFRVAAVNVHGTRGFTAPSKHFRSSKDPSAPPAPANL
RLANSTVNSDGSVTVTIVWDLPEEPDIPVHHYKVFWSWMVSSKSLVPTKKKRRKTTDGFQNSVILEKLQPDCDY
VVELQAITYWGQTRLKSAKVSLHFTSTHATNNKEQLVKTRKGGIQTQLPFQRRRPTRPLEVGAPFYQDGQLQVK
VYWKKTEDPTVNRYHVRWFPEACAHNRTTGSEASSGMTHENYIILQDLSFSCKYKVTVQPIRPKSHSKAEAVFF
TTPPCSALKGKSHKPVGCLGEAGHVLSKVLAKPENLSASFIVQDVNITGHFSWKMAKANLYQPMTGFQVTWAEV
TTESRQNSLPNSIISQSQILPSDHYVLTVPNLRPSTLYRLEVQVLTPGGEGPATIKTFRTPELPPSSAHRSHLK
HRHPHHYKPSPERY
```

B.

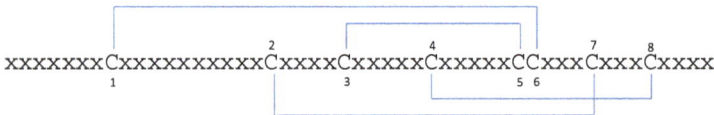

Fig. 2. Panel A. depicts the amino acid sequence of full length Anosmin-1. Residues, 127-176, (shown in red) spans the a-WAP domain with eight conserved cysteines involved in the intra FDSC(four-disulphide core) motif that are highlighted in blue.
Panel **B**. shows the highly conserved intramolecular disulphide bonding pattern in the a-WAP domain.

Some of the mutations in the a-WAP domain are believed to lead to the loss of functions of Anosmin-1. Missense mutations of C172R or C163Y in the KS patients are predicted to result in the disruption of the conserved disulphide bonds and consequently affecting the stability and folding of the protein (Hu et al., 2004). The present study is aimed at understanding the structure-function relationship of the a-WAP domain of Anosmin-1. Therefore, significant quantities of protein(s) are required to embark on detailed structure-function relationship studies. With the advent of the recombinant DNA technology, over-expression of recombinant proteins in heterologous hosts has been rendered easy. Among the expression hosts, *Escherichia coli* (*E.coli*) is an apt choice because it is one of the best studied model systems which is not only easy to handle but is also known to produce recombinant proteins in high yields.

In general, proteins of interest are cloned into a suitable expression vector and overexpressed under selected induction conditions. One of the most successful bacterial expression vectors available is the pGEX system. We have quite successfully cloned a number of genes/DNA segments into this vector and expressed them as soluble GST fusion proteins in milligram quantities from a liter of bacterial culture. In this chapter, we summarize the procedure used to clone, overexpress and purify the a-WAP domain of Anosmin-1. The method described can be applied to most other recombinant proteins that are difficult to be expressed in the soluble fraction.

2. Materials and methods

2.1 Cloning
- pGEX-KG' vector (Genentech, USA).
- BL21 (DE3) (New England Bio labs)
- DH5α (New England Bio labs)
- Mini prep plasmid isolation kit (QIAGEN, USA)
- Phusion PCR Master mix Kit (New England Bio Labs)
- Agarose gel electrophoresis buffer: 0.5X TAE buffer: Tris-acetate (20 mM; pH 8.0), EDTA (0.5 mM).

2.2 Protein expression
- LB (Luria-Bertani) medium (EMD Chemicals Inc.) for bacterial cultures: Dissolve 25g in 1L deionized water and autoclave at standard conditions.
- Ampicillin (Cellgro®): stock solution of 100 mg/ml was prepared and stored at -20 °C.
- Isopropyl β-D-thiogalactoside (IPTG) (Research Products International Corp.): 1M stock solution was prepared and stored at -20 °C.
- SDS PAGE electrophoresis buffer: 1X Tris-glycine buffer – trizma base (25 mM), glycine (50 mM), SDS (0.1%), final pH pH 8.3.
- Staining solution for SDS PAGE: Coomassie Brilliant Blue R-250 (0.25%) in destaining solution.
- Destaining solution for SDS PAGE: methanol (30%), glacial acetic acid (10%).
- 4X SDS gel-loading Buffer: Tris.HCl (100 mM; pH 6.8), SDS (4%), bromophenol blue (0.2%), glycerol (20%), β-Mercapthoethanol (200 mM).

2.3 Protein purification
- Phosphate Buffered Saline (1xPBS): Na_2HPO_4 (10 mM), KH_2PO_4 (2 mM), NaCl (137 mM), KCl (2.7 mM), final pH 7.2.
- GSH Sepharose® beads (GE Health care) are used as affinity based chromatography.
- Thrombin (Sigma) is dissolved in 2mL of 1xPBS pH 7.2 to obtain a concentration of 1U/µL.
- Thermo scientific FH 100 Peristaltic Pump was used for loading the buffers on the column.
- Purification was monitored using UV detector from BIORAD.

2.4 Cloning of a-WAP into the pGEX-KG expression vector

The pGEX-KG (Figure 3) vector from GE Healthcare was used for the expression of the gene of interest as the fusion protein with an N-terminal sequence of glutathione S-transferase (GST). The fusion protein can be cleaved with thrombin (cleavage sequence Leu-Val-Pro-Arg-↓-Gly-Ser) to obtain the recombinant protein with a 15 extra amino acids at the N-terminus.

The multiple cloning site (MCS) of this pGEX-2T contains only three sites for cloning: BamH1, SmaI, and EcoRI. To enhance the MCS, the pGEX-2T vector was linearized with EcoRI and was ligated to a cassette containing an expanded multiple cloning site (MCS) to produce a vector termed pGEX-KG'. The sequence of the MCS is shown (Figure 4). Therefore, the sequence beyond the cleavage site for thrombin in the pGEX-KG' vector will code for: Leu-Val-Pro-Arg-↓-Gly-Ser-Pro-Gly-Ile-Ser-Gly-Gly-Gly-Gly-Gly-Ile-Asp-Ser-Met-Gly-Arg-Leu-Glu-Leu-Lys- (continues further along). The NcoI site cuts between the Ser-Met residues and the XhoI site cuts between the Leu-Glu residues. The NcoI and XhoI sites were used for cloning the a-WAP gene of Anosmin-1.

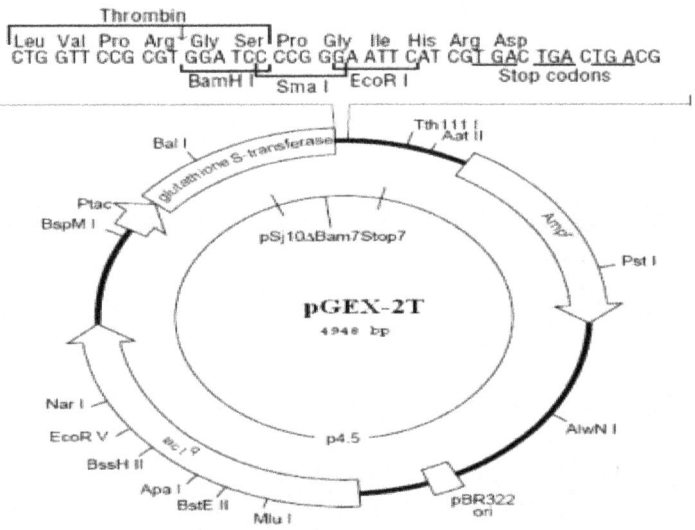

pGEX-2T vector carries ampicillin resistance, pBR322 origin of replication, Lac promoter and GST tag at the 5' end of the MCS.

Fig. 3. Vector Map of pGEX-2T

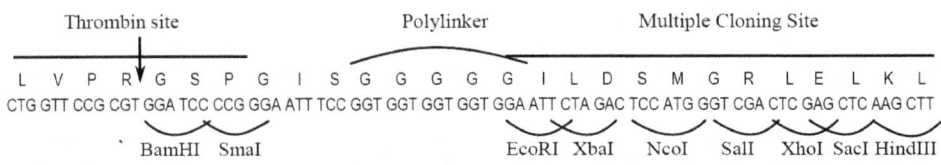

Fig. 4. MCS of the pGEX-KG vector

2.5 PCR-based cloning of a-WAP

The gene of interest was amplified from the optimized *E.coli* codon human full-length Anosmin-1 stabilized in pUC19 (GeneArt, Life technologies, USA) by using gene-specific primers a-WAP-FP (5'ACT GCCATGGTGCTGGTGAAACAGG3') and a-WAP-RP (5' ACTGCTCGAGTTTCGGAACCTGACAG 3'). The PCR conditions were as described in the manufacturer's protocol with the following modifications: DNA (50 ng -200 ng), forward and reverse primers (0.5μM each), 2xPhusion Flash PCR Master Mix (1x) (New England Biolabs) were added to a 20 μl reaction mixture. The samples were incubated in a Master cycler Gradient (Eppendorf) for an initial denaturation (10sec.; 98°C) followed by 30 cycles each including denaturation (2sec.; 98°C), annealing (5 sec.; 62°C), extension (15sec.; 72°C), and the final extension (1min.; 72°C). The PCR product was checked by agarose gel electrophoresis.

The PCR product was purified using a Qiagen PCR purification column according to the manufacturer's instructions to get rid of the unused primers, nucleotides and other material. The purified PCR product and pGEX-KG vector were subjected to double digestion with NcoI and XhoI. The digested PCR product and the vector were gel purified and were subjected to ligation reaction using the Rapid Ligation kit from MBI Fermentas. Conditions for agarose gel resolution, recovery of insert from gel slices, quantification, and ligation were as per standard cloning protocols. Ligation mixtures were used for transformation of DH5α chemical competent cells (New England Biolabs). Recombinant plasmids were purified from bacterial colonies and subjected to restriction analyses and DNA sequencing to confirm their identity.

2.6 Expression and purification of recombinant GST-a-WAP fusion protein in *E.coli*

The plasmid containing recombinant a-WAP was transformed into *E. coli* BL21 (DE3) cells. A single colony was inoculated into 10 ml LB broth containing ampicillin (100μg/ml) and incubated at 37 °C overnight. For large scale expression, LB broth containing ampicilin (100μg/ml) was inoculated with 5% (v/v) overnight culture and incubated at 37 °C and 250 rpm. Once the OD_{600} reached 0.8, 1 mM IPTG was added to the cells, which were further incubated for four hours. Cells were harvested at 6000 rpm for 20 minutes at 4 °C using Beckman JA-10 rotor. The pellets were washed using 1x PBS and either were used immediately or stored at -20 °C.

E. coli cells containing recombinant GST-a-WAP protein were resuspended in 20 mL of 1xPBS. The suspension was lysed using French press at a pressure of 1000 psi. Cell debris was removed by centrifugation at 20000 rpm for 20 min. at 4°C using Beckman JA-20 rotor.

The supernatant was loaded on to the pre-equilibrated (1xPBS) GSH-Sepharose column at a flow rate of 1ml/min. After loading the supernatant on the column, the column was washed with 1xPBS until a flat baseline was reached. Quantity of thrombin to be used for complete cleavage was standardized based on experiments involving in-solution cleavage of GST-a-WAP. For all further purification of a-WAP, thrombin was added at the ratio of 1U/250μg of recombinant GST-a-WAP protein onto the column for an on-column cleavage, which was incubated on the rocker (VWR) at slow speed (45-60 rpm) at room temperature for 24hrs. a-WAP was eluted using 1xPBS and concentrated using Amicon or Millipore concentrators.

2.7 Determination of protein concentration

The concentration of the protein was estimated by measuring absorbance at 280nm (ε_{280nm} = 1.303) using Agilent spectrophotometer.

2.8 MALDI-MS analysis of the a-WAP

Prior to MALDI-TOF (Bruker Daltonics) analysis, recombinant a-WAP (~50-100 µg) sample was desalted by passing through "ZIP™" tips (C-18 matrix). The theoretical molecular weight of a-WAP was calculated using ProtParam tool from Expasy and was found to be 7317.3 Da which was compared with the experimental value obtained from MS-Analysis.

2.9 Circular Dichorism (CD) analysis

CD data were recorded as an average of fifteen accumulations at room temperature using a Jasco J-720 spectropolarimeter. Far UV CD spectrum of a-WAP (166 µM) in 1xPBS pH 7.2 was recorded using a quartz cell of 0.1 mm pathlength in the standard sensitivity mode with a scan speed of 50 nm per minute. Appropriate blank corrections were made in the CD spectrum. The CD data are expressed as molar ellipticity (deg.cm^2.dmol^{-1}).

2.10 Differential scanning calorimetry

Heat capacity of a-WAP was measured as a function of temperature at pH 7.2 using NANO DSCIII with a ramping temperature of 1^{0}C/min from 10^{0}C to 90^{0}C. Thermal denaturation scans were performed using a protein (a-WAP) concentration of ~ 160 µM. The protein solution was degassed in 1xPBS, pH 7.2 prior to acquisition of DSC data. Both the heating and cooling cycles were recorded to examine the reversibility of the thermal unfolding process.

3. Results

3.1 PCR amplification and cloning of a-WAP

Human full length Anosmin-1 gene nucleotide sequence was codon optimized with respect to *E.coli*. The Anosmin-1 gene was cloned in the pET-20b vector. The expression yield(s) of the protein was not as expected and therefore each individual structural domain of the Anosmin-1 gene was cloned and overexpressed in *E.coli*. The significantly different structural and also functional properties of the a-WAP domain of Anosmin-1, as compared to the WAP domains in other proteins, directed our interest to embark on the cloning, overexpression, purification, and biophysical characterization of the a-WAP domain. The actual size of the a-WAP domain is 150 bp but after the PCR amplification the product size increased to 170 bps due to the addition of 10bps at the 5′ and 3′end that includes the desired restriction sites. This 170 bp NcoI – XhoI fragment was cloned into pGEX-KG′ expression vector (please see methods for a description of this expression vector) so that the coding sequence of a-WAP would be in-frame with the GST coding sequence of this vector. This would allow the expression of a GST-a-WAP fusion protein from the pTac promoter (T7 and Lac promoter) of the vector. The colonies obtained after transformation were checked for the positive clones using colony PCR, and plasmid isolated from the positive clones were confirmed by DNA sequencing.

3.2 Overexpression of a-WAP

E. coli BL-21 (DE3) cells transformed with pGEX-KG′-a-WAP were grown in the presence of 1mM IPTG to induce the overexpression of the GST-a-WAP fusion protein. A protein of ~32 kDa appeared to be induced in cultures grown in the presence of IPTG (Figure 5). The cells obtained after induction were collected to check for the presence of a-WAP in either the soluble fraction or in the pellet as inclusion bodies. Therefore, the bacterial pellet obtained from induced cells was subjected to cell lysis and both the supernatant and sonicated pellet fractions collected after centrifugation was resolved under reduced denaturation conditions, showed that the GST-a-WAP was totally expressed in the soluble fraction (Figure 5).

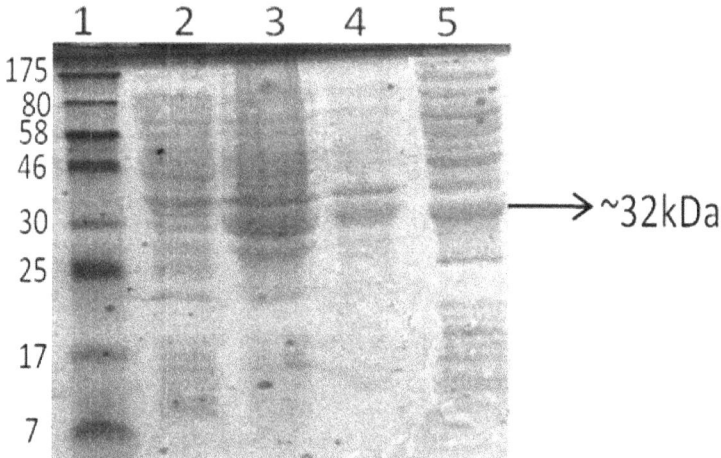

Fig. 5. Induction and purification of recombinant a-WAP. Cell lysates obtained before and after induction with IPTG were resolved on 15% SDS PAGE and was stained with Coomassie Brilliant Blue – R 250. Lane 1: Broad range molecular marker; Lane 2: lysate of uninduced culture; Lane 3: lysate of induced culture showing a overexpressed band ~32kDa; Lane 4: Pellet of the induced cells after the sonication followed by centrifugation; Lane 5: Supernatant obtained from induced cells after the sonication followed by centrifugation containing GST-a-WAP.

3.3 One step purification of a-WAP

a-WAP was purified to homogeneity by a single-step procedure using glutathione-sepharose affinity chromatography (Figure 6). The expression yield of GST-a-WAP was ~16mg/L of the induced *E.coli* culture. The GST-a-WAP fusion protein recovered from the affinity column was subjected to in-solution thrombin cleavage. It was observed that 1U of thrombin cleaved 250 μg of GST-a-WAP. The yield of recombinant a-WAP after thrombin cleavage was ~4mg/L of the induced *E.coli* culture(s). The purified a-WAP migrated as a single band corresponding to a molecular mass of ~7 kDa on SDS-PAGE under reducing conditions.

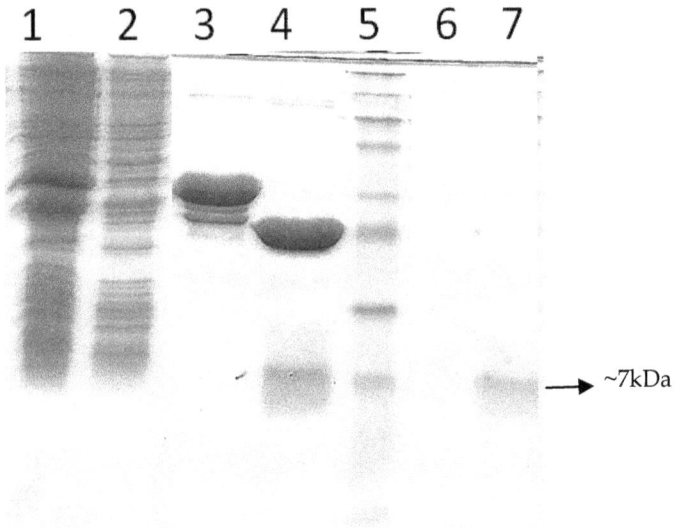

Fig. 6. Purification profile of a-WAP: Samples after each step of purification were resolved on 15% SDS-PAGE stained with Coomassie Brilliant Blue – R 250 Lane 1: Supernatant obtained after sonication; Lane 2: Flow through obtained after binding of the supernatant with 1xPBS pH 7.2; Lane 3: Sample collected upon elution with 10mM GSH(reduced); Lane 4: Sample of in-solution thrombin cleaved GST-a-WAP containing bands corresponding to 26kDA GST and also ~7kDA of a-WAP; Lane 5: Molecular weight marker ; Lane 6 : Empty lane ; Lane 7: Sample obtained by eluting the on-column thrombin cleaved (Incubated at room temperature for 24hrs) a-WAP using 1x PBS at pH 7.2.

3.4 MALDI-MS analysis of a-WAP

This is one of the most reliable and accurate method for the determination of molecular mass of a protein based on the ion mass-to-charge ratio. The protein obtained after purification was in 1X PBS buffer pH 7.2. The protein sample needs to be desalted prior to mass analysis because inorganic salts are known to significantly interfere with the ionization process in the mass spectrometry experiment(s). The monoisotopic mass of purified a-WAP from MALDI-MS analysis was observed to be 7312 Da (Figure 7). Theoretical molecular weight of a-WAP was calculated from the PROTPARAM tool from EXPASY was found to be 7317.3Da.

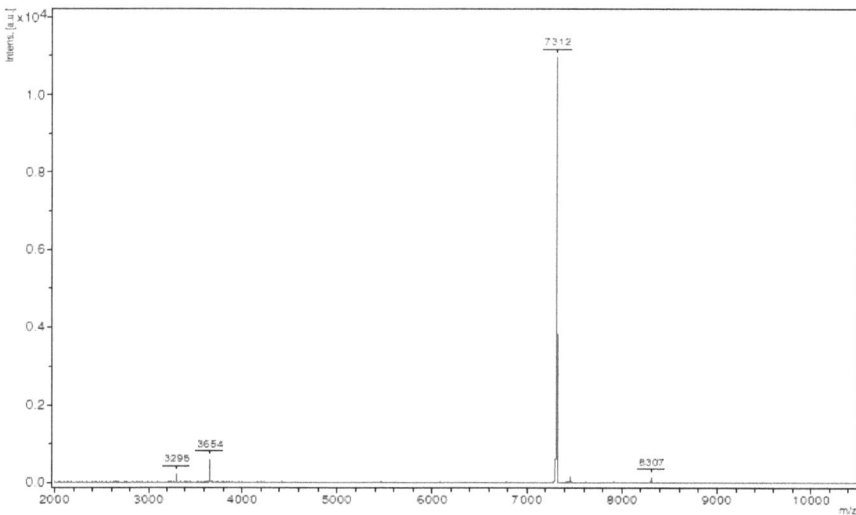

Fig. 7. MALDI-MS analysis of the a-WAP domain. a-WAP MALDI-MS analysis of the desalted a-WAP domain showed a peak that corresponds to a monoisotopic mass of 7312 Da.

3.5 CD analysis of a-WAP

Far UV Circular Dichorism spectroscopy is a powerful technique that is used for assessing the secondary structural elements in proteins. Far UV CD spectrum of the a-WAP showed the characteristic double minima centered at 208 nm and 222 nm suggesting that the backbone of the protein is predominantly in a helix conformation. Interestingly, the aromatic amino acid contributions to the far UV CD spectrum, in the wavelength region of 225 nm to 230 nm, are conspicuously missing. This observation corroborates well with the absence of tryptophan and tyrosine residues in the a-WAP domain of Anosmin-1(Figure 8).

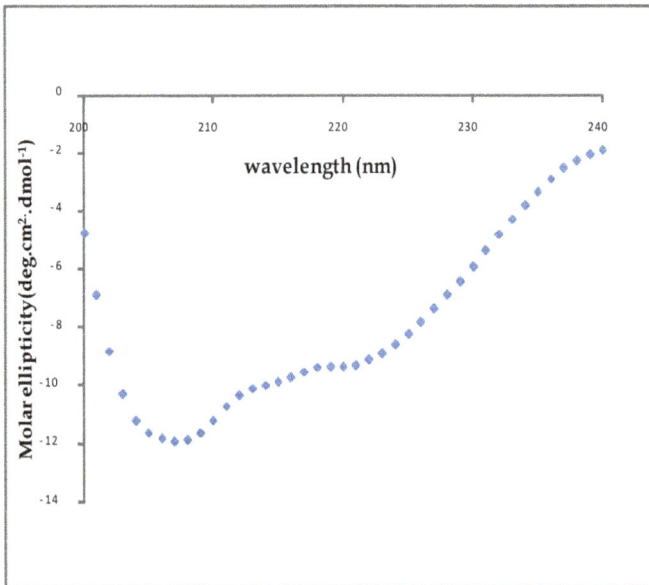

Fig. 8. CD Spectral Analysis of a-WAP. Far-UV CD profile of 166 μM a-WAP was recorded in 1xPBS at pH 7.2 at 25⁰C showed that the majority of the protein shows helicity.

3.6 DSC Analysis of a-WAP

Folding and unfolding of proteins are accompanied by heat effects which are largely exothermic in nature. Heat of unfolding measures the enthalpy of the process. Direct measurement of heat of unfolding was examined by differential scanning calorimetry. The change in enthalpy as function of temperature yields the specific heat associated with protein unfolding. It can be observed that the T_m (the temperature at which 50% of the protein population exists in the denatured state(s) is ~ 75.8 C. Interestingly, the thermogram representing the unfolding is broad suggesting that the two or more states of the protein are in equilibrium with each other (Figure 9). Such thermal transitions are not uncommon for proteins, such as the a-WAP domain of Anosmin-1, which are rich in disulfide bonds.

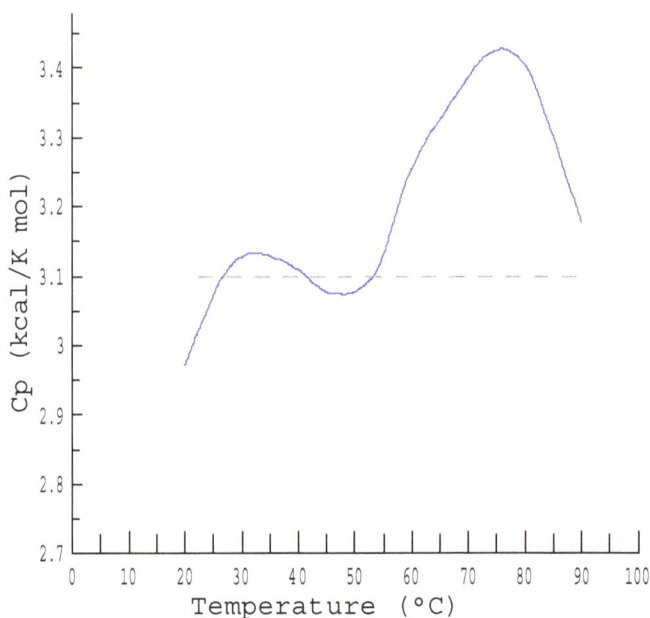

Fig. 9. DSC profile of a-WAP. DSC profile obtained for 166 μM a-WAP in 1x PBS pH 7.2. The T_m for the transition from the folded to the denatured state(s) of the a-WAP is 75.8°C. The change in enthalpy and entropy for the denaturation of the a-WAP are 7.5292kcal/mol and 0.0216kcal/(K mol), respectively.

4. Discussion

Recently, there is an increasing interest in understanding the structure and function of proteins. This endeavor warrants the design of avenues that facilitate cost effective production of recombinant proteins. To-date, E. coli has been the most suitable and commonly used heterologous host for production of recombinant proteins. However, its use requires two steps: introduction of DNA in to the host cell and efficient expression of the target protein. Therefore, an appropriate choice of the vector needs to be made to overexpress the recombinant protein(s) of interest. To-date, Anosmin-1 has been expressed only in eukaryotic hosts. To reach an abundant and inexpensive expression of Anosmin-1, for the first time an attempt was made to express the protein in E.coli. As the expression of the full length gene did not result in satisfactory yields of the protein in soluble fraction, the cloning and expression of independent structural domains of the protein was attempted in a quest to obtain viable yields. Of all the structural domains of Anosmin-1, the N-terminal a-WAP domain has been suggested to play an important role in the function of the protein. In this context, in the present study, the a-WAP domain was cloned and overexpressed in E.coli. All the WAP family proteins exhibit proteolytic activity but the a-WAP domain of Anosmin-1 shows anti-protease activity (Hu et al., 2004). Interestingly, mutations of the conserved cysteine residues in the a-WAP domain result in the Kallman syndrome. Therefore, detailed characterization of the a-WAP domain of Anosmin-1 is crucial for understanding the molecular basis underlying the Kallman syndrome.

To-date, a-WAP domain from Anosmin-1 has been expressed in S2 cells of Drosophila with a yield only sufficient for detection by Western blots. The very meager expression yields obtained therein precludes complete structural characterization of the protein using biophysical techniques such as X-ray crystallography or multidimensional NMR techniques which require substantially high protein concentrations. In this context, the high yields (~ 4 mg- 5mg/L of the bacterial culture) of the a-WAP domain obtained using our cloning and overexpression strategy is significant. Most importantly, the recombinant a-WAP domain, overexpressed and purified using our method, is well-folded with the backbone of the protein predominantly in the helical conformation.

In general, overexpression and purification of small cysteine-rich proteins in heterologous bacterial hosts such as E.coli is considered very challenging. However, the results of this study clearly show that optimization of the codons and proper choice of the protein affinity tag are crucial for the high expression yields of small and cysteine-rich proteins such as the a-WAP-domain of Anosmin-1. In our experience, the GST-tag considerably increases the expression yield and most importantly helps in the circumvention of problems generally associated due to lack of elaborate post-translational machinery in E.coli. In our opinion, the GST tag has a 'chaperoning' effect on the folding of the cysteine-rich a-WAP domain. This aspect is quite significant because most of the mammalian proteins are known to be produced as inclusion bodies when expressed in E.coli. In addition, from a structural biology point-of-view, production of GST-fused small molecular weight proteins is advantageous because the GST-tag does not significantly interfere with the different steps involved in the determination of 3D solution structures of proteins using multidimensional NMR experiments. We believe that the cloning and expression strategy employed in this study will be generally applicable to other small molecular weight, cysteine-rich proteins.

5. Conclusions and future trends

Anosmin-1 is known to play a crucial role in the regulation of the FGF signaling process. However, the exact mechanism by which Anosmin-1 regulates the FGF-induced signaling is still not understood. This lacuna in knowledge is primarily due to non-availability of an overexpression system to produce Anosmin-1 or any of its structural domains in high yields. The strategy for cloning and overexpression of the a-WAP domain of Anosmin-1, discussed in this study, is expected to provide the necessary impetus for future structural studies aimed at understanding the role of the a-WAP domain of Anosmin-1 in the regulation of the FGF receptor activation. This is a significant accomplishment because the a-WAP domain of Anosmin-1 is believed to modulate the FGF-receptor interaction due to its high affinity to bind to heparin. Understanding the structure-function of the a-WAP domain of Anosmin-1 is expected to provide important clues for the rational design of therapeutic principles against a multitude of FGF-mediated pathogenesis. In general, the cloning and overexpression strategy reported in this study is expected to be extremely useful for many other small mammalian proteins.

6. Acknowledgement

Financial support through the following funding agencies NIH COBRE (NCRR P20RR15569), NSF (NSF IOS084397), DOE (DE-FG02-01ER15161), and Arkansas Bioscience Institute to TKSK is kindly acknowledged. In addition, this work was also supported by a NIH grant (RO1 - 2 R01 HL035627) to IP and TKSK. BK gratefully acknowledges the financial help from the Atlantis program of FIPSE (US Dept of Education, FY 2007-058-MS).

7. References

Andrenacci, D., Grimaldi, M. R., Panetta, V., Riano, E., Rugarli, E. I., and Graziani, F. (2006) *BMC Genet* 7, 47

Andrenacci, D., Le Bras, S., Rosaria Grimaldi, M., Rugarli, E., and Graziani, F. (2004) *Gene Expr Patterns* 5, 67-73

Ardouin, O., Legouis, R., Fasano, L., David-Watine, B., Korn, H., Hardelin, J., and Petit, C. (2000) *Mech Dev* 90, 89-94

Ayari, B., and Soussi-Yanicostas, N. (2007) *Dev.Genes.Evol.* 217, 169-175

Bribian, A., Barallobre, M. J., Soussi-Yanicostas, N., and de Castro, F. (2006) *Mol Cell Neurosci* 33, 2-14

Bulow, H. E., Berry, K. L., Topper, L. H., Peles, E., and Hobert, O. (2002) *PNAS* 99, 6346-6351

Cariboni, A., Pimpinelli, F., Colamarino, S., Zaninetti, R., Piccolella, M., Rumio, C., Piva, F., Rugarli, E. I., and Maggi, R. (2004) *Hum. Mol. Genet.* 13, 2781-2791

Detillieux, K.A., Sheikh, F., Kardami, E., and Cattini, P.A. (2003) *Cardiovasc. Res.* 57, 8-19

Ernest, S., Guadagnini, S., Prevost, M. C., and Soussi-Yanicostas, N. (2007) *Gene Expr Patterns* 7, 274-281

Freeman, K.W., Gangula, R. D., Welm, B. E., Ozen, M., Foster. B. A., Rosen, J. M., Ittmann, M., Greenberg, N. M., and Spencer, D. M. (2003) *Cancer Res.* 63, 6237-6243

Gonzalez-Martinez, D., Kim, S.-H., Hu, Y., Guimond, S., Schofield, J., Winyard, P., Vannelli, G. B., Turnbull, J., and Bouloux, P.-M. (2004) *J. Neurosci.* 24, 10384-10392

Hardelin, J. P., Julliard, A. K., Moniot, B., Soussi-Yanicostas, N., Verney, C., Schwanzel-Fukuda, M., Ayer-Le Lievre, C., and Petit, C. (1999) *Dev Dyn* 215, 26-44

Hu, Y., Gonzalez-Martinez, D., Kim, S. H., and Bouloux, P. M. (2004) *Biochem J* 384, 495-505

Hu, Y., Guimond, SE., Travers, P., Cadman, S., Hohenester, E., Turnbull, JE., Kim, SH., Bouloux, PM. (2009) *J.Biol.Chem* 284 (43), 29905-20

Hu, Y., Bouloux, PM. (2010) *Trends in Endocrinology and Metabolism* 21 (6), 385-93

Kallmann, F. J., Schoenfeld, W. A., and Barrera, S. E. (1944) *Am J Ment Defic* XlVIII, 203-236

Maestre De San Jaun, A. (1856) *El Siglo Medico, Madrid* 3, 211

Murcia-Belmonte, V., Esteben, PF., Garcia-Gonzalez, D., De Castro, F. (2010). *J. Neurochem* 115 (5), 1256-65

Okubo, K., Sakai, F., Lau, E. L., Yoshizaki, G., Takeuchi, Y., Naruse, K., Aida, K., and Nagahama, Y. (2006) *Endocrinology* 147, 1076-1084

Robertson, A., MacColl, G. S., Nash, J. A., Boehm, M. K., Perkins, S. J., and Bouloux, P. M. (2001) *Biochem J* 357, 647-659

Rugarli, E. I., Ghezzi, C., Valsecchi, V., and Ballabio, A. (1996) *Hum Mol Genet* 5, 1109-1115

Soussi-Yanicostas, N., de Castro, F., Julliard, A. K., Perfettini, I., Chedotal, A., and Petit, C. (2002) *Cell* 109, 217-228

Soussi-Yanicostas, N., Faivre-Sarrailh, C., Hardelin, J. P., Levilliers, J., Rougon, G., and Petit, C. (1998) *J. Cell Sci.* 111, 2953-2965

Soussi-Yanicostas, N., Hardelin, J. P., Arroyo-Jimenez, M. M., Ardouin, O., Legouis, R., Levilliers, J., Traincard, F., Betton, J. M., Cabanie, L., and Petit, C. (1996) *J. Cell Sci.* 109, 1749-1757

Whitlock, K. E., Smith, K. M., Kim, H., and Harden, M. V. (2005) *Development* 132, 5491-5502

Yanicostas, C., Ernest, S., Dayraud, C., Petit, C., and Soussi-Yanicostas, N. (2008) *Dev Biol* 320, 469-479

4

Subcloning and Expression of Functional Human Cathepsin B and K in *E. coli*: Characterization and Inhibition by Flavonoids

Lisa Wen et al.*

Department of Chemistry, Western Illinois University,
USA

1. Introduction

1.1 Cathepsins

Cathepsins, originally identified as lysosomal proteases, play a fundamental role in intracellular protein turnover in lysosomes. However, several cathepsins and variants of cathepsins can also be found on the cell membrane, in the cytosol, nucleus, mitochondria, and extracellular space. These cathepsins are involved in a variety of important physiological and pathological processes [reviewed in: (Brix et al., 2008; Frlan and Gobec, 2006; Lutgens et al., 2007; Mohamed and Sloane, 2006; Nomura and Katunuma, 2005; Obermajer et al., 2008; Reiser et al., 2010; Stoka et al., 2005; Turk et al., 2001; Vasiljeva et al., 2007; Victor and Sloane, 2007)]. Cathepsins are classified mechanistically into groups which include serine (cathepsins A and G), aspartic (cathepsins D and E), and cysteine cathepsins (cathepsins B, C, F, H, L, K, O, S, V, W, and X). This classification is based on the nucleophilic residues present on their active sites responsible for proteolytic cleavage (Rawlings et al., 2006; Turk et al., 2001).

Cathepsins are synthesized as zymogens composed of a signal peptide, a propeptide, and mature protein of distinct length and substrate specificity for individual cathepsins (Rawlings et al., 2006). The signal peptide is cleaved in the Endoplasmic Reticulum and the pro-protein is activated by proteolytic removal of the N-terminal pro-peptide either by autocatalysis in acidic environments, or by other proteases. The pro-peptide region of the cathepsin plays multiple roles. It can act as an inhibitor to block access to the active site that regulates cathepsin activity. In addition the propeptide can act as an intramolecular chaperone that assists in protein folding, or as a trafficking signal that targets the protein to its destination (Turk et al., 2002). Cathepsins exhibit a broad range of functions and tissue expression (Brix et al., 2008; Turk et al., 2001). Some of the cathepsins are ubiquitously expressed and others are tissue or cell-type specific.

Cathepsins have been shown to be involved in the process of tumor invasion and metastasis (Białas and Kafarski, 2009; Lindeman et al., 2004; Nomura and Katunuma, 2005; Obermajer

*Soe Tha, Valerie Sutton, Keegan Steel, Franklin Rahman, Matthew McConnell, Jennifer Chmielowski, Kenneth Liang, Roxana Obregon, Jessica LaFollette, Laura Berryman, Ryan Keefer, Michael Bordowitz, Alice Ye, Jessica Hunter, Jenq-Kuen Huang and Rose M. McConnell
Department of Chemistry, Western Illinois University, USA

et al., 2008) and have been linked to many types of cancer including melanoma (Matarrese et al., 2010; Quintanilla-Dieck et al., 2008), hepatocellular carcinoma (Leto et al., 1997; Leto et al., 1996), breast cancer (Foekens et al., 1998; Laurent-Matha et al., 1998; Masson et al., 2011; Vashishta et al., 2007), lung cancer (Ledakis et al., 1996; Schweiger et al., 2000), prostate cancer (Brubaker et al., 2003; Kishore Kumar et al., 2010; Podgorski et al., 2009; Podgorski et al., 2007; Steffan et al., 2010), nasopharyngeal cancer (Cheng et al., 2008; Xu et al., 2009), thyroid cancer (Mikosch et al., 2008; Tedelind et al., 2010), bone cancer (Podgorski et al., 2009; Podgorski et al., 2007) as well as osteoporosis (Bone et al., 2010; Deal, 2009; Stoch and Wagner, 2007; Yasuda et al., 2005), rheumatoid arthritis (Skoumal et al., 2005; Skoumal et al., 2008), Alzheimer's Disease (Hook et al., 2007; Hook et al., 2009; Urbanelli et al., 2008), cardiovascular disease (Bengtsson et al., 2008; Lutgens et al., 2007), and obesity (Li et al., 2010; Naour et al., 2010; Podgorski et al., 2007; Yang et al., 2008).

1.2 Cysteine cathepsins

Cysteine cathepsins belong to papain-like enzyme family sharing similar amino acid sequences and tertiary structures (Turk et al., 2001). Eleven cysteine cathepsins B, C, F, H, L, K, O, S, V, W, and X have been identified in the human genome (Turk et al., 2001). Cysteine cathepsins have been documented to play a vital role in a variety of biological processes and pathological processes (Brix et al., 2008; Joyce et al., 2004; Lecaille et al., 2002; Obermajer et al., 2008; Reiser et al., 2010; Stoka et al., 2005; Turk et al., 2001; Victor and Sloane, 2007). The correct sorting and trafficking of the members of cysteine cathepsins are critical in their proteolytic actions to maintain homeostasis (Brix et al., 2008). The dysregulation of protease activity has resulted in numerous pathologies. Several cysteine cathepsins have been recognized as relevant drug targets in the development of many disease therapies (Deal, 2009; Le Gall et al., 2008; Mohamed and Sloane, 2006; Podgorski, 2009; Turk, 2006; Vasiljeva et al., 2007).

1.2.1 Cathepsin B and K

Human cathepsin B precursor is a protein of 339 amino acids that consists of a signal peptide at amino-terminal end of 17 amino acids (1-17). The proprotein is then activated as a single-chain form of 254 amino acids (80-333) or double-chain form of 47 amino acids (80-126) and 205 amino acids (129-333) (Pungerčar et al., 2009; Rozman et al., 1999). Unlike other cysteine proteases cathepsin B is unique due to its ability to act both as an endopeptidase and a peptidyldipeptidase. It contains a unique occluding loop which is characterized by two adjacent histidine residues (His 110 and His 111) and are responsible for the dipeptidityl carboxypeptidase activity (Illy et al., 1997). Cathepsin B is the most abundant in all of the cysteine proteases (Kirschke et al., 1995).

Cathepsin B participates in many diverse cellular processes including protein degradation, antigen processing (Zhang et al., 2000), and apoptosis (Bien et al., 2010; Chwieralski et al., 2006; Roberts et al., 1999). It has been implicated in a variety of diseases including cancer invasion and metastasis (Ledakis et al., 1996; Matarrese et al., 2010; Roshy et al., 2003; Sinha et al., 2001; Szpaderska and Frankfater, 2001; Yan et al., 1998), angiogenesis (Im et al., 2005; Kruszewski et al., 2004; Malla et al., 2011; Sinha et al., 1995), inflammation (Hashimoto et al., 2001; Kakegawa et al., 2004), and Alzheimer's Disease (Gan et al., 2004; Hook, 2006; Hook et al., 2008). Thus, cathepsin B is a promising target for anti-cancer drug design (Lim et al., 2004; Palermo and Joyce, 2008) and a potential target for Alzheimer's Disease (Hook et al., 2008).

Human cathepsin K precursor is a protein of 329 amino acids that consists of an amino-terminal signal peptide of 15 amino acids (1-15), a propeptide of 99 amino acids (16-114), and a catalytic region of 215 amino acids (115-329) (Lecaille et al., 2008).

Cathepsin K is highly expressed in osteoclasts (Drake et al., 1996) but also occurs in lung epithelia cells (Bühling et al., 1999), cultured primary neonatal skin fibroblasts activated chondrocytes, and in synovial fibroblasts of patients suffering from rheumatoid arthritis (Lecaille et al., 2008; Ruettger et al., 2008; Skoumal et al., 2005). Because of its strong collagenolytic activity cathepsin K has been described as the major enzyme responsible for the degradation of organic bone matrix, and is believed to play a fundamental role in bone resorption (Gowen et al., 1999; Saftig et al., 1998; Salminen-Mankonen et al., 2007; Stoch and Wagner, 2007). Cathepsin K is also involved in lung matrix homeostasis (Bühling et al., 2004), dermal extracellular matrix homeostasis (Rünger et al., 2007), and atherosclerotic plaque remodeling (Guo et al., 2009). Circulating serum cathepsin K has been found to play a significant role in both prostate cancer and breast cancer related bone metastasis (Tomita et al., 2008; Valta et al., 2008).

Cathepsin K has become an established drug target for osteoporosis (Deal, 2009; Stoch and Wagner, 2007; Vasiljeva et al., 2007) . Two cathepsin K inhibitors have progressed to Phase II and Phase III clinical trials for osteoporosis. Ono's ONO-5334 (Eastell et al., 2011; Manako, 2011) is in Phase II while Merck's odanacatib (Pérez-Castrillón et al., 2010) is in Phase III clinical trials.

Cathepsin K has also been found to play roles in atherosclerosis (Guo et al., 2009), inflammation (Asagiri et al., 2008; Lecaille et al., 2008), and obesity (Podgorski et al., 2007; Yang et al., 2008). These findings suggested that cathepsin K may be one of the common biological links connecting low bone density to cardiovascular disease (Lutgens et al., 2007; Podgorski, 2009; Podgorski et al., 2007). Thus pointing to possible future anti-cathepsin K drug applications toward dual therapy for skeletal disease and atherosclerosis (Podgorski, 2009)

1.3 Research objectives

Specific proteinase inhibitors are useful in investigations of the mechanisms and pathways of intracellular protein degradation and could lead to the development of therapeutic agents for treatment of many types of carcinomas, skeletal disease and atherosclerosis, as well as Alzheimer's Disease. In the present communication, we report the successful production of functional human recombinant cathepsin B and K. The active enzymes have been used successfully in screening flavonoids for effective inhibitors against human cathepsins B and K enzymes.

2. Experimental

2.1 Materials

E. coli strains JM109 and BL21(DE3)pLysS were used as host cells. The pET-15b was used for the expression vector. Antibiotics ampicillin and chloramphenicol were purchased from Sigma and Fisher Scientific, respectively. Yeast extract and bactotryptone were from BD Biosciences. Isopropyl thio-β-galactoside (IPTG) was from Calbiochem. FideliTaq™ PCR master mix was from United States Biochemical Corp and PCR master mix (2x) from Promega corporation. XhoI, alkaline phosphatase, and T4 DNA ligase were from New

England Biolabs. HisTrap FF column was from GE Healthcare. Z-Phe-Arg-pNA (Carbobenzoxycarbonyl-L-Phenylalanyl-L-Arginine para-nitroanilide) was from Enzo Life Sciences. Oligonucleotides were synthesized by Integrated DNA Technology. Glutathione (reduced) and GSSG (oxidized) were from Sigma–Aldrich. Precision Plus Protein Kaleidoscope Standards were from Bio-Rad Laboratories.

2.2 Subcloning of recombinant human procathepsin B and K

Full-length cDNA encoding of the human pre-pro-cathepsin B (GenBank accession number BC095408) and the human pre-pro-cathepsin K (GenBank accession number BC016058) were purchased from Open Biosystems. Plasmid DNA isolated from each clone was used as templates for amplification of coding region of procathepsin B and K (excluding signal peptide) using gene-specific primer pairs (see Table 1). Each primer was appended with XhoI restriction enzyme recognition site (underlined) to facilitate cloning. The XhoI restriction site was chosen because both cathepsin genes lack XhoI recognition site. The extra six nucleotides (selected at random) at the 5'end of each primer were to facilitate cleavage by restriction enzymes.

	Primers
ProCatB forward primer	ATA TAA CTC GAG CGG AGC AGG CCC TCT TTC C
ProCatB reverse primer	ATA TAA CTC GAG TTA GAT CTT TTC CCA GTA CTG ATC GGT G
ProCatK forward primer	ATG CGA CTC GAG CTG TAC CCT GAG GAG ATA CTG G
ProCatK reverse primer	ATG CGA CTC GAG TCA CAT CTT GGG GAA GCT GG

Table 1. Synthetic primers for amplification of coding sequences of procathepsin B and K.

Each PCR solution (50 µl) consisted of 25 µl FideliTaq™ PCR Master Mix (2x), 1 µM of each forward and reverse primer, and 10 ng plasmid template. The reaction was pre-denatured at 94 °C for 5 min. Then 25 cycles were conducted, which consisted of denaturation at 94 °C for 45 sec, annealing at 55±10 °C for 45 sec, and extension at 68 °C for 1.5 min. A final extension was done at 68 °C for 10 min. The cycling process was accomplished by the Eppendorf Mastercycler. The PCR amplified DNA fragments were extracted once with equal volume of phenol/chloroform (1/1), once with chloroform/isoamyl alcohol (24/1), and DNA precipitated with ethanol (Sambrook and Russell, 2001). Each PCR fragments of procathepsin B or K was digested with XhoI to generate sticky ends and the digested products were separated on a 1.2% agarose gel. The desired bands were purified from agarose gel slice (Kim, 1992) and ligated to pET-15b which had been treated with XhoI and alkaline phosphatase. Each ligation reaction was then transformed into JM109 competent cells according to the method of Chung et al. (Chung et al., 1989) and plated on ampicillin containing plates. The transformants were screened to find recombinant DNA containing colonies by Quick Screening (Huang et al., 2004). In Quick Screening, a toothpick was used to isolate a colony and the cells were resuspended in 25 µl of STE solution (100 mM NaCl, 20 mM Tris–Cl pH 7.5, and 10 mM EDTA). The suspension was extracted with phenol/chloroform and the aqueous layer was analyzed on a 1% agarose gel. The chimeric DNA (with DNA insert) migrates slower than the vector alone. The selected chimeric plasmids were isolated using boiling method (Sambrook and Russell, 2001).

The orientation of the insert in the chimeric plasmids was checked by PCR using the pET 5' sequencing primer and the gene specific reverse primer. Promega's 2x PCR master mix was used because proofreading was not required in this case. PCR products are expected only for clones harboring an insert in the correct orientation. The correct recombinant plasmids (confirmed by DNA sequencing) were each transformed into BL21(DE3)pLysS host cells for protein expression.

2.3 Overexpression of recombinant human procathepsin B and K

The chosen transformants of BL21(DE3)pLysS harbored the chimeric plasmid (procathepsin B or K) in the correct orientation and reading frame was induced for protein expression in the presence of IPTG. Several growth conditions (varying growth media, induction temperatures, induction time, and inducer concentration) were tested and the one yielded the best results is described below. Each clone was grown in 10 mL terrific broth (1.2 % bacto-tryptone, 2.4 % bacto-yeast extract, 0.4 % glycerol, 0.017 M KH_2PO_4 and 0.072 M K_2HPO_4) containing 100 µg/ml ampicillin and 20 µg/ml chloramphenicol overnight on a shaker at 37 °C. Next day, the culture was diluted 1:20 in terrific broth containing ampicillin and chloramphenicol in a baffled Erlenmeyer flask. The culture was grown until A_{600} reached 0.8–1.2. At this point IPTG inducer was added to a final concentration of 0.5 mM along with 25% fresh terrific broth. The incubation temperature was dropped to 25 °C and cells were harvested 4-16 hours after IPTG induction. Overexpression of the target protein was checked by comparing total protein patterns before and after IPTG induction by SDS-PAGE.

2.4 Purification of recombinant procathepsin B

Recombinant procathepsin B was obtained from purified inclusion bodies of IPTG induced cells by a procedure described by Kuhej, et al, 1995 with minor modifications. The IPTG induced cell pellet (4.6 g from 500 mL culture) was re-suspended in 46 mL of 50 mM Tris-HCl, pH 8.0 containing 5 mM EDTA and 5% sucrose, and was sonicated. The homogenate was divided into two centrifuge tubes and centrifuged at 6,000 x g for 10 min. The pellets containing inclusion body were washed twice by homogenizing each in 15 mL of 50 mM Tris-HCl, pH 8.0 containing 5 mM EDTA and 0.1% Triton X-100 using a generator sawtooth followed by centrifugation at 6,000 x g for 10 min at 4 °C. Each inclusion body pellet was consecutively washed twice with 10 -15 mL of 50 mM Tris-HCl, pH 8.0 containing 5 mM EDTA and 2 M Urea. The content was centrifuged as before. The purified inclusion bodies were solubilized in approximately 10 mL of 8 M Urea/0.1 M Tris-HCl, pH 8.0/10 mM DTT by stirring for 1.5 h at room temperature. The protein concentration was approximately 10.8 mg/mL.

The procathepsin B in the solubilized inclusion body was refolded and re-oxidized by dilution and dialysis according to a reported procedure (Kuhelj et al., 1995) with minor modification. An aliquot of the solubilized inclusion body was diluted to approximately 30 µg/mL with 8 M urea in 0.1 M Tris-HCl, pH 8.0 containing 10 mM DTT and dialyzed against 4 L of 0.1 M sodium phosphate pH 7.0 containing 5 mM EDTA and 5 mM L-Cysteine overnight. (It took about 2 hours of occasional stirring with a glass rod to diffuse some urea so that the dialysis tubing could float). Dialysis was then carried out for approximately 24 hours. Fresh buffer was changed once during dialysis. The dialyzed procathepsin B solution was collected and centrifuged at 10,000 x g for 10 min to remove precipitated protein. The protein was stored at -70 °C until use. Protein purification progress and purity were checked by 12% SDS-PAGE.

2.5 Activation of the procathepsin B

Procathepsin B was autoactivated to its mature cathepsin B form by lowering the pH to 3.6 (Kuhelj et al., 1995) with 1 M formic acid. The reaction was terminated by increasing the pH to 6 with addition of 1 M sodium phosphate pH 6.0. The autoactivation was carried out at 4°C for various time intervals. The successful activation/cleavage was monitored by cathepsin B activity assay and 15% SDS-PAGE analysis.

2.6 Activity assay of cathepsin B

Cathepsin B was assayed using chromogenic substrate, Z-Phe-Arg-ρNA. The reaction consisted of 20 mM sodium acetate, pH 5.0, 1 mM EDTA, 5 mM L-Cysteine, 0.9-1.8 µg/mL cathepsin B, and 0.015-0.6 mM Z-Phe-Arg-ρNA in a total volume of 500 µl. The release of the para-nitroaniline (ρNA) chromophore was monitored at 405 nm at 25 °C for 3 min, and the reaction rate was calculated from the slope of the trace showing the increase in absorbance over time.

2.7 Purification of recombinant procathepsin K

The recombinant procathepsin K was purified from IPTG induced *E. coli* cells according to a procedure reported by Hwang and Chung (Hwang and Chung, 2002) with slight modifications. Briefly, the cell pellet (from 250 mL culture) was sonicated in 25 mL of buffer A (50 mM Tris-HCl, pH 8.0, 0.5 M NaCl). The homogenate was centrifuged at 10,000 g for 20 minutes at 4 °C. The pellet (containing inclusion body) was washed in buffer A plus 2 M urea. The content was centrifuged at 10,000 g for 20 minutes at 4 °C. The pellet was re-suspended in 12.5 mL of buffer B (50 mM Tris-HCl, pH 8.0, 0.5 M NaCl, 6 M guanidine hydrochloride, 5 mM imidazole) to solubilize the inclusion body. The suspension was homogenized using a generator sawtooth and stirred at 4 °C for 1 hour. This was centrifuged at 10,000 g for 30 minutes. The supernatant was filtered using a 0.8 µM syringe filter. The filtrate (approximately 12 mL) was loaded onto a Ni column (5 mL HisTrap™ FF from GE Healthcare) that had been equilibrated with buffer B. The column was washed with buffer B until A_{280} reached baseline. The bound protein was eluted with buffer B plus 1 M imidazole and ten 1.3 mL fractions were collected. Protein assay was performed using Bio-Rad protein binding assay. Fractions #4 and #5 which contained the most protein were used in refolding. The purified procathepsin K was observed on 12 % SDS-PAGE.

2.8 Refolding and activation of recombinant procathepsin K

The refolding of procathepsin K was performed by dilution and dialysis (Hwang and Chung, 2002). The eluted protein (2 mL) in guanidine-HCl was diluted in 100 mL of refolding buffer A (5 mM EDTA, 10 mM GSH, 1 mM GSSG, 0.7 M L-Arg, 1 % CHAPS). The dilution was achieved by adding the denatured protein drop by drop at 1 mL/minute into the refolding buffer followed by stirring overnight at 4 °C. The content was dialyzed (MWCO 8,000 – 14,000) against 4 L of 25 mM Tris-HCl, pH 8.0, 0.5 M NaCl overnight at 4 °C. Fresh buffer was changed once during dialysis. Dialyzed protein was clarified by centrifugation and then concentrated using Vivaspin 20 with MWCO 10,000. The volume was reduced to one third of the original volume. The procathepsin K protein was then activated in 0.2 M sodium acetate, pH 4.0, 5 mM DTT, 5 mM EDTA, and porcine pepsin (20 µg/mL). The progress of activation was monitored by activity assay and 15% SDS-PAGE. Activation was terminated by raising the pH to 5.5 with addition of sodium acetate (pH 5.5) to 0.1 M as pepsin is inactive at this pH (Tauber and Kleiner, 1934).

2.9 Cathepsin K activity assay

Cathepsin K was assayed using chromogenic substrate, Z-Phe-Arg-ρNA. The reaction mixture consisted of 100 mM sodium acetate buffer, pH 5.5, 2.5 mM EDTA, 2.5 mM DTT, various amounts of Z-Phe-Arg-ρNA, and appropriate amount of mature cathepsin K in a total volume of 500 μl using a procedure reported by Hwang and Chung (Hwang and Chung, 2002) with minor modifications. The release of the ρNA chromophore was monitored at 405 nm at 25 °C for 3 min, and the reaction rate was calculated from the slope of the trace showing the increase in absorbance over time.

2.10 Inhibition of active cathepsin B and K with flavonoids

Amentoflavones have been shown to inhibit cathepsin B and K (Pan et al., 2005; Zeng et al., 2006). Therefore, several different commercially available flavonoids were utilized, including baicalin, ametoflavone, celastrol, fisetin, kaempferol, luteolin, rutin, limonin, myricetin, and apigenin. Each flavonoid was tested for its activity as cathepsin B or K inhibitors. The enzyme reaction was monitored for three minute intervals by spectrophotometic assay at 405 nm using the chromogenic substrate described above.

3. Results and discussion

3.1 Subcloning of human procathepsin B and K

The full-length human cDNA clone encoding cathepsin B purchased from Open Biosystems was in pBluescriptR vector and the cathepsin K cDNA was in pOTB7 vector. Each of the procathepsin B or K genes was amplified by PCR as described in the Experimental section. The PCR products were analyzed by 1% agarose gel (Figure 1). The reaction worked well at all three annealing temperatures (57.1 oC, 54.4 oC, 51.7 oC for procathepsin B and 54.4 oC, 51.7 oC, 49.3 oC for procathepsin K) as a strong band of about 900 bp was observed in every lane.

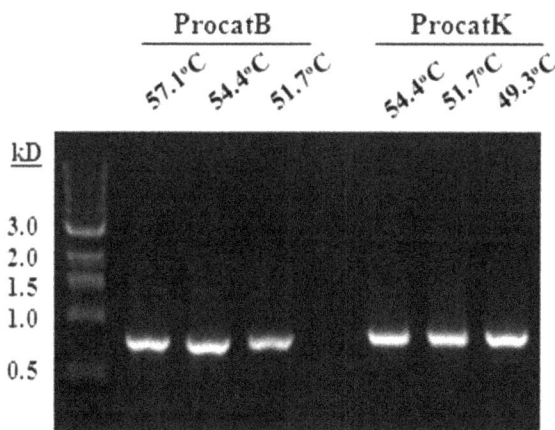

Fig. 1. Agarose gel electrophoresis of PCR products of procathepsin B and K DNA inserts. The plasmid pBluescriptR containing human cathepsin B cDNA was used as template and gene specific primer pair shown in Table 1 to amplify procathepsin B gene. Similarly, the plasmid pOTB7 containing human cathepsin K cDNA was used as template and its gene specific primer pair to amplified procathepsin K gene. PCR cycles were described in the Experimental section.

The XhoI-treated procathepsin B and K fragment from each gel was purified, inserted into the expression vector pET-15b, and transformed into JM109 competent cells. Transformants were treated with ampicillin to induce ampicillin resistance. All transformants were screened for the presence of a chimeric plasmid by the quick screening method as described in the Experimental. Since the cDNA was inserted into the vector at a single restriction site, XhoI, ligation of the insert is possible in either orientation. Only the chimeric plasmid that contains the procathepsin B or K cDNA insert in the correct orientation is useful. Therefore, several insert-containing clones were checked for the orientation of the insert by PCR. Plasmids isolated by boiling method from each selected colony were used as template in the PCR. When the insert is in the correct orientation, a PCR product of about 1050 bp appears with pET 5′ primer and procat B or K reverse primer (Figure 2). This PCR product cannot be formed unless the insert is in the correct orientation. A positive PCR control was included using Forward and Reverse primer pair of procatB or K which yields about a 900 bp PCR product regardless of the insert orientation. Clones 3, 12 and 13 of procathepsin B candidates appeared to contain the DNA insert in the correct orientation, while clone 5 contained a deletion and clones 1 and 14 contained the DNA insert in the wrong orientation (Figure 2 left panel). Clones 6, 7, 11, 12, 14 and 15 of procathepsin K candidates appeared to contain the DNA insert in the correct orientation, while clones 8, 10 and 13 had the DNA insert in the wrong orientation (Figure 2 right panel).

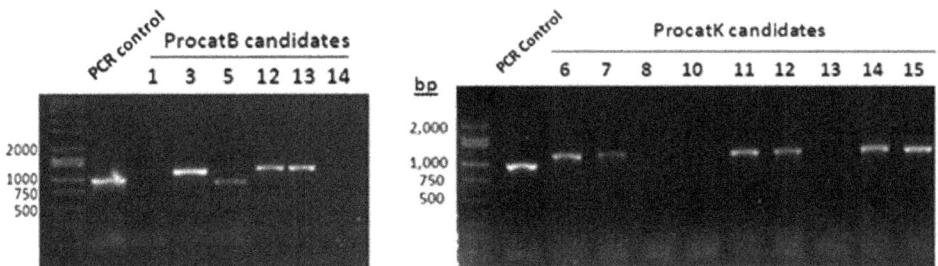

Fig. 2. Orientation check of putative pET-15b-procat B and K clones by PCR. Template used: isolated plasmids from clones #1, 3, 5, 12, 13, and 14 of procatB candidates; isolated plasmids from clones #6, 7, 8, 10, 11, 12, 13, 14, and 15 of procatK candidates. Primers used: for the PCR control, procatB or K forward and reverse primer pairs were used; for procatB candidate clones pET 5′ primer and procatB reverse primer were used; for procatK candidate clones pET 5′ primer and procatK reverse primer were used. Promega's 2x PCR mix was used. PCR cycles were described in the Experimental section.

3.2 Overexpression of recombinant procathepsin B and K

The chimeric plasmids with DNA inserts in the correct orientation and correct reading frame were transformed into BL21(DE3)pLysS for protein expression. Overexpression of the procathepsin B had been achieved according to the SDS-PAGE which showed overexpression of the proenzyme at about 35 kDa (Figure 3. left panel). Expression level of the recombinant procathepsin B after 4 h IPTG induction was greater than 40% of total bacterial protein and the procathepsin B level remained high after overnight induction. Overexpression of the recombinant procathepsin K was clearly visible after 4 h IPTG induction and the expression level increased with increasing induction time up to 22 hours

(Figure 3. Right panel). The recombinant procathepsin K showed an approximate molecular
weight of 38 kD on SDS-PAGE gel.

3.3 Purification of recombinant procathepsin B and K
3.3.1 Purification of recombinant procathepsin B
The recombinant procathepsin B was expressed largely as insoluble inclusion bodies as very
little procathepsin B was found in the soluble protein fraction (Figure 4. lane 3).
Procathepsin B was purified from the inclusion bodies as described in the Experimental
section. The progress of purification was monitored by SDS-PAGE (Figure 4). As shown in
Figure. 4 lane 4, the inclusion bodies contained greater than 90% of procathepsin B. Overall
yield of the procathepsin B enzyme was about 1.5 mg per 100 ml of bacterial culture.

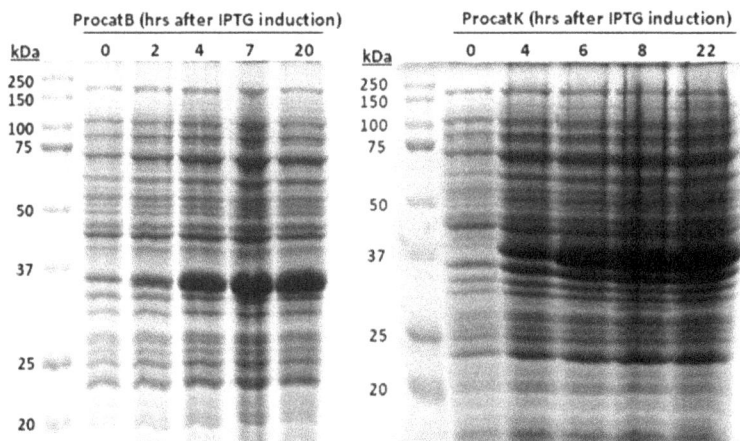

Fig. 3. SDS-PAGE showing induction progress of recombinant procathepsin B and K after
IPTG addition. BL21 (DE3)pLysS cells containing pET-15b-procathepsin B or pET-15b-
procathepsin K were cultured and protein expression induced by IPTG as described in the
Experimental section. An aliquot of each culture was collected at various time intervals after
IPTG addition. The cells were lysed in SDS-PAGE loading buffer by repeated vortexing and
heating. The cell lysate was analyzed by 12% SDS-PAGE.

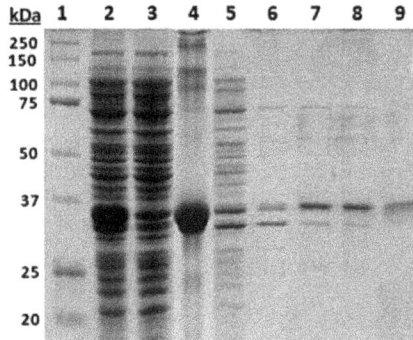

Fig. 4. SDS-PAGE showing the progress of purification of procathepsin B. The gel contained 12% acrylamide. Lane 1: Precision Plus Protein Kaleidoscope Standards; lane 2: Total IPTG induced *E. coli* cell lysate; Lane 3: Soluble proteins in the cell-free extract; Lane 4: Crude inclusion bodies before being subjected to washes; Lanes 5 & 6: First and second washes of the inclusion bodies with 50 mM Tris-HCl pH 8.0 containing 5 mM EDTA and 0.1% Triton X-100; lane 7 & 8: First and second washes of the inclusion bodies with 50 mM Tris-HCl, pH 8.0 containing 5 mM EDTA and 2 M Urea; Lane 9: Purified and refolded procathepsin B.

3.3.2 Purification of recombinant procathepsin K

The procathepsin K was also expressed as insoluble proteins in inclusion bodies as the procathepsin K band was missing from the soluble protein fraction (Figure. 5. lane 5). His-tagged procathepsin K protein was purified from the solubilized inclusion bodies by affinity chromatography on a HisTrap™ FF column to near homogeneity. The protein was then refolded and concentrated (Figure. 5 lane 7). Overall yield of the procathepsin K enzyme was about 1 mg per 100 ml of bacterial culture.

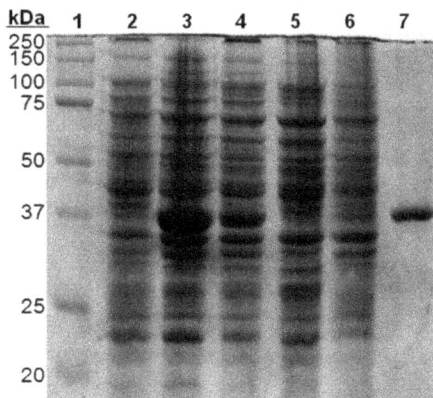

Fig. 5. SDS-PAGE showing overexpression of the human procathepsin K and the purification process. The gel contained 12% acrylamide. Lane 1: Precision Plus Protein Kaleidoscope Standards; lane 2: total lysate of the uninduced cells; lane 3: total lysate of 8 hours IPTG induced cells; lane 4: homogenate of the IPTG induced sample; lane 5: soluble proteins in cell-free extract of the IPTG induced sample; lane 6: Ni column wash; lane 7: purified procathepsin K.

Subcloning and Expression of Functional Human Cathepsin B and K in E. coli: Characterization and
Inhibition by Flavonoids

61

3.4 Activation of the procathepsin B and K proteins

3.4.1 Activation of procathepsin B and functional cathepsin B assay

The recombinant procathepsin B activation was carried out by autoactivation as described in the Experimental section. The success of the activation was demonstrated by the cathepsin B activity assay and SDS-PAGE. As observed in the SDS-PAGE (Figure 6) clear bands are observed and the proteolytic cleavage was evident (lane 2).

The cathepsin B activity was demonstrated by its ability to cleave the Z-Phe-Arg-pNA. These results show a linear relationship between the formation pNA products with increasing cathepsin B levels from 0.024 to 0.19 µM (Figure 7).

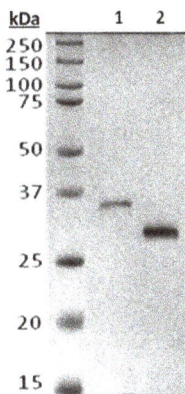

Fig. 6. SDS-PAGE showing autoactivation of procathepsin B. The gel contained 15% acrylamide. Lane 1: purified procathepsin B protein; Lane 2: autoactivated cathepsin B protein.

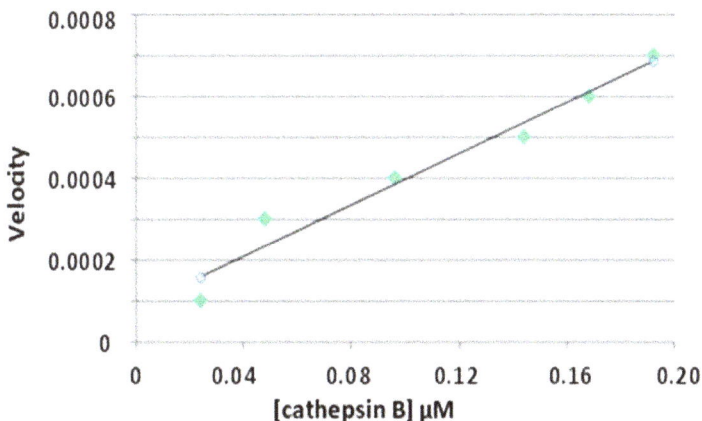

Fig. 7. Cathepsin B demonstrated enzyme concentration-dependent activity. The product formation linearity was verified by the use of increasing amounts of cathepsin B from 0.024 to 0.19 µM. The activity was assayed using a chromogenic substrate, Z-Phe-Arg-pNA, and the rate was monitored by the increase in absorption at 405 nm due to the release of the pNA chromophore.

3.4.2 Activation of procathepsin K and functional cathepsin K assay

The refolded procathepsin K protein was activated to mature cathepsin K with the aid of porcine pepsin. Figure 8 shows successful cleavage of the procathepsin K (~38 kDa) into 2 fragments: mature cathepsin K (~26.5 kDa) and propeptide fragment (~11.5 kDa).

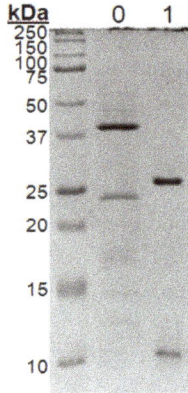

Fig. 8. SDS-PAGE showing activation of procathepsin K with pepsin. The gel contained 15% acrylamide. Lane 0: procathepsin K untreated; Lane 1: procathepsin K treated with pepsin for 1 hr. After 1 hr of pepsin activation, the procathepsin K (~38 kDa) was successfully cleaved into 2 fragments: mature cathepsin K (~26.5 kDa) and propeptide fragment (~11.5 kDa).

The cathepsin K activity is demonstrated by its ability to cleave the Z-Phe-Arg-pNA. The results show a linear relationship between the formation pNA chromophore product with increasing cathepsin K levels from 0.09 to 0.72 μM (Figure 9).

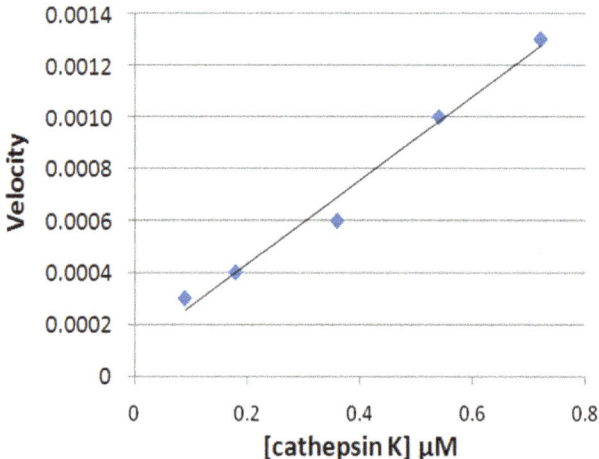

Fig. 9. Cathepsin K demonstrated enzyme concentration-dependent activity. The product formation linearity was verified by the use of increasing amounts of cathepsin K from 0.09 to 0.72 μM. The activity was assayed spectrophotometrically by monitoring the increase in absorption at 405 nm as described in the text.

3.5 Screening of flavonoid inhibitors of cathepsin B and K

Flavonoids, polyphenolic compounds predominantly found in colorful fruits and vegetables, have been regarded of as possessing "antiviral, anti-allergic, antiplatelet, anti-inflammatory, antitumor, and antioxidant (Arora et al., 1998; Rice-Evans and Miller, 1996)" properties. Flavonoids have been known to have anti-cancer effects and could potentially inhibit cathepsin B. In particular, amentoflavones have been shown to inhibit cathepsin B and K (Pan et al., 2005; Zeng et al., 2006). Therefore a search for other flavonoid inhibitors of cathepsins was undertaken. Several commercially available flavonoids were tested (amentoflavone, apigenin, baicalin, celastrol, fisetin, kaempferol, limonin, luteolin, myricetin, and rutin), for their ability to inhibit cathepsin B or K. The inhibition of cathepsin B activity by each of the flavonoid at 40 µM is shown in Figure 10. Ametoflavone, celastrol, and luteolin demonstrated the best flavonoid inhibitors of cathepsin B of those screened. The IC_{50} value for celastrol was 125 µM.

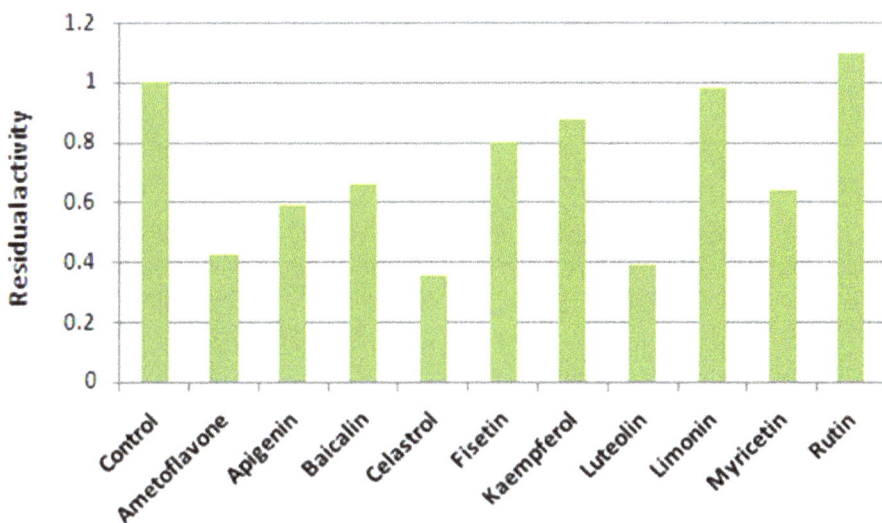

Fig. 10. Inhibition data of cathepsin B by flavonoids. Cathepsin B was incubated with 40 µM flavonoid. The residual activity was determined using chromogenic substrate, Z-Phe-Arg-pNA. The enzymatic reaction was monitored by the increase in absorption at 405 nm due to the release of the pNA chromophore.

Most of these flavonoids were also screened for their inhibition of cathepsin K activity (Figure 11). Results show that apigenin, celastrol, and myricetin are among the strongest inhibitors of cathepsin K of those tested. Their IC_{50} values of the cathepsin K inhibitors were determined to be 8.7 µM, 135.5 µM, and 100.5 µM for apigenin, celastrol, and myricetin, respectively.

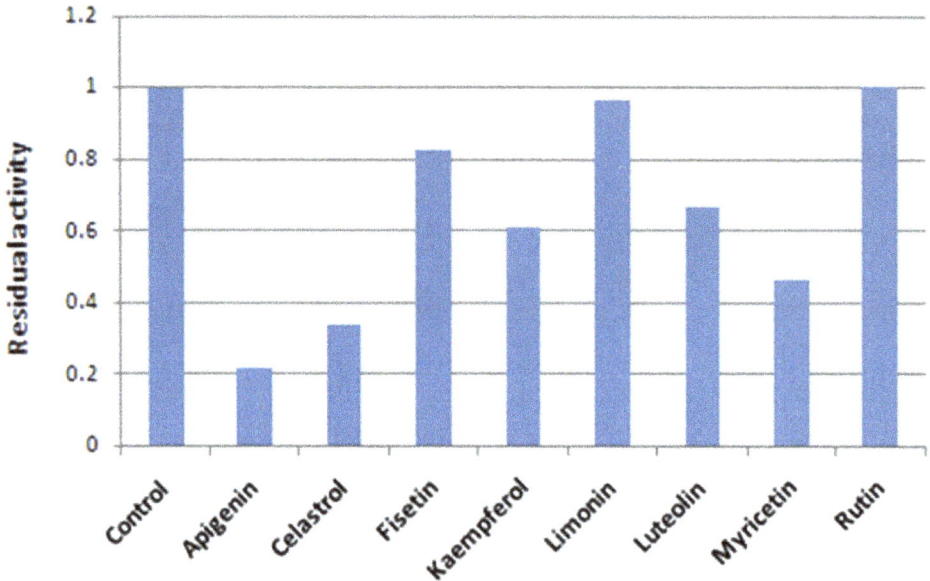

Fig. 11. Inhibition data of cathepsin K by flavonoids. Cathepsin K was incubated with 40 µM flavonoid. The residual enzyme activity was determined using chromogenic substrate, Z-Phe-Arg-pNA, by monitoring the increase in absorption at 405 nm due to the release of the pNA chromophore.

4. Conclusion

In conclusion, the subcloning, expression, and purification of recombinant procathepsins B and K has been accomplished. Each of the proenzyme has been activated and shown to be functional as demonstrated by the enzyme's ability to cleave the chromogenic substrate, Z-Phe-Arg-pNA. The enzymes have been used successfully to screen potent flavonoid inhibitors. The availability of substantial amounts of active cathepsin B and K will greatly facilitate the identification of small molecule inhibitors. The information obtained from studies such as this will greatly facilitate drug discovery programs in the efforts to develop effective therapies for osteoporosis, cancer, and Alzheimer's Disease.

5. Acknowledgment

This work was supported by National Cancer Institute at NIH (Grant No. 3R15CA086933-04 and 3R15CA086933-04A2S1) and the College of Arts and Sciences, Western Illinois University.

Subcloning and Expression of Functional Human Cathepsin B and K in E. coli: Characterization and
Inhibition by Flavonoids

65

6. References

Arora, A., Nair, M.G., and Strasburg, G.M. (1998). Structure-Activity Relationships for Antioxidant Activities of a Series of Flavonoids in a Liposomal System. Free Radical Biology and Medicine 24, 1355-1363.

Asagiri, M., Hirai, T., Kunigami, T., Kamano, S., Gober, H.-J., Okamoto, K., Nishikawa, K., Latz, E., Golenbock, D.T., Aoki, K., et al. (2008). Cathepsin K-Dependent Toll-Like Receptor 9 Signaling Revealed in Experimental Arthritis. Science 319, 624-627.

Bengtsson, E., Nilsson, J., and Jovinge, S. (2008). Cystatin C and cathepsins in cardiovascular disease. Frontiers in Bioscience 13, 5780-5786.

Białas, A., and Kafarski, P. (2009). Proteases as Anti-Cancer Targets - Molecular and Biological Basis for Development of Inhibitor-Like Drugs Against Cancer. Anti-Cancer Agents in Medicinal Chemistry 9, 728-762.

Bien, S., Rimmbach, C., Neumann, H., Niessen, J., Reimer, E., Ritter, C.A., Rosskopf, D., Cinatl, J., Michaelis, M., Schroeder, H.W.S., et al. (2010). Doxorubicin-induced cell death requires cathepsin B in HeLa cells. Biochemical Pharmacology 80, 1466-1477.

Bone, H.G., McClung, M.R., Roux, C., Recker, R.R., Eisman, J.A., Verbruggen, N., Hustad, C.M., DaSilva, C., Santora, A.C., and Ince, B.A. (2010). Odanacatib, a cathepsin-K inhibitor for osteoporosis: A two-year study in postmenopausal women with low bone density. Journal of Bone and Mineral Research 25, 937-947.

Brix, K., Dunkhorst, A., Mayer, K., and Jordans, S. (2008). Cysteine cathepsins: Cellular roadmap to different functions. Biochimie 90, 194-207.

Brubaker, K.D., Vessella, R.L., True, L.D., Thomas, R., and Corey, E. (2003). Cathepsin K mRNA and Protein Expression in Prostate Cancer Progression. Journal of Bone and Mineral Research 18, 222-230.

Bühling, F., Gerber, A., Hackel, C., Kruger, S., Kohnlein, T., Bromme, D., Reinhold, D., Ansorge, S., and Welte, T. (1999). Expression of Cathepsin K in Lung Epithelial Cells. American Journal of Respiratory Cell and Molecular Biology 20, 612-619.

Bühling, F., Röcken, C., Brasch, F., Hartig, R., Yasuda, Y., Saftig, P., Brömme, D., and Welte, T. (2004). Pivotal Role of Cathepsin K in Lung Fibrosis. The American Journal of Pathology 164, 2203-2216.

Cheng, A.-L., Huang, W.-G., Chen, Z.-C., Zhang, P.-F., Li, M.-Y., Li, F., Li, J.-L., Li, C., Yi, H., Peng, F., et al. (2008). Identificating Cathepsin D as a Biomarker for Differentiation and Prognosis of Nasopharyngeal Carcinoma by Laser Capture Microdissection and Proteomic Analysis. Journal of Proteome Research 7, 2415-2426.

Chung, C.T., Niemela, S.L., and Miller, R.H. (1989). One-step preparation of competent Escherichia coli: transformation and storage of bacterial cells in the same solution. Proceedings of the National Academy of Sciences of the United States of America 86, 2172-2175.

Chwieralski, C., Welte, T., and Bühling, F. (2006). Cathepsin-regulated apoptosis. Apoptosis 11, 143-149.

Deal, C. (2009). Future therapeutic targets in osteoporosis. Current Opinion in Rheumatology 21, 380-385

Drake, F.H., Dodds, R.A., James, I.E., Connor, J.R., Debouck, C., Richardson, S., Lee-Rykaczewski, E., Coleman, L., Rieman, D., Barthlow, R., et al. (1996). Cathepsin K, but Not Cathepsins B, L, or S, Is Abundantly Expressed in Human Osteoclasts. Journal of Biological Chemistry *271*, 12511-12516.

Eastell, R., Nagase, S., Ohyama, M., Small, M., Sawyer, J., Boonen, S., Spector, T., Kuwayama, T., and Deacon, S. (2011). Safety and efficacy of the Cathepsin K inhibitor, ONO-5334, in postmenopausal osteoporosis - the OCEAN study. Journal of Bone and Mineral Research, *26*, 1303-1312.

Foekens, J.A., Kos, J., Peters, H.A., Krasovec, M., Look, M.P., Cimerman, N., Meijer-van Gelder, M.E., Henzen-Logmans, S.C., van Putten, W.L., and Klijn, J.G. (1998). Prognostic significance of cathepsins B and L in primary human breast cancer. Journal of Clinical Oncology *16*, 1013-1021.

Frlan, R., and Gobec, S. (2006). Inhibitors of cathepsin B. Current Medicinal Chemistry *13*, 2309-2327.

Gan, L., Ye, S., Chu, A., Anton, K., Yi, S., Vincent, V.A., von Schack, D., Chin, D., Murray, J., Lohr, S., *et al.* (2004). Identification of Cathepsin B as a Mediator of Neuronal Death Induced by Aβ-activated Microglial Cells Using a Functional Genomics Approach. Journal of Biological Chemistry *279*, 5565-5572.

Gowen, M., Lazner, F., Dodds, R., Kapadia, R., Feild, J., Tavaria, M., Bertoncello, I., Drake, F., Zavarselk, S., Tellis, I., *et al.* (1999). Cathepsin K Knockout Mice Develop Osteopetrosis Due to a Deficit in Matrix Degradation but Not Demineralization. Journal of Bone and Mineral Research *14*, 1654-1663.

Guo, J., Bot, I., de Nooijer, R., Hoffman, S.J., Stroup, G.B., Biessen, E.A.L., Benson, G.M., Groot, P.H.E., Van Eck, M., and Van Berkel, T.J.C. (2009). Leucocyte cathepsin K affects atherosclerotic lesion composition and bone mineral density in low-density lipoprotein receptor deficient mice. Cardiovascular Research *81*, 278-285.

Hashimoto, Y., Kakegawa, H., Narita, Y., Hachiya, Y., Hayakawa, T., Kos, J., Turk, V., and Katunuma, N. (2001). Significance of cathepsin B accumulation in synovial fluid of rheumatoid arthritis. Biochemical and Biophysical Research Communications *283*, 334-339.

Hook, V., Kindy, M., and Hook, G. (2007). Cysteine protease inhibitors effectively reduce in vivo levels of brain β-amyloid related to Alzheimer's disease. Biological Chemistry *388*, 247-252.

Hook, V.Y.H. (2006). Unique neuronal functions of cathepsin L and cathepsin B in secretory vesicles: biosynthesis of peptides in neurotransmission and neurodegenerative disease. Biological Chemistry *387*, 1429-1439.

Hook, V.Y.H., Kindy, M., and Hook, G. (2008). Inhibitors of Cathepsin B Improve Memory and Reduce β-Amyloid in Transgenic Alzheimer Disease Mice Expressing the Wild-type, but Not the Swedish Mutant, β-Secretase Site of the Amyloid Precursor Protein. Journal of Biological Chemistry *283*, 7745-7753.

Hook, V.Y.H., Kindy, M., Reinheckel, T., Peters, C., and Hook, G. (2009). Genetic cathepsin B deficiency reduces [beta]-amyloid in transgenic mice expressing human wild-type amyloid precursor protein. Biochemical and Biophysical Research Communications *386*, 284-288.

Huang, J.-K., Tsai, S., Huang, G.H., Sershon, V.C., Alley, A.M., and Wen, L. (2004). Molecular Cloning of Bovine eIF5A and Deoxyhypusine Synthase cDNA. DNA Sequence - The Journal of Sequencing and Mapping 15, 26-32.

Hwang, H.-S., and Chung, H.-S. (2002). Preparation of active recombinant cathepsin K expressed in bacteria as inclusion body. Protein Expression and Purification 25, 541-546.

Illy, C., Quraishi, O., Wang, J., Purisima, E., Vernet, T., and Mort, J.S. (1997). Role of the Occluding Loop in Cathepsin B Activity. Journal of Biological Chemistry 272, 1197-1202.

Im, E., Venkatakrishnan, A., and Kazlauskas, A. (2005). Cathepsin B Regulates the Intrinsic Angiogenic Threshold of Endothelial Cells. Molecular Biology of the Cell 16, 3488-3500.

Joyce, J.A., Baruch, A., Chehade, K., Meyer-Morse, N., Giraudo, E., Tsai, F.-Y., Greenbaum, D.C., Hager, J.H., Bogyo, M., and Hanahan, D. (2004). Cathepsin cysteine proteases are effectors of invasive growth and angiogenesis during multistage tumorigenesis. Cancer Cell 5, 443-453.

Kakegawa, H., Matano, Y., Inubushi, T., and Katunuma, N. (2004). Significant accumulations of cathepsin B and prolylendopeptidase in inflammatory focus of delayed-type hypersensitivity induced by Mycobacterium tuberculosis in mice. Biochemical and Biophysical Research Communications 316, 78-84.

Kim, S.-S. (1992). Expression in Escherichia coli and site-directed mutagenesis of a pumpkin seed protein inhibitor of trypsin and bloof coagulation factor XIIa. In Department of Chemistry (Macomb, Western Illinois University), pp. 140.

Kirschke, H., Barrett, A.J., and Rawlings, N.D. (1995). Proteinases 1: lysosomal cysteine proteinases. Protein Profile 2, 1581-1643.

Kishore Kumar, G.D., Chavarria, G.E., Charlton-Sevcik, A.K., Arispe, W.M., MacDonough, M.T., Strecker, T.E., Chen, S.-E., Siim, B.G., Chaplin, D.J., Trawick, M.L., et al. (2010). Design, synthesis, and biological evaluation of potent thiosemicarbazone based cathepsin L inhibitors. Bioorganic & Medicinal Chemistry Letters 20, 1415-1419.

Kruszewski, W.J., Rzepko, R., Wojtacki, J., Skokowski, J., Kopacz, A., and Drucis, K. (2004). Overexpression of cathepsin B correlates with angiogenesis in colon adenocarcinoma. Neoplasma 51, 38-43.

Kuhelj, R., Dolinar, M., Pungerčar, J., and Turk, V. (1995). The Preparation of Catalytically Active Human Cathepsin B from Its Precursor Expressed in Escherichia coli in the Form of Inclusion Bodies. European Journal of Biochemistry 229, 533-539.

Laurent-Matha, V., Farnoud, M.R., Lucas, A., Rougeot, C., Garcia, M., and Rochefort, H. (1998). Endocytosis of pro-cathepsin D into breast cancer cells is mostly independent of mannose-6-phosphate receptors. Journal of Cell Science 111, 2539-2549.

Le Gall, C., Bonnelye, E., and Clézardin, P. (2008). Cathepsin K inhibitors as treatment of bone metastasis. Current Opinion in Supportive and Palliative Care 2, 218-222

Lecaille, F., Brömme, D., and Lalmanach, G. (2008). Biochemical properties and regulation of cathepsin K activity. Biochimie 90, 208-226.

Lecaille, F., Kaleta, J., and Brömme, D. (2002). Human and Parasitic Papain-Like Cysteine Proteases: Their Role in Physiology and Pathology and Recent Developments in Inhibitor Design. Chemical Reviews 102, 4459-4488.

Ledakis, P., Tester, W.T., Rosenberg, N., Romero-Fischmann, D., Daskal, I., and Lah, T.T. (1996). Cathepsins D, B, and L in malignant human lung tissue. Clinical Cancer Research 2, 561-568.

Leto, G., Tumminello, F.M., Pizzolanti, G., Montalto, G., Soresi, M., and Gebbia, N. (1997). Lysosomal cathepsins B and L and Stefin A blood levels in patients with hepatocellular carcinoma and/or liver cirrhosis: potential clinical implications. Oncology 54, 79-83.

Leto, G., Tumminello, F.M., Pizzolanti, G., Montalto, G., Soresi, M., Ruggeri, I., and Gebbia, N. (1996). Cathepsin D serum mass concentrations in patients with hepatocellular carcinoma and/or liver cirrhosis. European Journal of Clinical Chemistry and Clinical Biochemistry 34, 555-560.

Li, X., Wu, K., Edman, M., Schenke-Layland, K., MacVeigh-Aloni, M., Janga, S.R., Schulz, B., and Hamm-Alvarez, S.F. (2010). Increased Expression of Cathepsins and Obesity-Induced Proinflammatory Cytokines in Lacrimal Glands of Male NOD Mouse. Investigative Ophthalmology & Visual Science 51, 5019-5029.

Lim, I.T., Meroueh, S.O., Lee, M., Heeg, M.J., and Mobashery, S. (2004). Strategy in Inhibition of Cathepsin B, A Target in Tumor Invasion and Metastasis. Journal of the American Chemical Society 126, 10271-10277.

Lindeman, J.H.N., Hanemaaijer, R., Mulder, A., Dijkstra, P.D.S., Szuhai, K., Bromme, D., Verheijen, J.H., and Hogendoorn, P.C.W. (2004). Cathepsin K Is the Principal Protease in Giant Cell Tumor of Bone. The American Journal of Pathology 165, 593-600.

Lutgens, S., Cleutjens, K.B., Daemen, M.J., and Heeneman, S. (2007). Cathepsin cysteine proteases in cardiovascular disease. The FASEB Journal 21, 3029-3041.

Malla, R.R., Gopinath, S., Gondi, C.S., Alapati, K., Dinh, D.H., Gujrati, M., and Rao, J.S. (2011). Cathepsin B and uPAR knockdown inhibits tumor-induced angiogenesis by modulating VEGF expression in glioma. Cancer Gene Therapy 18, 419-434

Manako, J. (2011). New approach for osteoporosis treatment: cathepsin K inhibitor, ONO-5334. Clinical Calcium 21, 64-69.

Masson, O., Prébois, C., Derocq, D., Meulle, A., Dray, C., Daviaud, D., Quilliot, D., Valet, P., Muller, C., and Liaudet-Coopman, E. (2011). Cathepsin-D, a Key Protease in Breast Cancer, Is Up-Regulated in Obese Mouse and Human Adipose Tissue, and Controls Adipogenesis. PLoS ONE 6, e16452.

Matarrese, P., Ascione, B., Ciarlo, L., Vona, R., Leonetti, C., Scarsella, M., Mileo, A.M., Catricalà, C., Paggi, M.G., and Malorni, W. (2010). Cathepsin B inhibition interferes with metastaticpotential of human melanoma: an in vitro andin vivo study. Molecular Cancer 9, 207-220.

Mikosch, P., Kerschan-Schindl, K., Woloszczuk, W., Stettner, H., Kudlacek, S., Kresnik, E., Gallowitsch, H.J., Lind, P., and Pietschmann, P. (2008). High Cathepsin K Levels in Men with Differentiated Thyroid Cancer on Suppressive 1-Thyroxine Therapy. Thyroid 18, 27-33.

Subcloning and Expression of Functional Human Cathepsin B and K in E. coli: Characterization and
Inhibition by Flavonoids

69

Mohamed, M.M., and Sloane, B.F. (2006). Cysteine cathepsins: multifunctional enzymes in cancer. Nature Reviews, Cancer 6, 764-775.

Naour, N., Rouault, C., Fellahi, S., Lavoie, M.-E., Poitou, C., Keophiphath, M., Eberle, D., Shoelson, S., Rizkalla, S., Bastard, J.-P., et al. (2010). Cathepsins in Human Obesity: Changes in Energy Balance Predominantly Affect Cathepsin S in Adipose Tissue and in Circulation. The Journal of Clinical Endocrinology & Metabolism 95, 1861-1868.

Nomura, T., and Katunuma, N. (2005). Involvement of cathepsins in the invasion, metastasis and proliferation of cancer cells. The Journal of Medical Investigation 52, 1-9.

Obermajer, N., Jevnikar, Z., Doljak, B., and Kos, J. (2008). Role of Cysteine Cathepsins in Matrix Degradation and Cell Signalling. Connective Tissue Research 49, 193-196.

Palermo, C., and Joyce, J.A. (2008). Cysteine cathepsin proteases as pharmacological targets in cancer. Trends in Pharmacological Sciences 29, 22-28.

Pan, X., Tan, N., Zeng, G., Zhang, Y., and Jia, R. (2005). Amentoflavone and its derivatives as novel natural inhibitors of human Cathepsin B. Bioorganic & Medicinal Chemistry 13, 5819-5825.

Pérez-Castrillón, J.L., Pinacho, F., De Luis, D., Lopez-Menendez, M., and Laita, A.D. (2010). Odanacatib, a New Drug for the Treatment of Osteoporosis: Review of the Results in Postmenopausal Women. Journal of Osteoporosis 2010, Article ID 401581, 401585 pages.

Podgorski, I. (2009). Future of anticathepsin K drugs: dual therapy for skeletal disease and atherosclerosis? Future Medicinal Chemistry 1, 21-41.

Podgorski, I., Linebaugh, B.E., Koblinski, J.E., Rudy, D.L., Herroon, M.K., Olive, M.B., and Sloane, B.F. (2009). Bone Marrow-Derived Cathepsin K Cleaves SPARC in Bone Metastasis. The American Journal of Pathology 175, 1255-1269.

Podgorski, I., Linebaugh, B.E., and Sloane, B.F. (2007). Cathepsin K in the bone microenvironment: link between obesity and prostate cancer? Biochemical Soceity Transactions 35, 701-703.

Pungerčar, J.R., Caglič, D., Sajid, M., Dolinar, M., Vasiljeva, O., Požgan, U., Turk, D., Bogyo, M., Turk, V., and Turk, B. (2009). Autocatalytic processing of procathepsin B is triggered by proenzyme activity. FEBS Journal 276, 660-668.

Quintanilla-Dieck, M.J., Codriansky, K., Keady, M., Bhawan, J., and Rünger, T.M. (2008). Cathepsin K in Melanoma Invasion. Journal of Investigative Dermatology 128, 2281-2288.

Rawlings, N.D., Morton, F.R., and Barrett, A.J. (2006). MEROPS: the peptidase database. Nucleic Acids Research 34, D270-D272.

Reiser, J., Adair, B., and Reinheckel, T. (2010). Specialized roles for cysteine cathepsins in health and disease. The Journal of Clinical Investigation 120, 3421-3431.

Rice-Evans, C.A., and Miller, N.J. (1996). Antioxidant activities of flavonoids as bioactive components of food. Biochemical Society Transactions 24, 790-795.

Roberts, L., Adjei, P., and Gores, G. (1999). Cathepsins as effector proteases in hepatocyte apoptosis. Cell Biochemistry and Biophysics 30, 71-88.

Roshy, S., Sloane, B.F., and Moin, K. (2003). Pericellular cathepsin B and malignant progression. Cancer and Metastasis Reviews 22, 271-286.

Rozman, J., Stojan, J., Kuhelj, R., Turk, V., and Turk, B. (1999). Autocatalytic processing of recombinant human procathepsin B is a bimolecular process. FEBS Letters 459, 358-362.

Ruettger, A., Schueler, S., Mollenhauer, J.A., and Wiederanders, B. (2008). Cathepsins B, K, and L Are Regulated by a Defined Collagen Type II Peptide via Activation of Classical Protein Kinase C and p38 MAP Kinase in Articular Chondrocytes. Journal of Biological Chemistry 283, 1043-1051.

Rünger, T.M., Quintanilla-Dieck, M.J., and Bhawan, J. (2007). Role of Cathepsin K in the Turnover of the Dermal Extracellular Matrix during Scar Formation. Journal of Investigative Dermatology 127, 293-297.

Saftig, P., Hunziker, E., Wehmeyer, O., Jones, S., Boyde, A., Rommerskirch, W., Moritz, J.r.D., Schu, P., and von Figura, K. (1998). Impaired osteoclastic bone resorption leads to osteopetrosis in cathepsin-K-deficient mice. Proceedings of the National Academy of Sciences 95, 13453-13458.

Salminen-Mankonen, H.J., Morko, J., and Vuorio, E. (2007). Role of Cathepsin K in Normal Joints and in the Development of Arthritis. Current Drug Targets 8, 315-323.

Sambrook, J., and Russell, D.W. (2001). Molecular Cloning: A Laboratory Manual, 3rd edn (Cold Spring Harbor Laboratory).

Schweiger, A., Staib, A., Werle, B., Krasovec, M., Lah, T.T., Ebert, W., Turk, V., and Kos, J. (2000). Cysteine proteinase cathepsin H in tumours and sera of lung cancer patients: relation to prognosis and cigarette smoking. Journal of Cancer 82, 782-788.

Sinha, A., Gleason, D., Staley, N., Wilson, M., Sameni, M., and Sloane, B. (1995). Cathepsin B in angiogenesis of human prostate: an immunohistochemical and immunoelectron microscopic analysis. The Anatomical Record 241, 353-362.

Sinha, A.A., Jamuar, M.P., Wilson, M.J., Rozhin, J., and Sloane, B.F. (2001). Plasma membrane association of cathepsin B in human prostate cancer: Biochemical and immunogold electron microscopic analysis. The Prostate 49, 172-184.

Skoumal, M., Haberhauer, G., Kolarz, G., Hawa, G., Woloszczuk, W., and Klingler, A. (2005). Serum cathepsin K levels of patients with longstanding rheumatoid arthritis: correlation with radiological destruction. Arthritis Research & Therapy 7, R65-R70.

Skoumal, M., Haberhauer, G., Kolarz, G., Hawa, G., Woloszczuk, W., Klingler, A., Varga, F., and Klaushofer, K. (2008). The imbalance between osteoprotegerin and cathepsin K in the serum of patients with longstanding rheumatoid arthritis. Rheumatology International 28, 637-641.

Steffan, J.J., Williams, B.C., Welbourne, T., and Cardelli, J.A. (2010). HGF-induced invasion by prostate tumor cells requires anterograde lysosome trafficking and activity of Na+-H+ exchangers. Journal of Cell Science 123, 1151-1159.

Stoch, S.A., and Wagner, J.A. (2007). Cathepsin K Inhibitors: A Novel Target for Osteoporosis Therapy. Clinical Pharmacology & Therapeutics 83, 172-176.

Subcloning and Expression of Functional Human Cathepsin B and K in E. coli: Characterization and
Inhibition by Flavonoids

71

Stoka, V., Turk, B., and Turk, V. (2005). Lysosomal cysteine proteases: structural features and their role in apoptosis. IUBMB Life 57, 347-353.

Szpaderska, A.M., and Frankfater, A. (2001). An Intracellular Form of Cathepsin B Contributes to Invasiveness in Cancer. Cancer Research 61, 3493-3500.

Tauber, H., and Kleiner, I.S. (1934). The Inactivation of Pepsin, Trypsin, and Salivary Amylase by Proteases. Journal of Biological Chemistry 105, 411-414.

Tedelind, S., Poliakova, K., Valeta, A., Hunegnaw, R., Yemanaberhan, E.L., Heldin, N.-E., Kurebayashi, J., Weber, E., Kopitar-Jerala, N., Turk, B., et al. (2010). Nuclear cysteine cathepsin variants in thyroid carcinoma cells. Biological Chemistry 391, 923-935.

Tomita, A., Kasaoka, T., Inui, T., Toyoshima, M., Nishiyama, H., Saiki, H., Iguchi, H., and Nakajima, M. (2008). Human breast adenocarcinoma (MDA-231) and human lung squamous cell carcinoma (Hara) do not have the ability to cause bone resorption by themselves during the establishment of bone metastasis. Clinical and Experimental Metastasis 25, 437-444.

Turk, B. (2006). Targeting proteases: successes, failures and future prospects. Nature Reviews Drug Discovery 5, 785-799.

Turk, V., Turk, B., Guncar, G., Turk, D., and Kos, J. (2002). Lysosomal cathepsins: structure, role in antigen processing and presentation, and cancer. Advances in Enzyme Regulation 42, 285-303.

Turk, V., Turk, B., and Turk, D. (2001). Lysosomal cysteine proteases: facts and opportunities. The EMBO Journal 20, 4629-4633.

Urbanelli, L., Emiliani, C., Massini, C., Persichetti, E., Orlacchio, A., Pelicci, G., Sorbi, S., Hasilik, A., Bernardi, G., and Orlacchio, A. (2008). Cathepsin D expression is decreased in Alzheimer's disease fibroblasts. Neurobiology of Aging 29, 12-22.

Valta, M.P., Tuomela, J., Bjartell, A., Valve, E., Väänänen, H.K., and Härkönen, P. (2008). FGF-8 is involved in bone metastasis of prostate cancer. International Journal of Cancer 123, 22-31.

Vashishta, A., Ohri, S.S., Proctor, M., Fusek, M., and Vetvicka, V. (2007). Ribozyme-targeting procathepsin D and its effect on invasion and growth of breast cancer cells: An implication in breast cancer therapy. International Journal of Oncology 30, 1223-1230.

Vasiljeva, O., Reinheckel, T., Peters, C., Turk, D., Turk, V., and Turk, B. (2007). Emerging Roles of Cysteine Cathepsins in Disease and their Potential as Drug Targets. Current Pharmaceutical Design 13, 387-403.

Victor, B.C., and Sloane, B.F. (2007). Cysteine cathepsin non-inhibitory binding partners: modulating intracellular trafficking and function. Biological Chemistry 388, 1131-1140.

Xu, X., Yuan, G., Liu, W., Zhang, Y., and Chen, W. (2009). Expression of Cathepsin L in Nasopharyngeal Carcinoma and its Clinical Significance. Experimental Oncology 31, 102-105.

Yan, S., Sameni, M., and Sloane, B.F. (1998). Cathepsin B and human tumor progression. Biological Chemistry 379, 113-123.

Yang, M., Sun, J., Zhang, T., Liu, J., Zhang, J., Shi, M.A., Darakhshan, F., Guerre-Millo, M., Clement, K., Gelb, B.D., *et al.* (2008). Deficiency and Inhibition of Cathepsin K Reduce Body Weight Gain and Increase Glucose Metabolism in Mice. Arteriosclerosis, Thrombosis, and Vascular Biology *28*, 2202-2208.

Yasuda, Y., Kaleta, J., and Brömme, D. (2005). The role of cathepsins in osteoporosis and arthritis: Rationale for the design of new therapeutics. Advanced Drug Delivery Reviews *57*, 973-993.

Zeng, G.Z., Pan, X.L., Tan, N.H., Xiong, J., and Zhang, Y.M. (2006). Natural biflavones as novel inhibitors of cathepsin B and K. European Journal of Medicinal Chemistry *41*, 1247-1252.

Zhang, T., Maekawa, Y., Hanba, J., Dainichi, T., Nashed, B.F., Hisaeda, H., Sakai, T., Asao, T., Himeno, K., Good, R.A., *et al.* (2000). Lysosomal cathepsin B plays an important role in antigen processing, while cathepsin D is involved in degradation of the invariant chain in ovalbumin-immunized mice. Immunology *100*, 13-20.

5

Effects of Two Novel Peptides from Skin of *Lithobates Catesbeianus* on Tumor Cell Morphology and Proliferation

Rui-Li ZHAO[1,2], Jun-You HAN[3],
Wen-Yu HAN[2], Hong-Xuan HE[4] and Ji-Fei MA[1]

[1]*Department of Animal Science, Tianjin Agricultural University, Tianjin*
[2]*College of Animal Science and Veterinary Medicine, Jilin University, Changchun*
[3]*College of Plant Science, Jilin University, Changchun*
[4]*Research Center for Wildlife Borne Diseases, Institute of Zoology,*
Chinese Academy of Sciences,
China

1. Introduction

The number of resistant bacterial strains has continued to increase due to the overuse of antibiotics, necessitating continued development of novel antimicrobial agents. These resistant strains have made the care of infected patients with cancer more difficult [1, 2]. Indeed, infection control in patients with malignant tumors is one of the most important aspects of hospital management. Amphibian skin secretions contain a large number of antimicrobial peptides, including aurein 1.2, citropin 1.1, gaegurins, magainin, and magainin analogues, which have shown selective toxicity against human cancer cells [3, 4, 5]. There are now 105 antitumor peptides listed in the antimicrobial peptide database (APD, http://aps.unmc.edu/AP), 11 of which were isolated by Roaek [6] from the Australian bull frogs. Most of the remaining peptides have been isolated from the family magainins [7] or from Xenopus laevis.

Antimicrobial peptides rely on amphiphilic α-helix structures to associate with prokaryotic lipid bilayers, often forming transmembrane ion channels that cause cell leakage, mitochondrial failure, or the activation of apoptotic signaling cascades. Antimicrobial peptides have little effect on healthy eukaryotic cells, however, and tend to act selectively on prokaryotic cells and damaged eukaryotic cells. Thus, they have the capacity to kill bacteria, fungi, and viruses. In addition, antimicrobial peptides can damage cancer cells without damaging normal cells, possibly due to differences in the composition of the cell membrane and cytoskeleton. Whether the antimicrobial peptides kill bacteria and tumor cells through the same molecular mechanisms is uncertain [8]. In recent years, antimicrobial peptides have becoming an intense focus of research in zoology, physiology, and pharmacology [9, 10], as these peptide are high efficiency, low toxicity, broad-spectrum antimicrobial agents that may complement traditional antibiotics.

We isolated two novel polypeptides named temporin-La [11] and palustrin-Ca with strong antimicrobial and antitumor activities by screening the skin cDNA library of *Lithobates catesbeianus*. In order to explore the antitumor mechanism of temporin-La and palustrin-Ca, we investigated the influence of these two polypeptides on HeLa ultrastructure by transmission electron microscopy. These two antimicrobial peptides are potential candidates for clinical trials as antibiotics and anticancer drugs.

2. Materials and methods

2.1 Materials

A *Lithobates catesbeianus* skin cDNA library was constructed in the Department of Microorganisms and Immunity, College of Animal Science and Veterinary Medicine, Jilin University. Tumor cells used in the anticancer assays were the gastric tumor cell lines SGC7901 and MGC-803, and the liver tumor cell lines SMMC7721 and BEL-7402. The tumor cells were obtained from the First Hospital of Jilin University. The Gram-negative bacteria *Escherichia coli* (ATCC 25922), *Klebsislla pneumoniae* (ATCC700603), *Salmonella* (ATCC20020), and *Pseudomonas aeruginosa* (ATCC227853), the Gram-positive bacterium *Staphylococcus aureus* (ATCC25923), *Streptococcus suis 2* (CVCC606), *Listeria* (ATCC 54004), and *Bacillus subtilis* (ADB403), and the fungus *Candida albicans* (ATCC10231) were all obtained from China Institute of Veterinary drug Control.

2.2 Peptide synthesis

Temporin-La and palustrin-Ca were synthesized by GL Biochem (Shanghai, China) Ltd. according to the deduced amino acid sequences of mature peptides. High performance liquid chromatography and ESI-MS mass spectrometry confirmed that the purity of the synthetic peptides was higher than 95%. The synthetic peptides were dissolved in sterile water and used to evaluate antimicrobial and antitumor activities.

2.3 Antimicrobial assays

Microorganisms were incubated in LB broth at 37 °C to log-phase and then diluted in fresh LB broth to approximately 2×10^6 CFU/mL. To test the dose-response of the peptides, 50 µL samples of diluted microorganisms were mixed with 1:1 serial dilutions of the synthetic peptides in fresh LB in 96-well microtiter plates. Plates were incubated at 37 °C in a moist atmosphere for 16-18 h. After incubation, the absorbance of the plates at 600 nm was measured and recorded. The minimal inhibitory concentration (MIC) of the peptides was defined as the minimal concentrations at which no visible growth of the microorganisms was detected [12].

2.4 Hemolytic assays

The hemolytic effect of the toxins was tested using rabbit erythrocytes in liquid medium according to a method previously described [13]. Briefly, serially diluted peptides were incubated with washed rabbit erythrocytes at 37 °C for 30 min. Following incubation, the cells were centrifuged and the 595 nm absorbance of the supernatant was measured. Absorbance was compared to that of supernatant from cells completely lysed with 1% Triton X-100.

2.5 Transmission electron microscopy

Transmission electron microscopy was performed to study the potential antitumor mechanisms of temporin-La and palustrin-Ca on HeLa cells. Briefly, the tumor cells were digested by trypsin and incubated in 50 mL cell culture flasks with toxin for 2 h. Cells were then harvested and centrifuged. The cell pellets were embedded, stained, dehydrated, and embedded in white resin. The embedded cells were then sliced, stained, and imaged using a JEM−1200EXll microscope under standard operating conditions.

3. Results

3.1 cDNA cloning of temporin-La and palustrin-Ca

Two bacterial clones encoding the two novel antimicrobial peptides were identified separately. The first clone encoded a mature peptide named temporin-La and the second encoded palustrin-Ca (Figure 1). The cDNAs were deposited in GenBank under accession numbers FJ430082 and FJ830669.

Temporin-La

```
atgttcccccttgaagaaatccctgttactccttttttttccttgggaccatcaacttatct   60
 M  F  P  L  K  K  S  L  L  L  L  F  F  L  G  T  I  N  L  S
Ttttgtgaggaagagagagatgtcgatcaagatgaaagaagagatgatccaggtgaaagg    120
 F  C  E  E  E  R  D  V  D  Q  D  E  R  R  D  D  P  G  E  R
Aatgttcaagtggaaaaacgattgttacgacatgttgtaaagattctcgaaaaatatttg    180
 N  V  Q  V  E  K  R  L  L  R  H  V  V  K  I  L  E  K  Y  L
Ggaaaataaccagaaatgttgaaactttgaaaatggaattggaaatcatttgatgtggaa    240
 G  K  *
Tattatttggctaaatgctcaacagatgttttataaaaataaataaatatgttgcaaaaa    300
Aaaaaaaaaaaaaaaaaaaaaaaa                                        324
```

Palustrin-Ca

```
atgttcaccatgaagaaatccctgttgctccttttctttcttgggaccatctccttatct   60
 M  F  T  M  K  K  S  L  L  L  L  F  F  L  G  T  I  S  L  S
Ctctgtgagcaagagagagatgccgatggagatgaaggggaagttgaagaagtaaaaaga    120
 L  C  E  Q  E  R  D  A  D  G  D  E  G  E  V  E  E  V  K  R
Ggtttcctggatatcatcaaggatacggggaaggaatttgctgtgaaaattttgaataat    180
 G  F  L  D  I  I  K  D  T  G  K  E  F  A  V  K  I  L  N  N
Ttaaaatgtaaattggctggaggatgtccaccctgaatcagaagtcatctcatgtggaat    240
 L  K  C  K  L  A  G  G  C  P  P  *
Atcacttagctaaatctgtaatgtcttattaaaaaataaaaatatcacatgcaaaaaaaa    300
Aaaaaaaaaaaaaaaaaaaaaaaa                                        322
```

Fig. 1. Nucleotide and deduced amino acid sequences of temporin-La and palustrin-Ca precursors. Sequences of the mature peptides are underlined. The asterisks represent the termination codon.

The deduced amino acid sequences, net charges, molecular masses, and isoelectric points of temporin-La and palustrin-Ca are shown in Table 1. Temporin-La consists of 13 amino acid residues with a net charge of +3 at pH 7.0. The C-terminus of temporin-La is amidated. A NCBI-BLAST search revealed that the nucleotide sequences of the precursors of temporin-La exhibit 40% sequence identity with other temporin precursors from *Rana* species. Palustrin-Ca consists of 31 amino acid residues with a net charge of +2 at pH 7.0. The seven peptide "Rana box" was formed by the Cys^{23} and Cys^{29}. Palustrin-Ca demonstrated approximately 50% homology with brevinin-2TD from the European frog and with ranatuerin-2Ca, ranatuerin-3, and ranatuerin 2Cb from bullfrog.

		Net charge (pH7.0)	Mr (Da)	pI
temporin-La	LLRHVVKILEKYLamide	+3	1623.08	9.70
palustrin-Ca	GFLDIIKDTGKEFAVKILNN LKCKLAGGCPP	+2	3303.97	8.79

Table 1. Primary structure, net charge, molecular masses, and isoelectric points of temporin-La and palustrin-Ca. The isoelectric point values were calculated using the ExPASy MW/pI tool (http://www.expasy.ch/tools/pi_tool.html). The molecular masses were determined by ESI-MS mass spectrometry.

3.2 Antimicrobial and hemolytic activities of temporin-La and palustrin-Ca

Minimum inhibitory concentrations (MICs) were determined to characterize antimicrobial activity (Table 2). Temporin-La and palustrin-Ca were both more effective against Gram-positive than Gram-negative bacteria. Temporin-La exhibited stronger inhibition against *S.aureus* and *S.suis* than palustrin-Ca. The MIC of temporin-La against *Staphylococcus aureus* was 7.8 µg/mL and 15.6 µg/mL against *Streptococcus suis*, while MICs for palustrin-Ca against these microbes were 7.8 µg/mL and 31.25 µg/mL. The antifungal activity [14, 15] of temporin-La, as assed by the MIC was 31.25 µg/mL. In contrast, palustrin-Ca was not nearly as potent as an antifungal agent with an MIC > 100 µg/mL. Temporin-La showed no measurable hemolytic activity against rabbit erythrocytes at 250 µg/ml. Similarly, palustrin-Ca showed little hemolytic activity (0.25% at 250 µg/ml).

Bacterium	Antimicrobial activity (MIC µg/ml)	
	temporin-La	palustrin-Ca
Gram-positive bacteria		
Listeria ATCC 54004	ND	30
S. aureus ATCC 25923	2.5	7.8
S. suis 2 CVCC 606	15.6	31.25
Bacillus subtilis ADB403	>100	30
Candida albicans ATCC10231	31.25	>100
Gram-negative bacteria		
Salmonella ATCC 20020	15.6	>100
E. coli ATCC 25922	>100	ND
K. pneumoniae ATCC 700603	>100	60
P. aeruginosa ATCC 227853	60	30

Table 2. Antimicrobial activity of temporin-La and palustrin-Ca. The data represent mean values of three independent experiments. ND: not determined.

3.3 Temporin-La and palustrin-Ca possess strong anticancer activity

The antitumor activity of temporin-La and palustrin-Ca was evaluated by MTT assay (Table 3). The IC_{50} of temporin-La against SMMC7721 cells was 1.384 µg/mL, and palustrin-Ca was also a potent antitumor agent, with an IC_{50} of 0.951 µg/mL against SGC7901 cells.

Tumor cell	Antitumor activity IC_{50} (µg/mL)	
	temporin-La	palustrin-Ca
SGC7901	2.755	0.951
MGC803	3.937	1.572
SMMC7721	1.384	1.077
BEL7402	2.670	1.375
HeLa	4.685	1.202

Table 3. Antitumor activity of temporin-La and palustrin-Ca.

3.4 Electron microscopy

Transmission electron microscopy was performed to evaluate the possible cytotoxic mechanisms of temporin-La and palustrin-Ca (Fig. 2). After 24 h treatment with temporin-La or palustrin-Ca at 10×MIC, HeLa cells exhibited significant damage, with breached membrane integrity, cytoplasmic leakage, mitochondria vacuolization, cristae loss, and blurred boundaries.

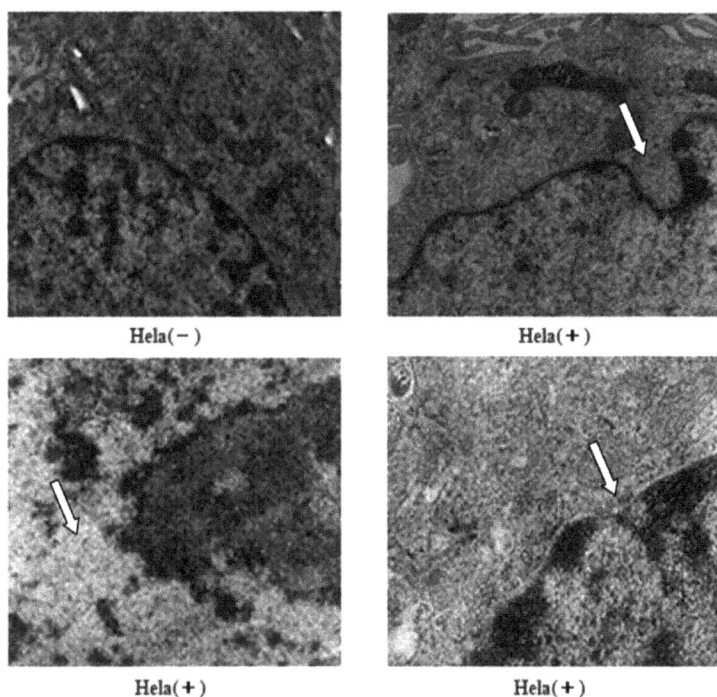

Hela(−)   Hela(+)

Hela(+)   Hela(+)

Fig. 2. The cytotoxic activity of temporin-La and palustrin-Ca against Hela cell *in vitro* (×20 000) Hela(-): untreated; Hela(+): temporin-La and palustrin-Ca treated. Peptides were used at 10 times the MIC. Arrows indicate the morphological changes.

4. Discussion

Antimicrobial peptides demonstrate selective toxicity against tumor cells [16, 17], but the cytotoxic mechanisms remain poorly understood. Recent studies suggest that the cytotoxic mechanisms of antimicrobial peptides against bacteria are distinct for these antitumor activities. The sensitivity of tumor cells to antimicrobial peptides is far higher than the normal cells, thus allowing selective antitumor effects. This selective cytotoxicity may be due to the higher acidic phospholipid content of tumor cell membranes and the reduced integrity of the cytoskeleton system. Tumor cells can be inhibited and killed by direct contact with antimicrobial peptides, suggesting that the effect on cancer cells may be due to insertion of antimicrobial peptides into the plasma membrane. Membrane insertion causes bilayer melting and membrane puncture, leading to leakage of intracellular contents, mitochondria vacuolization, cristae loss, blurred boundaries, chromosomal DNA breakage, inhibition of chromosomal DNA synthesis, and other detrimental events associated with membrane damage. Another possible mechanism of antimicrobial peptides is the induction of apoptosis. Mai *et al* [18, 19] showed that local injection of the peptide DP1 into mice quickly induced tumor cell apoptosis. On the other hand, the protective immune response can be enhanced by antimicrobial peptides, as shown by the resistance to the invaded cancer cells conferred by humoral immunity. Temporin-La and palustrin-Ca demonstrated significantly different antimicrobial and antitumor efficacies against the tested battery of bacteria, fungi, and tumor cell lines, suggesting that the cytotoxic mechanisms may be at least partially distinct for different targets.

Temporin-La had higher antibacterial activity against *Staphylococcus aureus, Streptococcus suis 2* and *Salmonella* than palustrin-Ca, with an MIC for *Staphylococcus aureus* of 2.5 µg/mL. The IC_{50} against the gastric cancer cell line SMMC7721 was only 1.384 µg/mL, and the IC_{50} values measured for all other tumor cells tested were lower than 5 µg/mL. This is far below the concentration required for mammalian red blood cell hemolysis, indicating that temporin-La can inhibit the growth of some Gram-positive bacteria and cancer cells while not damaging the membranes of other healthy mammalian cells. Palustrin-Ca had stronger antibacterial activity against *Staphylococcus aureus, Listeria,* and *Pseudomonas aeruginosa* than temporin-La. The MIC against *Staphylococcus aureus* was 7.8 µg/mL, and palustrin-Ca also exhibited promising inhibition against all the tumor cells studied, with IC_{50} values lower than 2 µg/mL. The IC_{50} against the liver cancer cell line SGC7901 was only 0.951 µg/mL, underscore the potential antitumor efficacy of palustrin-Ca. Temporin-La and palustrin-Ca are potential candidate antimicrobial and anticancer compounds that warrant further study.

5. Acknowledgements

This work was supported by the Chinese National Natural Science Foundation (30571416)

6. References

[1] Biedler JL, Riehm H .Cellular resistance to actinomycind in Chinese hamster cells in vitro: cross resistance, radioautographic and cytogenetic studies[J]. Cancer Res 1970; 30:1174–84.

[2] Sunkyu Kima, Sukwon S. Kima, Yung-Jue Bangb. In vitro activities of native and designed peptide antibiotics against drug sensitive and resistant tumor cell lines[J]. Peptides, 2003, 24: 945-953.

[3] J. Doyle, C.S. Brinkworth, K.L. Wegener, nNOS inhibition,antimicrobial and anticancer activity of the amphibian skin peptide,citropin 1.1 and synthetic modifications. The solution structure of a modified citropin 1.1[J], Eur. J. Biochem. 2003, 27:1141–1153.

[4] H.-S. Won, M.-D. Seo, S.-J. Jung.Structural determinants for the membrane interaction of novel bioactive undecapeptides derived from gaegurin 5[J], J. Med. Chem. 2006, 49: 4886–4895.

[5] L. Cruz-Chamoro, M.A. Puertollano, E. Puertollano.In vitro biological activities of magainin 1 alone or in combination with nisin[J], Peptides .2006,27 :1201–1209.

[6] Roaek T, Wegener KL , Bowie J H. The antibiotic and anticancer active aurein peptides from the Australian bell frogs Litoria aurea and Litoria raniformis [J]. European Journal of Biochemistry , 2000 , 267 : 5 330～5 341.

[7] M. Zasloff, Magainins, a class of antimicrobial peptides from Xenopus skin: isolation, characterization of two active forms, and partial DNA sequence of a precursor[J], Proc. Natl. Acad. Sci. U. S. A. 1987,84 :5449–5453.

[8] David W. Hoskin , Ayyalusamy Ramamoorthy. Studies on anticancer activities of antimicrobial peptides[J]. Biochimica et Biophysica Acta, 2008,1778 :357–375.

[9] Rozek T, Wegener KL, Bowie JH, et al. The antibiotic and anticancer active aurein peptides from the Australian Bell Frogs Litoria aurea and Litoria raniformis[J]. Eur J Biochem,2000, 267: 5330−5331.

[10] Xu ND, Zhong L, Zhang W, et al. The relationship between structure and antibacterial activity of cationic antimicrobial peptides [J]. Chin J Public Health, 2005, 21(9): 1143.

[11] Zhao RL, Han JY, Han WY. Molecular Cloning of Two Novel Temporins From Lithobates catesbeianus and Studying of Their Antimicrobial Mechanisms[J].Progress in Biochemistry and Biophysics .2009, 36(8): 1064～1070.

[12] G ill S C, von Hippel PH. Calculation of protein extinction coefficients from amino acid sequence data[J]. Anal Biochem. 1989, 182(2): 319-26.

[13] Bignami GS. A rapid and sensitive hemolysis neutralization assay for palytoxin. Toxicant 1993; 31: 817–20.

[14] Yashuhara T, Nakajima T, Erspamer V, Falconieri-Erspamer G, Tukamoto Y, Mori M. Isolation and sequential analysis of peptides in Rana Erythraea skin. In: Y. Kiso, Editor, Peptide Chemistry 1985, Protein Research Foundation, Osaka (1986), pp. 363-368.

[15] Simmaco M, Mignogna G, Canofeni S, Miele R, Mangoni ML, Barra D. Temporins, antimicrobial peptides from the European red frog Rana temporaria. Eur J Biochem 1996; 242:788-92.

[16] Matsuzaki K, Harada M, Funakoshi S, Fuji N, Miyajima K. Physicochemical determinants for the interactions of magainins 1 and 2 with acidic lipid bilayers[J]. Biochem. Biophys. Acta. 1991, 1063: 162-170.

[17] Frank Schweizer. Cationic amphiphilic peptides with cancer-selective toxicity[J].European Journal of Pharmacology, 2009,625:190-194.

[18] Mai JC, *et al*. A proapoptotic peptide for the treatment of solid tumors [J]. Cancer Res, 2001, 61: 7709-7712

[19] Mi Z, et al. Identification of a synovial fibroblast-specific protein transduction domain for delivery of apoptotic agents to hyperplastic synovium [J]. Mol Ther, 2003, 8(2): 295-305.

Part 3

Immunology/Hematology

6

Molecular Cloning of Immunoglobulin Heavy Chain Gene Translocations by Long Distance Inverse PCR

Takashi Sonoki
Hematology/Oncology
Wakayama Medical University
Japan

1. Introduction

B-cell functions as a key player in the humoral immunity producing immunoglobulin protein in mammalian. To produce wide-ranged and antigen-specific antibody, the B-cell undergoes genetic rearrangement of immunoglobulin genes during its maturation process. The genetic rearrangements in immunoglobulin genes require unstable steps for genome; breakage and re-ligation of the double strand DNA. In the unstable steps, misconduct can be occurred; leading chromosome translocations involving immunoglobulin genes.

Chromosome translocations involving the immunoglobulin heavy chain (IGH) locus on chromosome 14q32 are found in various B-lymphoid malignancies, including B-cell precursor lymphoblastic leukemia (BCP-ALL), B-cell non-Hodgkin's lymphoma (B-NHL), and myeloma. Chromosome translocations involving IGH locus are often associated to disease entities among the B-lymphoid malignancies; thus, identification of certain translocation is clinically important information for definitive diagnosis. Moreover, molecular cloning of the IGH translocation breakpoint allows the identification of genes of physiologically or pathologically importance in B-cells (Willis and Dyer, 2000; Küppers and Dalla-Favera, 2001; Siebert et. al., 2001). The consequence of IGH translocations is the deregulated expression of the target gene controlled by potent B-cell-specific transcriptional enhancers within the IGH locus, resulting from physically close apposition with the IGH locus. The vast majority of target genes of IGH translocations play a fundamental role in B-cell biology, such as cell growth, differentiation, apoptosis and signal transduction (Wills and Dyer, 2000). Thus, deregulated expression of those genes alters B-cell fate and may initiate malignant transformation of the affected B-cell. To date, many IGH translocation breakpoints have been molecularly cloned and the target genes have been identified; however, several recurrent breakpoints remain to be cloned (Heim and Miltelman, 1995). Molecular cloning of IGH translocation breakpoints would reveal the involvement of either genes of unknown biological functions or the unsuspected oncogenic potential of known genes. Moreover, some target genes of IGH translocation have been found to be deregulated by gene amplification or unknown genetic mechanisms rather than IGH translocation, thereby contributing to disease development or progression (Dyer et. al, 2010).

2. Structure of IGH and its genetic modifications during B-cell differentiation

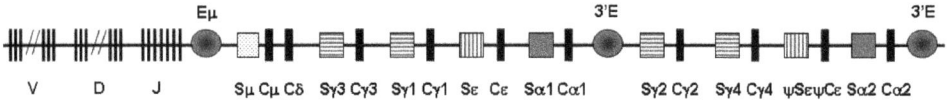

The human IGH locus consists of 51 functional variable (V) segments, 27 diverse (D) segments, six joining (J) segments and nine constant (C) segments in noncontiguous fashion starting from the telomeric end of 14q32 and spans 1.4 megabases (Mb) (Honjo and Alt, 1995) (Fig. 1). Each C segment, except Cδ, is preceded by repetitive DNA sequences named switch (S) segments, which play a role in class switch recombination, as discussed below. During the B-cell maturation process, the IGH gene undergoes three major genetic modifications: VDJ recombination, somatic hypermutation, and class switch recombination. Generally, the former occurs in the bone marrow, whereas the latter two occur in the germinal center of the lymph nodes. As a consequence of these genetic modifications, the IGH gene can produce part of the immunoglobulin protein (antibody), which can bind to a specific antigen with distinct effecter functions.

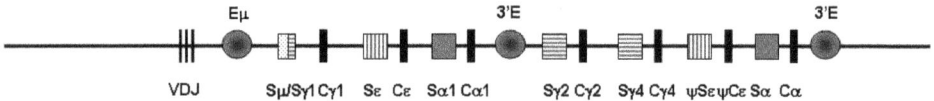

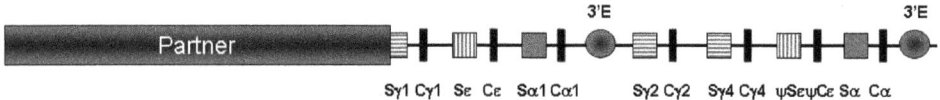

Fig. 1. A, Schematic diagram of human IGH structure. The coding exons lie on 14q32 as non-contiguous fashion. There are three potent enhancers within the IGH locus (Eμ and two 3'E). B, Schematic diagram of functionally rearranged IGH allele. In this schema, VDJ recombination and switch recombination of IgM to IgG1 are shown. Note Sμ and Sγ1 are combined. C, Schematic diagram of IGH translocation involving Joining region. The target gene on the partner chromosome is deregulated by Eμ enhancer. D, Schematic diagram of IGH translocation involving Switch region. The target gene on the partner chromosome is deregulated by 3'E enhancer.

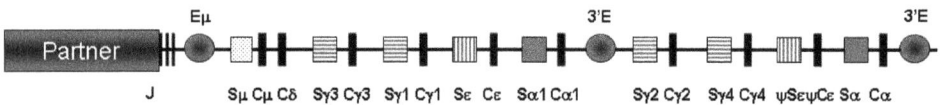

Each B-progenitor cell forms a unique VDJ complex by selecting one V, one D, and one J segment randomly, and this process is called VDJ recombination. The VDJ complex therefore encodes part of the variable domain of immunoglobulin heavy chain protein. After encountering an antigen in the germinal center of a lymph node, VDJ complex is further introduced by nucleotide substitution, deletion or duplication, a process called somatic hypermutation, resulting in increased affinity with the respective antigen. In the germinal center, some B-cells exchange constant segments from Cμ to others by the class switch recombination process, which results in the exchange effecter domain of immunoglobulin protein (Fc-domain) retaining the antigen specific binding domain. This process involves the recombination of S regions preceding the Cμ segment and other C segment, and the intervening DNA segment is looped out. When class switching from IgM to IgG1 occurs, for example, Sμ and Sγ1 are recombined with deletion of the intervening DNA, thereafter, Cγ1 segment positions at 3′ flanking the VDJ complex thus produce VDJ-Cγ1 transcript (Figure 1). All VDJ recombination, somatic hypermutation and class switch recombination involve DNA double-strand cleavage and re-ligation; errors in each process may result in chromosome translocation targeted to the IGH locus (Küppers and Dalla-Favera, 2001). The IGH translocations seen in B-cell malignancies commonly take place in either the joining (IGHJ) or the switch (IGHS) segments.

3. Disease entities and IGH translocations

IGH translocation is a common genetic aberration in B-cell malignancies; however, the incidence various according to the disease entity. Precursor B-cell acute lymphoblastic leukemia shows a low frequency of IGH translocation, but recent studies have revealed diverse target genes of IGH translocation in this disease (Dyer et. al., 2010). Examples include cytokine receptor genes (Russell et.. al., 2009), the CCAAT enhancer-binding protein gene family (Chapiro et. al. 2006; Akasaka et. al., 2007), and microRNA 125b-1 (Sonoki et. al 2005; Chapiro et al, 2010, Tassano et al., 2010). Non-Hodgkin lymphoma often shows specific IGH translocations corresponding to the disease entity. t(14;18)(q32;q21) is seen in ~85% of follicular lymphoma and t(8;14)(q24;q32) is seen in ~90% of Burkitt lymphoma. Myeloma, which arises from terminal differentiated B-cells, shows extensive IGH translocation; however, the IGH translocation seen in myeloma defines distinct molecular subtypes of myeloma and is associated with clinical behavior (Bergsagel and Kuehl, 2005).

The vast majority of IGH translocation breakpoints are clustered in joining or switch regions. Since most IGH translocation results from physiological genetic modification errors, the breakpoints correspond to the differentiation stage of the tumor cells. IGH translocation breakpoints seen in B-cell precursor acute lymphoblastic leukemia are clustered in the joining region, whereas those seen in myeloma are clustered in the switch region. In B-NHL, IGH translations are seen in joining as well as switch regions. Follicular lymphoma and Mantle cell lymphoma show breakpoints in the joining region of IGH breakpoints suggesting that IGH translocations occur in the early stage of B-cell differentiation in bone marrow. Diffuse large B-cell lymphoma shows switch region breakpoints, suggesting that the translocation occurs in the late stage of B-cell differentiation in lymph nodes. Table 1 summarizes known partner genes of IGH translocation seen in B-cell malignancies.

Disease entity	Chromosome translocation	Involved gene	Biological function in B-cell
Precursor B-ALL	t(14;X)(q32;p22) or t(14;Y)(q32;p11)	CRFL2*	Cytokine Receptor
	t(5;14)(q31;q32)	IL3	Cytokine
	t(1;14)(q21;q32)	BCL9*	Modification of WNT signaling
	t(6;14)(p21;q32)	ID4*	Unkown
	t(8;14)(q11;q32)	CEBPD*	Transcription factor
	t(14;19)(q32;q13)	CEBPA*	Transcription factor
	t(14;14)(q11;q32)	CEBPE*	Transcription factor
	t(14;20)(q32;q13)	CEBPB*	Transcription factor
	t(14;17)(q32;q21)	IGF2BP1*	unkown
	t(11;14)(q24;q32)	microRNA125 b-1*	Anti-apoptosis
Follicular lymphoma	t(14;18)(q32;q21)	BCL2	Anti-apoptosis
Mantle cell lymphoma	t(11;14)(q13;q32)	Cyclin D1	Regulation of cell cycle
Burkitt Lymphoma	t(8;14)(q24;q32)	C-MYC	Cell proliferation and metabolism
Diffuse Large B-cell lymphoma	t(3;14)(q27;q32)	BCL6	Transcriptional repressor
	t(1;14)(q21;q32)	ITRA	B-cell signaling
	t(11;14)(p13:q32)	CD44	Cell migration or proliferation
Marginal cell lymphoma	t(6;14)(p21;q32)	Cyclin D3*	Regulation of cell cycle
B-CLL	t(2;14)(p13;q32)	BCL11A*	Unkown
MALT	t(1;14)(p21;q32)	BCL10*	Modification of NF-kB signaling
	t(14;18)(q32;q21)	MALT1	Modification of NF-kB signaling
Lymphoplasmacytic lymphoma	t(9;14)(p13;q32)	PAX-5	Transcription factor

Disease entity	Chromosome translocation	Involved gene	Biological function in B-cell
Myeloma	t(11;14)(q13;q32)	Cyclin D1	Regulation of cell cycle
	t(4;14)(p16;q32)	FGFR3/MMSET	Unkown
	t(8;14)(q24;q32)	C-MYC	Cell proliferation and metabolism
	t(14;16)(q32;q23)	C-MAF	Unkown
	t(6;14)(p21;q32)	Cyclin D3	Regulation of cell cycle
	t(14;20)(q32;q12)	MAFB	Unkown

Table 1. Disease entity and IGH chromosome translocation. *, Genes identified by LDI-PCR.

4. Molecular cloning of IGH translocation breakpoints using long distance inverse-PCR

Inverse PCR is a useful method to define unknown nucleotide alignments flanking known nucleotide sequences, such as retroviral DNA integration sites into host chromosome DNA (Takemoto et. al. 1994). The principle of inverse PCR uses self-ligated circular DNA as a PCR template and primers setting in a known sequence in the opposite direction. Inverse PCR is easily applied for determination of unknown partner sequence combined with IGH resulting from chromosome translocation. After thermostable DNA polymerase that can amplify long target DNA is available, the combination of inverse PCR and long target thermostable DNA polymerase (long distance inverse PCR; LDI-PCR) is a powerful tool for molecular cloning of an unknown sequence from IGH. Molecular cloning by LDI-PCR requires a small amount of genomic DNA and only couple of days to determine the translocation breakpoints involving IGH translocation. Using this method, many novel partner genes have been identified as listed in Table 1.

Since PCR may yield non-specific products, Southern blot analysis using IGH-specific probes is informative to interpret LDI-PCR results. The LDI-PCR product can be estimated by rearranging the band detected in Southern blot analysis; however, cloning of IGHJ translocations is simple and clear, and can be performed without reference to Southern blot data (Willis et al. 1997; Akasaka et al., 2007). In contrast, IGHS might undergo complex rearrangements, including insertion, deletion, and inversion (Bergsagel et. al. 1999) To distinguish translocation or other rearrangements, Southern blot assays using several probes are required. Thus, it is necessary to have large amounts of high-molecular-weight DNA available to allow Southern blot analysis before performing LDI-PCR for switch regions (Sonoki et al, 2001).

The detailed protocol of LDI-PCR for molecular cloning of the immunoglobulin heavy chain gene has been published elsewhere (Karran et al. 2005). A brief principle is shown in Figure 2. After complete digestion of high molecular genomic DNA with the restriction enzyme, the DNA is reacted with T4-ligase at low DNA concentration to form self-ligated circular DNA. The circular DNA is then purified by a silica-based column and subjected to PCR using primer pairs set at the IGH sequence. When the PCR product of interest is obtained, the nucleotide alignment can be determined directly. Although genomic DNA can be isolated from paraffin-embedded materials, the DNA is fragmented, and thus is not suitable for LDI-PCR.

LDI-PCR allows the rapid cloning of vast majority of translocations targeting IGH in B-cell malignancies. The benefits of LDI-PCR are first the speed and ease with which this cloning can now be performed and second that cloning can be performed when only small amounts of material are available (Willis TG, et al. 1997). The principle for LDI-PCR can be applied for other cloning purpose than IGH translocations.

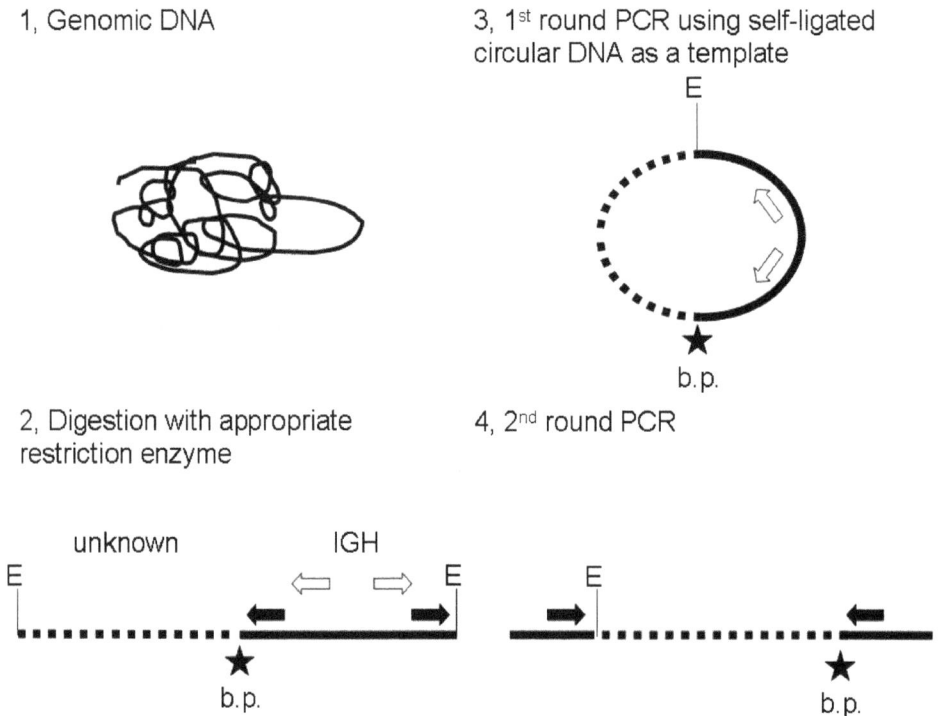

Fig. 2. Schematic diagram of LDI-PCR for cloning IGH translocation breakpoint. Using long distant DNA polymerase, 4~5 kb PCR products are able to be amplified routinely. E; restriction enzyme site, star; translocation breakpoint.

5. References

Akasaka T, Balasas T, Russell LJ, Sugimoto KJ, Majid A, Walewska R, Karran EL, Brown DG, Cain K, Harder L, Gesk S, Martin-Subero JI, Atherton MG, Bruggemann M, Calasanz MJ, Davies T, Haas OA, Hagemeijer A, Kempski H, Lessard M, Lillington DM, Moore S, Nguyen-Khac F, Radford-Weiss I, Schoch C, Struski S, Talley P, Welham MJ, Worley H, Strefford JC, Harrison CJ, Siebert R, Dyer MJ. Five members of the CEBP transcription factor family are targeted by recurrent IGH translocations in B-cell precursor acute lymphoblastic leukemia (BCP-ALL). Blood. 2007 Apr 15;109(8):3451-61.

Chapiro E, Russell L, Radford-Weiss I, Bastard C, Lessard M, Struski S, Cave H, Fert-Ferrer S, Barin C, Maarek O, Della-Valle V, Strefford JC, Berger R, Harrison CJ, Bernard OA, Nguyen-Khac F. Overexpression of CEBPA resulting from the translocation t(14;19)(q32;q13) of human precursor B acute lymphoblastic leukemia. Blood. 2006 Nov 15;108(10):3560-3.

Chapiro, E., L. Russell, S. Struski, H. Cave, I. Radford-Weiss, V. Valle, J. Lachenaud, P. Brousset, O. Bernard, C. Harrison, and F. Nguyen-Khac. A new recurrent translocation t(11;14)(q24;q32) involving IGH@ and miR-125b-1 in B-cell progenitor acute lymphoblastic leukemia. Leukemia 2010 24:1362-1364.

Dyer MJ, Akasaka T, Capasso M, Dusanjh P, Lee YF, Karran EL, Nagel I, Vater I, Cario G, Siebert R. Immunoglobulin heavy chain locus chromosomal translocations in B-cell precursor acute lymphoblastic leukemia: rare clinical curios or potent genetic drivers? Blood. 2010 Feb 25;115(8):1490-9.

Heim S and Mitelman F Cancer Cytogenetics, The second edition, Wiley-Liss, 1995

Honjo T and Frederick WA ed. Immunoglobulin genes, The second edition., Academic Press, 1995

Karran EL, Sonoki T, Dyer MJ. Cloning of immunoglobulin chromosomal translocations by long-distance inverse polymerase chain reaction. Methods Mol Med. 2005;115:217-30.

Küppers R, Dalla-Favera R. Mechanisms of chromosomal translocations in B cell lymphomas. Oncogene. 2001 Sep 10;20(40):5580-94.

Russell LJ, Capasso M, Vater I, Akasaka T, Bernard OA, Calasanz MJ, Chandrasekaran T, Chapiro E, Gesk S, Griffiths M, Guttery DS, Haferlach C, Harder L, Heidenreich O, Irving J, Kearney L, Nguyen-Khac F, Machado L, Minto L, Majid A, Moorman AV, Morrison H, Rand V, Strefford JC, Schwab C, Tonnies H, Dyer MJ, Siebert R, Harrison CJ. Deregulated expression of cytokine receptor gene, CRLF2, is involved in lymphoid transformation in B-cell precursor acute lymphoblastic leukemia. Blood. 2009 Sep 24;114(13):2688-98.

Siebert R, Rosenwald A, Staudt LM, Morris SW. Molecular features of B-cell lymphoma. Curr Opin Oncol. 2001 Sep;13(5):316-24.

Sonoki T, Willis TG, Oscier DG, Karran EL, Siebert R, Dyer MJ. Rapid amplification of immunoglobulin heavy chain switch (IGHS) translocation breakpoints using long-distance inverse PCR. Leukemia. 2004 Dec;18(12):2026-31.

Takemoto S, Matsuoka M, Yamaguchi K, Takatsuki K. A novel diagnostic method of adult T-cell leukemia: monoclonal integration of human T-cell lymphotropic virus type I provirus DNA detected by inverse polymerase chain reaction. Blood. 1994 Nov 1;84(9):3080-5.

Tassano E, Acquila M, Tavella E, Micalizzi C, Panarello C, Morerio C. MicroRNA-125b-1 and BLID upregulation resulting from a novel IGH translocation in childhood B-Cell precursor acute lymphoblastic leukemia. Genes Chromosomes Cancer. 2010 Aug;49(8):682-7.

Willis TG, Dyer MJ. The role of immunoglobulin translocations in the pathogenesis of B-cell malignancies. Blood. 2000 Aug 1;96(3):808-22.

Willis TG, Jadayel DM, Coignet LJ, Abdul-Rauf M, Treleaven JG, Catovsky D, Dyer MJ. Rapid molecular cloning of rearrangements of the IGHJ locus using long-distance inverse polymerase chain reaction. Blood. 1997 Sep 15;90(6):2456-64.

7

Molecular Cloning, Characterization, Expression Analysis and Chromosomal Localization of the Gene Coding for the Porcine αIIb Subunit of the αIIbβ3 Integrin Platelet Receptor[1,2]

Gloria Esteso, Ángeles Jiménez-Marín, Gema Sanz,
Juan José Garrido and Manuel Barbancho
Unidad de Genómica y Mejora Animal, Departamento de Genética,
Universidad de Córdoba
Spain

1. Introduction

Integrins are a long family of heterodimeric transmembrane glycoproteins consisting of multiple combinations of noncovalently linked α and β chains, which generate different complex receptors with different expression patterns and ligand binding profiles. The integrins bind to extracellular matrix (ECM) or to cell-surface ligands, regulating numerous downstream pathways (Hynes, 2002).

Each integrin binds to only a limited series of ligands, ensuring that cell adhesion and migration are precisely regulated. The α subunit mainly determines the substrate specificity with extracellular matrix molecules (ECM) (Yamada, 1991), while the intracytoplasmic tail of the β chain is predominantly responsible for the integrin interaction with the cell cytoskeleton by binding to vinculin, talin and α-actin (Isenberg, 1991). Thus, this heterodimeric association between α and β subunits allows the integrins to act as bidirectional signaling molecules in the different tissues and cell types in which they are widely distributed, mediating a variety of biological processes so diverse as embryogenesis, haemostasis, tissue repair, migration, cell polarity, immune response and metastatic diffusion of tumor cells (Hynes, 1987, 1992; Hemler et al., 1994).

Mammalian integrins have been divided into subfamilies according to their β subunit. The most important β integrin subfamilies are β_1, β_2 and β_3. Within a subfamily, the same β subunit can associate with different α subunits. To date, 18 α and 8 β chains -whose combinations provide up to 24 different integrins- have been described in mammal species (Hynes et al., 2002; Alam et al, 2007).

---

[1]The experiments and results showed in this chapter belong to the PhD by G. Esteso directed by J.J. Garrido.
[2]Sequence data from this article has been deposited with the GenBank Data Libraries under Accession N° **JF808665**.

The α_{IIb} chain only associates with the β_3 chain providing the $\alpha_{IIb}\beta_3$ (CD41/CD61) integrin receptor which is the most abundant one in platelets. The main role of the $\alpha_{IIb}\beta_3$ receptor is the binding of fibrinogen to the surface of the activated platelet, thereby resulting in the platelet aggregation with significant consequences in the thrombosis and the homeostasis attainment (Clark & Brugge, 1995; Schwartz et al., 1995). Although for a long time it was thought that the α_{IIb} integrin expression was limited to platelets and their precursors (the megakaryocytes), several studies have revealed that the α_{IIb} chains are also expressed in myeloid and in hematopoietic cells (Ody et al., 1999; Corbel and Salaun, 2002). In addition, $\alpha_{IIb}\beta_3$ integrin plays an important role in the progression and invasion of tumors (Chen et al., 1992; 1997) and in the differentiation of cells from the myeloid lineage in bone marrow (Chen et al, 1997; Wall et al, 1997). In humans, both subunits of the $\alpha_{IIb}\beta_3$ integrin show a high level of polymorphism resulting in some cases in clinically important hemorrhagic disorders (Weiss et al., 1996).

Consequently, α_{IIb} has been involved in many, different and important functions related with platelet activation and tumor progression. However, most studies related to α_{IIb} integrin have been carried out in humans, and little is known about the expression of α_{IIb} subunit in porcine tissues and cell types, although pig is generally accepted as an optimal experimental model which is used in different areas as immunology, xenotransplantation, artherosclerosis, cancer or cardiovascular disease because of its similarity to humans (Misdorp, 2003; Lunney, 2007).

In the present study we describe the cloning and molecular characterization of a cDNA encoding the porcine α_{IIb} (CD41) integrin, and the expression pattern of the α_{IIb} mRNAs in a variety of porcine cells and tissues. In parallel, we use immunohistochemistry and flow cytometry to accurately locate the porcine α_{IIb} integrin protein in the same tissues and cell types. For this, we produce a monoclonal antibody against a porcine recombinant α_{IIb} protein. We also study if any change is produced in the level of α_{IIb} transcripts in thrombin stimulated platelets. Additionally, we identify the chromosomal localization of the porcine *CD41* gene.

2. Material and methods

2.1 Tissues and cells

Fresh pig blood from approximately 1 year old healthy pigs was collected at the slaughterhouse into sodium citrate to final concentration of 10% v/v of the anticoagulant. Platelets isolation was carried out according to García et al., 2005. Porcine platelets were pelleted from plateled-rich plasma (PRP) obtained by centrifugation at 200g for 20 min of blood after addition of ACD solution (117mMsodium citrate, 282mMglucose and 78mMcitric acid) to a concentration of 7%v/v. The upper third of the PRP was centrifuged again after addition of prostacyclin (final concentration 2.5 mM) to avoid platelets activation. For platelets activation, the cells were stimulated by the addition of 1 U of thrombin for 3min at 37°C. Porcine tissues were recovered from adult pigs immediately after slaughtering at the local abattoir and frozen in liquid nitrogen until use. Peripheral blood mononuclear cells (PBMC) were isolated from heparinized whole blood using Ficoll-Hypaque (density 1077 g/ml, Sigma) centrifugation at 900 g for 30 min. Mononuclear cells (lymphocytes and monocytes) and granulocytes were collected by aspiration from their respective gradient interphases and washed twice in PBS.

Molecular Cloning, Characterization, Expression Analysis and Chromosomal Localization of the Gene Coding for
the Porcine αIIb Subunit of the αIIbβ3 Integrin Platelet Receptor

93

2.2 RNA isolation, RT-PCR and RACE

Total RNA from platelets, cells or tissues was purified according to the *M-MLV Reverse Transcriptase system* (Invitrogene) using the random primers pd(N)$_6$-5'-PO$_3$NA$^+$Salt (Pharmacia Biotech). RNA samples were kept at –80°C after controlling the quality on a denaturing agarose gel. 5 μg RNA, resuspended in 9.5 μl water, were heated for 3 min at 65°C in the presence of random hexamers (7.5 μM final concentration), and then cooled in ice. RNA was reverse transcribed using 1 μl Moloney murine leukemia virus reverse transcriptase (200 units/μl) (GibcoBRL) for 1 h at 42°C in a final volume of 20 μl containing 4 μl of 5X reverse transcriptase buffer, 0,5 μl ribonuclease inhibitor (50 U/μl) (Roche), 1 μl 20 mM dNTP (Pharmacia) and 2 μl 0.1 M dithiothreitol. After 10 min at room temperature, 1 h at 42 ° C, and 10 min at 95°C, DEPC H$_2$O were added until a final volume of 100 μl. 2 μl of this mixture were subjected to PCR using 1μl *Tth* DNA polymerase (1U/μl) (Biotools) and 2,5 μl each CD41-specific primer (20μM) (see Table 1) in a final volume of 50 μl containing 5 μl 10x buffer, 2 μl MgCl$_2$ (50 mM), and 8 μl dNTP MIX (1,25 mM each) (Biotools). The amplification consisted in 35 cycles of PCR and each cycle consisted of incubations at 94°C for 1 min, Tm°C for 1 min, and 72°C for 1 min. The amplifications were electrophoresed on 1% agarose/1X TAE gel. RT-PCR on RNA18S cDNA was used as a control. For RACE (Rapid Amplification of cDNA Ends), 1 μg total RNA from platelet was used to reverse-transcribe using 1 μl Moloney murine leukemia virus reverse transcriptase (200 units/μl) (GibcoBRL) for 1 h at 42°C in a final volume of 20 μl containing 4 μl of 10X reverse transcriptase buffer, 1,0 μl ribonuclease inhibitor (50 U/μl) (Roche), 4 μl 2,5 mM dNTP (Pharmacia) and 2 μl 3' RACE ADAPTER (20 μM) in a final volume of 20 μL. 3' CD41 cDNAs were obtained by PCR using a specific porcine CD41 primer and the anchor primer provided in the kit (Table 1).

2.3 DNA sequencing and sequences analysis

Sequencing was performed using *ABI PRISM BigDye Terminator Cycle Sequencing Ready Reaction Kit* (Applied Biosystems, Foster City, CA, USA) on a thermal DNA cycler GeneAmp PCR System 2400 (Applied Biosystems, Foster City, CA, USA), according to the instructions of the manufacturer, and analysed on an ABI PRISM 3100 Sequencer (Applied Biosystems, Foster City, CA, USA). Porcine CD41 sequence has been deposited at GenBank under number **JF808665**. Sequences were analyzed using the analysis software LaserGene (DNAstar, Londo, UK) and the analysis tools provided by the expasy web site <http://www.expaxy.org>. Primers design was performed with Oligo 6 (MBI, Cascade, CO, USA) and Amplify 3 (<http://www.engels.genetics.wisc.edu/amplify/>). Multiple alignment among CD41 peptide sequences from *Sus scrofa* (GenBank accession no. **JF808665**), *Homo sapiens* (GenBank accession no. **AAI26443**), *Bos taurus* (GenBank accession no. **NP_001014929**), *Mus musculus* (GenBank accession no. **NP_034705**), *Rattus norvegicus* (GenBank accession no. **XP001063315**), *Canis familiaris* (GenBank accession no. **NP_001003163**), *Equus caballus* (GenBank accession no. **NP_001075262**), *Oryctolagus cuniculus* (GenBank accession no. **Q9TUN4**), *Danio rerio* (GenBank accession no. **AAQ82784**) and *Xenopus laevis* (GenBank accession no. **Q5XH72**) was performed by using MUSCLE program (Edgar, 2004).

2.4 Recombinant CD41 protein (rpCD41) expression and purification

DNAs encoding extracellular domains of the porcine CD41 were amplified by PCR. Primers used for amplification contained restriction sites enabling ligation into the expression vector *pET28b* (Novagen) following digestion of the PCR product and the vector with *Bam*HI and *Hind*III. Two different pairs of primers were used: F1rp-*EcoRI*/R1r-p*XhoI* and F2rp-*Bam*HI/R2rp-*Hind*III (Table1). PCR product was ligated into the expression vector *pET28b* and used to transform *Escherichia coli* strain *BL21 (DE3)* (Novagen). Recombinant proteins (rpCD41-F1R1 and rpCD41-F2R2), expression and purification were carried out following previously procedures described by us (Jiménez-Marín et al., 2000).

2.5 Antibodies production

A monoclonal antibody, GE2B6, against rpCD41-F2R2 and two polyclonal antibodies, anti-rpCD41-F1R1 and anti- rpCD41-F2R2, were produced using previously described immunization and cells fusion procedures (Arce et al., 2002; Jiménez-Marín et al., 2000). Briefly, female BALB/c mice were immunized with 50 µg of rpCD51. Spleen cells from immune mice were fused with Sp2/0 myeloma cells. Hybridoma clones were selected on the basis of binding secreted antibody to rpCD61 by indirect ELISA. Antibody-producing hybridomas reacting positively were cloned at least twice by limiting dilution. Immunoglobulin classes and subclasses were determined in solid-phase ELISA using rabbit antisera specific for mouse heavy and light chains and a peroxidase-conjugated goat anti-rabbit immunoglobulin (Sigma).

2.6 Electrophoresis and immunoblottings

Platelets (10^8/sample) were lysated in NP-40 lysis buffer with PMSF 2 mM with vigorous shaking for 1 h at 4°C, and then centrifuged at 12,000 rpm, 20 min. 100 µl supernatant were mixed with 100 µl of sample treatment buffer, and 100 µl were loaded in the gel. Electrophoresis was carried out in 5%-15% gradient polyacrilamide gels. For the 2D electrophoresis, the platelet proteins pellet was resuspended in lyses buffer (7 M urea, 2 M thiourea, 4% CHAPS, 1% DTT, 0,8% ampholytes). Immobilized pH gradient strips (17 cm, 5-8 linear pH gradient, Bio-Rad) were rehydrated with 300 µl (300 µg) of the protein solution for 16 h, and focused in a PROTEAN IEFcell (Bio-Rad). Second dimension was performed on 10% SDS-PAGE. For the immunoblottings, proteins were transferred from gels to PVDF Inmobilon P membranes (Milipore). Membranes were blocked and washed three times in PBS-T, and then incubated with 3 ml antibody or PBS as negative control, overnight at 4°C in shaking. After three washing ups in PBS-T, the membranes were incubated with rabbit anti-immunoglobulin-peroxidase (Sigma). Afterwards, they were washed up three times in PBS-T, and finally reactions were detected with the ECL™ detection system (Amersham) following the manufactures instructions.

2.7 Immunoprecipitation of platelet CD41 proteins

Platelets (10^8/sample) were incubated with 0.4 mg sulfobiotin (Pierce) with gently shaking for 15 min at 4°C, and then centrifuged at 3,000 rpm, 15 min. Pellet was washed three times in PBS, and then resuspended in lyses buffer (500 µl/sample) and PMSF 2 mM. After incubation in dark with vigorous shaking for 1 h at 4°C, it was centrifuged at 13,000 rpm, 20 min, and the supernatant was collected. 50 µl of protein G-Sepharose (Pharmacia) were added per ml of supernatant and incubated with shaking overnight at 4°C, and then centrifuged at 2,000 rpm, 5 min. 500 µl lysate were incubated with 1 ml of the anti-porcine CD41 antibody for 2 h at room temperature. At the same time, when monoclonal antibody was going to be used, to increase its binding ability, the G protein is recovered with an anti-mouse immunoglobulin rabbit serum (Pierce) 1/10 in lyses buffer, for 2 h at room temperature. Then, the G protein is washed three times in lyses buffer, and centrifuged at 2,000 rpm, 2 min. This step was not needed when polyclonal antibodies were used. 50 μl of the antibody recovered G-Sepharose were added to the lysate containing the anti-CD41 antibody and incubated for 1 h at room temperature with shaking, and then centrifuged at 2,000 rpm, 5 min. The supernatant was collected and washed three times in lyses buffer, the first being in buffer and sucrose. Finally, supernatant was subjected to SDS-PAGE in 5%-15% gels in reducing or not reducing conditions. After electrophoresis, the proteins were transferred to PVDF, Inmobilon P membranes, as described before, and, after be blocked, incubated with a solution of Streptavidin-HRP (Amersham) solution 1/500 in PBS for 1 h in dark. Then, membranes were washed three times in PBS-T and revealed with the ECL™ detection system (Amersham) following the manufactures instructions.

2.8 Immunohistochemistry

Expression of CD41 protein from healthy animals was studied following previously procedures described by us (Jiménez-Marín et al., 2008) using monoclonal antibody GE2B6 supernatant or polyclonal antibodies (1/3000 dilution in PBS) or an irrelevant mAb (as negative control). Briefly, all tissue specimens were fixed in Bouin liquid for 16 hours. Tissues were dehydrated in ascending concentrations of ethanol and xylene and embedded in paraffin. Sections of 5 μm were placed on slides coated with Vectabound (Vector Laboratiores, Inc.). The tissue slides were kept at 55°C for 45 min to improve the adherence of sections to glass. The sections were deparaffinized and rehydrated in xylene and descending concentrations of ethanol, respectively. Endogenous peroxidase activity was inhibited with 3% hydrogen peroxidase. The sections were incubated with normal goat serum (1:10 dilution in PBS) (Vector) for 30 min at room temperature. After removing the serum, anti-porcine CD41 antibodies or an irrelevant mAb (as negative control) were added for 18 hours at 4°C in a wet chamber. The sections were incubated with biotinylated anti-mouse Ig (Dako) diluted 1/50 in PBS for 30 min at room temperature. Tissue sections were covered with avidin-biotin-peroxidase complex (Sigma) diluted 1/50 with PBS for 1 h in a wet chamber at room temperature, washed and then developed with 3, 3'-diaminobenzidine (Sigma) (5 μg in 10 ml PBS). Sections were counterstained with Mayer hematoxylin and mounted with Eukitt.

2.9 Flow cytometry

100 μl of platelets (10^6 cells/ml) and 100 μl of the antibody (or PBS as a control) were incubated 30 min at 4°C. After washing with PBS, tubes were centrifuged at 3,000 rpm, 6 min, and the platelets resuspended in 50 μl of a rabbit FITC-anti-immunoglobulin (1/160 in PBS) (Sigma). After incubation at 4°C 30 min in dark, the platelets were washed three times in PBS and the fixed in 1% PFA/PBS. Samples were analyzed in a FACsort cytometer (Bencton Dickinson) equipped with a CellQuest v 1.2software.

2.10 Chromosome localization

The INRA somatic cell hybrid panel (Yerle et al., 1996) was screened with porcine primers (VARP1 and VARP2), which specifically amplify a 212 bp fragment (Table 1). For genotyping of the hybrid panel, 10 ng of DNA from each cell line and control sample (pig, hamster, and mouse) were amplified. PCR products were evaluated on a 1% agarose gel and individual cell lines were evaluated for the presence or absence of a fragment of the correct size. Statistical calculations of the assignment were performed using the software developed by Chevalet et al (1997) (<http://www.inra.toulouse.fr>). The INRA-Minesota porcine radiation hybrid (IMpRH) panel (Yerle et al., 1998; Hawken et al., 1999) was screened with the same porcine specific primers in the same PCR conditions (Table 1). Statistical calculations of the assignment were performed using the IMpRH mapping tools (<http://www.imprh.toulouse.inra.fr>).

2.11 Quantitave real time RT-PCR

CD41 cDNA was quantified by real time quantitative PCR (RT-Q-PCR) relative to β-actin cDNA reverse transcribed from total RNA from platelets. The PCR reaction was carried out with 0.5 μl of each VARP1/VARP2 and β–actinF/ β-actinR primers (Table 1) (20μM), 12.5 μl of iQ™ SYBR1 Green Supermix (Bio-Rad), and 1.5 μl of the cDNA sample. The PCR conditions included 40 cycles of 30s at 94°C, 30s at 60°C and 30s at 72°C. All experiments were performed three times to confirm accuracy and reproducibility of real-time PCR. The efficiency of the primers (E) was calculated according to the equation (1).

$$E = 10^{[-1/p]} \tag{1}$$

being p the slope of the standard curve log(fluorescence)/Ct.

The relative abundance of *CD41* gene expression was determined by the ratio (R) equation (2)

$$R = 2^{[\Delta Ct \, (target) - \Delta Ct \, (control)]} \tag{2}$$

being Ct = threshold cycle (cycle at with PCR amplification reaches a significant value).

3. Results

3.1 Cloning and sequence analysis of the porcine CD41 cDNA

The porcine full length *CD41* cDNA was obtained by a combination of PCR and RACE (Rapid Amplification of cDNA Extremes). A partial sequence that lacked the 3′ cDNA extreme, including part of the coding sequence, was deposited in GenBank (**AF170526**). So, we first amplified the pig *CD41* 5′ cDNA using forward and reverse primers designed from this sequence. Three pairs of primers were used P3/R4, P1/R2 and F1/R1 (Table1).

Three 1352, 870 and 677 bp long overlapping fragments were produced, respectively. Altogether, the three fragments provided a 2701 bp long sequence that belongs to the 5′ extreme of the *CD41* cDNA (Figure1). To obtain the remaining 3′ sequence of the *CD41* cDNA we carried out a RACE by using the RACE-out and the RACE-P5 primers shown in Table1. This allowed us to obtain an additional 622 bp long 3′ sequence (Figure1). Finally, we obtained the full length *CD41* cDNA molecule amplifying RNA from platelets by RT-PCR by using the FcDNA5 and RcDNA3 pair of primers (Table1 and Figure1) and MBLong polimerase.

Primers	Primer sequences 5'- 3'	Template, localization (5'-5')	Product size (bp)	Tm (°C)
FP3	TGTGGAAGAAGGAAGATGG	cDNA, 2-1353	1352	58.1
RP4	GCAGAGCCTGCGGCAAAGG			
FP1	GGGCCAAGTATCGGTGTTC	cDNA, 1264-2133	870	60.0
RP2	TGGGTACAGATGAGCCTCTCTAAG			
F1	CCCCAGGTGCTCACTACA	cDNA, 2066- 2742	677	60.0
R1	TCGGCAGCTCAGGAGAATTGGA			
RACE-*adapter*	GCGAGCACAGAATTAATACGACTCACT ATAGGT$_{12}$			
RACE-*outer*	GCGAGCACAGAATTAATACGACT	cDNA, 2638-3'	622	60.0
RACE-P5	CTCCCCTGTGTACCCAGCTCATCA			
FcDNA5	CCTAAGCTTAAGATGGCCAGAGCTTTGT GT	cDNA, 13-3125	3113	59.2
RcDNA3	GCAAAGCTTTCACTCCTCCTCTTCATCA GA			
VARP1	GAGGCATGACCTCTTGGTGG	Genomic DNA cDNA, 1012-1223	212	59.0
VARP2	CATTGTAGCCATCCCGGTTC			
F1rp-*Eco*RI	CGAC<u>GAATTC</u>CCCCAGGTGCTCACTAC A	cDNA, 2066- 2742	677+20	60.0
R1rp-*Xho*I	CGA<u>CTCGAG</u>GCAGCTCAGGAGAATTGG A			
F2rp-*Bam*HI	GGTC<u>GGATCC</u>TTGAACCTGGACCCAGT GCAT	cDNA, 110-1105	996+20	59.0
R2rp-*Hind*III	GGT<u>AAGCTT</u>CTGCAGGAACAAGTAAAC ACG			

Table 1. Primers used in PCRs and RACE.

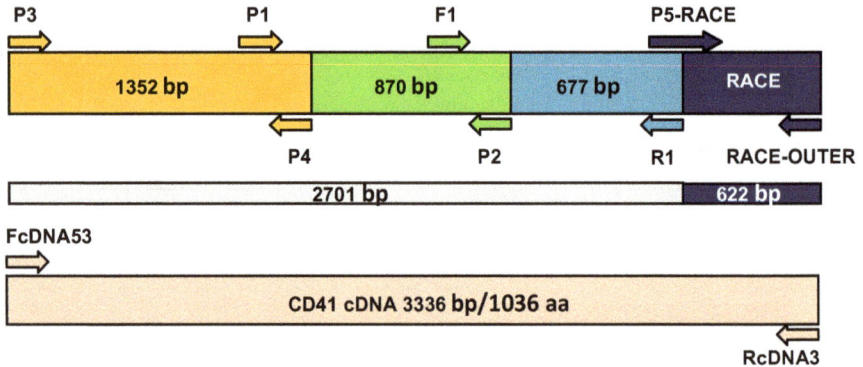

Fig. 1. Strategy and localization of the primers used for cloning the porcine full length *CD41* cDNA.

The full *CD41* cDNA was 3336 bp long and contained an open reading frame of 3111 bp (Figure2) which encodes a CD41 polypeptide of 1036 amino acid residues, and a 198-bp long untranslated 3' flanking region. The nucleotide sequence encoding the full length CD41 pig cDNA was submitted to GenBank (Accession number JF808665).

The first 31 amino acid residues of CD41 are predominantly hydrophobic and correspond to the signal peptide sequence. So, the pig mature pre-CD41 molecule consists of 1005 amino acid residues, and, as this amino acid sequence has a proteolytic cleavage site (KR/D) located between amino acids residues 899 and 900 in pre-CD41 (Takada et al., 1989), the mature porcine CD41 polypeptides –lacking the signal peptide- must be composed by two different chains (914 and 91 amino acid residues) linked by disulfide bridges, similar to those reported in homologous CD41 integrins. Other sequences and structural domains contained in other CD41 proteins are also presents in the porcine CD41 chains (Figure2).

The seven FG-GAP tandem repeats are shown as W with arrows marking their initial and final limits. An α helix is shown in red. The long extracellular domain of the porcine CD41 integrin consists of 869 amino acids residues. It contains 8 consensus N-glycosylation sites (Asn-X-Ser/Thr, where X is not Pro) identified by the NetNGlyc 1.0 program (www.expasy.org), and 18 –from 19- cystein residues. As in other α integrins, the extracellular domain of the porcine CD41 contains four Ca++ binding domains (DX(D/N)XDGXXD) and seven FG-GAP tandem repeats -which are identified by the SABLE 2.0 program (http://sable.cchmc.org/)- each one containing four helixes similar to those previously described (Springer, 1997; Xiong et al., 2001). The secondary structure of CD41 molecule is shown in Figure3 and the tertiary one, obtained with the Swismodel (www.expasy.org), is shown in Figure4.

The stretch sequence of 26 hydrophobic amino acid residues located in the carboxy-terminal portion of the polypeptide must constitute the transmembrane domain. Following it there is a short 20-amino acid sequence that must represent the cytoplasmic domain of the molecule. It contains a GFFKR (1019-1023 in pre-CD41) domain, which is conserved in all human α integrin chains and is involved in the link of both α and β chains of the heterodimeric complex (Rojiani et al., 1991). It also contains a β–like turn (PPLEE) (1026-1030), that in comparison to the $α_v$ chain (PPREE) could aid in the ligand interaction of fibronectin and vitronectin with the intact $\alpha_{IIb}\beta_3$ heterodimer which is essential for various transductional processes during mammalian organogenesis (Filardo & Cheresh, 1994).

```
P 3
TTGTGGAAGAAGGAAGATGCCCAGAGCTTTGTGTCTACTTCATGCCCTCTGGCTTCTGGAGTGGGTGCAACTGCTCTTG  79
                 M  A  R  A  L  C  L  L  H  A  L  W  L  L  E  W  V  Q  L  L  L   21
              F 2
GGACCCGGTGCTGCCCCTCCAACCTGGGCCTTGAACCTGGACCCAGTGCATCTCACCATCTACACAGGCCCCAATGGCAGCCACTTTGGG 169
 G  P  G  A  A  P  P  T  W  A  L  N  L  D  P  V  H  L  T  I  Y  T  G  P  N  G  S  H  F  G   51
TTTTCATTGGACTTCTACAAGAAGAGCCATGGCAGCGTATCCATCGTGGTGGGGGCCCCGCGGACCTTGGGCCGCAACCTGGAGGAGACC 259
 F  S  L  D  F  Y  K  K  S  H  G  S  V  S  I  V  V  G  A  P  R  T  L  G  R  N  L  E  E  T   81
GGCGGCGTTTTCCTGTGTCCCTGGAAGGCCAAGAGCGTCCAGTGCGTCGCGCTGTCCTTCAACCTCGATGATGAGACGCGAAACGTAGGC 349
 G  G  V  F  L  C  P  W  K  A  K  S  V  Q  C  V  A  L  S  F  N  L  D  D  E  T  R  N  V  G  111
GCCCAAACTTTCCAAACCTTCAAGGCCCGTCAAGGACTAGGGGCGTCGGTCTTAACTTGGAGAGACAACGTTGTGGCCTGCGCCCCCTGG 439
 A  Q  T  F  Q  T  F  K  A  R  Q  G  L  G  A  S  V  L  T  W  R  D  N  V  V  A  C  A  P  W  141
CAGCACTGGAACGTCCTAGAAAAGAACGAGGAGGCTGAGAAGACACCTGTAGGTGGCTGCTTCGTGGCTCAGTCCAGAACAGCGGCCGC 529
 Q  H  W  N  V  L  E  K  N  E  E  A  E  K  T  P  V  G  G  C  F  V  A  Q  L  Q  N  S  G  R  171
GCGGAGTTCTCGCCCTGTCGGGCCAACCTCTTGAGCCTGGTTTACGTGGAAAGTAAATTCAATGACAGGCGCTATTGCGAGGTCGGCTTC 619
 A  E  F  S  P  C  R  A  N  L  L  S  L  V  Y  V  E  S  K  F  N  D  R  R  Y  C  E  V  G  F  201
AGCTCTGCGGTCACTCAGGCTGGGGAGCTGGTGCTTGGGGCTCCTGGGGGCTACTATTTCTTAAGTCTCATGGCACGGGCTCCAATTGCC 709
 S  S  A  V  T  Q  A  G  E  L  V  L  G  A  P  G  G  Y  Y  F  L  S  L  M  A  R  A  P  I  A  231
GATATCATCTCGAGTTACCGCCCAGGCACCCTCTTGTGGCACGTGCCCACCCAGAAGCTCACCTTCGACTCCGACCTTCCTGAGTACTAC 799
 D  I  I  S  S  Y  R  P  G  T  L  L  W  H  V  P  T  Q  K  L  T  F  D  S  D  L  P  E  Y  Y  261
GAGAGCTACTTGGGGTACTCGGTGGCTGTGGGCGAGTTCGACCGGAATCCCAACACCACAGAGTACATCCTCGGTGGCCCCACCTGGAGC 889
 E  S  Y  L  G  Y  S  V  A  V  G  E  F  D  R  N  P  N  T  T  E  Y  L  G  G  P  T  W  S  291
ATGACCCTGGGAGCGGTGGAAATTTTTACCTCGAAACACCAGAGGCTGCACCTGCTGCAGGGAGAGCAGGTGGCTTCATATTTCGGGCAT 979
 M  T  L  G  A  V  E  I  F  T  S  K  H  Q  R  L  H  L  L  Q  G  E  Q  V  A  S  Y  F  G  H  321
TCAGTGGCCGTCACCGACGTCAACGGGGACGGGAGGATGGACCTCTTGGTGGGAGCGCCACTGTACATGGAGAGCCGTGCTGACCACAG 1069
 S  V  A  V  T  D  V  N  G  D  G  R  M  D  L  L  V  G  A  P  L  Y  M  E  S  R  A  D  H  K  351
                                          R 2
CTGGCCGAGGTGGGGCGTGTTTACTTGTTCCTGCAGTCTCTCGAGGTCACCACTCGCTGGGCACCCCCAGCCTCCTGCTGACAGGCACACAG 1159
 L  A  E  V  G  R  V  Y  L  F  L  Q  S  R  G  H  H  S  L  G  T  P  S  L  L  L  T  G  T  Q  381
CTCTATGGACGATTTGGCTCAGCCATCGCGCCTCTGGGCGACTTGAACCGGGATGGCTACAATGATGTTGCCGTGGCCGCCCCCTATGGG 1249
 L  Y  G  R  F  G  S  A  I  A  P  L  G  D  L  N  R  D  G  Y  N  D  V  A  V  A  A  P  Y  G  411
                                                                               P 1
GGTCCCACCGGTCAGGGCCAAGTATCGGTGTTCCTGGGTCAGAGTGAGGGGCTTAACTCGCAGCCCTCCCAGGTCCTGCACAGCCCCTTT 1339
 G  P  T  G  Q  G  Q  V  S  V  F  L  G  Q  S  E  G  L  N  S  Q  P  S  Q  V  L  H  S  P  F  441
P 4
GCCGCAGGCTCTGCCTTTGGCTTCTCCCTTCGAGGTGCCACAGACATCGATGACAATGGATACCCAGACCTGCTGGTAGGAGCTTACGGG 1429
 A  A  G  S  A  F  G  F  S  L  R  G  A  T  D  I  D  D  N  G  Y  P  D  L  L  V  G  A  Y  G  471
GCCGACAAGGTTGTCGTGTACCGAGCTCAGCCGGTGGTGACCGCCATGTCCAGCTGATGGTGCAAGAATCCCTGAATCCTGCTGTGAAG 1519
 A  D  K  V  V  V  Y  R  A  Q  P  V  V  T  A  T  V  Q  L  M  V  Q  E  S  L  N  P  A  V  K  501
AATTGTGTCCAGCCCCAAACCAAGACACCAGTGAGCTGCTTTACCATCCAGATGTGTGTGGGAGCCACTGGGCACAACATTCCTGAGAAG 1609
 N  C  V  Q  P  Q  T  K  T  P  V  S  C  F  T  I  Q  M  C  V  G  A  T  G  H  N  I  P  E  K  531
CTGCGCCTAAATGCCGAGCTGCAGCTGGACCGGCAGAAACCCCACCAGAGCCGGCGGGTGCTGCTGCTGGCTTCCAACAGGCGAGCACT 1699
 L  R  L  N  A  E  L  Q  L  D  R  Q  K  P  H  Q  S  R  R  V  L  L  L  A  S  Q  Q  A  S  T  561
GTCCTGGACGTGGATCTGACCTGGAGGCAGAGCCCCACCTGCCACAACACCACGGCCTTCCTCCGGGATGAGGCTGATTTCCGGGACAAG 1789
 V  L  D  V  D  L  T  W  R  Q  S  P  T  C  H  N  T  T  A  F  L  R  D  E  A  D  F  R  D  K  591
CTGAGCCCCATCGTGCTCAGCTTCAATGTGTCCCTGCAGCCGGAGAAGAATGGAGACGCCCTCACGTTCATGCTGCATGGAGACACCCAC 1879
 L  S  P  I  V  L  S  F  N  V  S  L  Q  P  E  K  N  G  D  A  L  T  F  M  L  H  G  D  T  H  621
GTTCAGGAACAGACCCGCATCATCCTGGACTGTGGGGAAGACCAAGTGTGCGTGCCAAAGCTCCAGCTCTCGGCCAACACGACAGGCTCC 1969
 V  Q  E  Q  T  R  I  I  L  D  C  G  E  D  Q  V  C  V  P  K  L  Q  L  S  A  N  T  T  G  S  651
CCACTCCTAGTTGGAGCTGATAATGTGCTGGAGCTGCACGTGGTCGCGGCCAATGAGGGCGAGGGAGCCCATGAGGCCGAGCTGGTTGTG 2059
 P  L  L  V  G  A  D  N  V  L  E  L  H  V  V  A  A  N  E  G  E  G  A  H  E  A  E  L  V  V  681
          F 1                                                                P 2
CACCTGCCCCCCAGGTGCTCACTACATGCAGGCCCTCAGCAACACCAAGAGCTTCGAGAGGCTCATCTGTACCCAGAAGAAGGAGAATGAG 2149
 H  L  P  P  G  A  H  Y  M  Q  A  L  S  N  T  K  S  F  E  R  L  I  C  T  Q  K  K  E  N  E  711
ACCAAAGTGGTGCTGTGTGAGCTGGGCAACCCCATGAAGGGGGACACCCAGATAGAAATCACGATGTTGGTGAGTGTGGGGAACCTGGAA 2239
 T  K  V  V  L  C  E  L  G  N  P  M  K  G  D  T  Q  I  E  I  T  M  L  V  S  V  G  N  L  E  741
GAGGCTGGGGAGCACGTGTCCTTCCGGCTGCAGATCAGGAGCAAGAACAGCCAGAATCCAAACAGCGAGACAGTGGTGCTGGATGTGCAA 2329
 E  A  G  E  H  V  S  F  R  L  Q  I  R  S  K  N  S  Q  N  P  N  S  E  T  V  V  L  D  V  Q  771
```

```
GTCCGGGCAGAGGCCCACCTGGAGCTGAGAGGGAACTCCTTTCCAGCCTCCCTGTTGATGGCGGCAGAAGGCGATTGGGAGAACAGCTCG 2419
 V  R  A  E  A  H  L  E  L  R  G  N  S  F  P  A  S  L  L  M  A  A  E  G  D  W  E  N  S  S  801

GACAATTTGGGCCCCAAAGTGGAGCACACCTATGAGCTCCACAACAATGGCCCTAGTACCGTAAGTGGCCTCCACCTCAGGCTCCACCTG 2509
 D  N  L  G  P  K  V  E  H  T  Y  E  L  H  N  N  G  P  S  T  V  S  G  L  H  L  R  L  H  L  831

CCCGGCCAGTCCCAGACCTCCGACCTGCTCTACATCGTGGACATACAGACCCAGGGAGGCCTTCAGTGCTCCCCGCAGCCCTCTCCCAAC 2599
 P  G  Q  S  Q  T  S  D  L  L  Y  I  V  D  I  Q  T  Q  G  G  L  Q  C  S  P  Q  P  S  P  N  861
                                    P 5 RACE
CCCCGCCAGCTGGACTGGGGGCTGCACAGCCCCACACCCTCCCCTGTGTACCCAGCTCATCACAAGCGGAATCGCAGACAGGCAGTCCTG 2689
 P  R  Q  L  D  W  G  L  H  S  P  T  P  S  P  V  Y  P  A  H  H  K  R  N  R  R  Q  A  V  L  891
                                          R 2
CCAGGGCAGAAGCAGCCGTTGAGTCTTCAGGATCCAATTCTCCTGAGCTGCGACTCGGCGCCCTGCACCGTGGTGCACTGTGACCTTCCG 2779
 P  G  Q  K  Q  P  L  S  L  Q  D  P  I  L  L  S  C  D  S  A  P  C  T  V  V  H  C  D  L  P  921
                        ▲
GAGATGGCGCGTGGGCAGCGGGCCATGGTCACCGCGCAGGCTTTCCTGTGGCTGCCCAGCCTCCGCCAGAGACCCCTGGATCAGTTTGTG 2869
 E  M  A  R  G  Q  R  A  M  V  T  A  Q  A  F  L  W  L  P  S  L  R  Q  R  P  L  D  Q  F  V  951

CTGCGGTCGCACGCTTGGTTCAACGTCTCTTCCCTTCCCTACGCCGTGCCCGCCCTCAGCCTGCCCAGCGGGGAAGCTCAGGTGCAGACA 2959
 L  R  S  H  A  W  F  N  V  S  S  L  P  Y  A  V  P  A  L  S  L  P  S  G  E  A  Q  V  Q  T  981

CAGGTGCTTCGGGTCTTGGAGGACAGAGAGGTTCCACTCTGGTGGATGCTGGTAGGCGTGCTGGGCGGCCTGCTGTTGCTCACACTGCTG 3049
 Q  V  L  R  V  L  E  D  R  E  V  P  L  W  W  M  L  V  G  V  L  G  G  L  L  L  L  T  L  L  1011

GTCCTGGCCATGTGGAAGTGTGGCTTCTTCAAGCGGAATCGGCCACCCCTGGAAGAATCTGATGAAGAGGAGGAGTGACGCTGCAGCCCA 3139
 V  L  A  M  W  K  C  G  F  F  K  R  N  R  P  P  L  E  E  S  D  E  E  E  E  *            1036
CACATCTCTGTTCCTGCAGAAAGGCTGCATGTGGTGACTGCCCTGTCCCTCTTCTCCAGCCAGTGCCCTGCGAGCTCTGCTCGCCCATGA 3229
ATTGGAGCTGTTCCCTGGAGGCCTCCCGATGTCCTCCTGCTCCCATCAGAGCTGGCTCACCCCCCTTCCTGCTGCCTAATAAAGAGGCTG 3319
AACCCCAAAAAAAAAAA                                                                          3336
```

Fig. 2. The nucleotide and deduced amino acid sequences of pig *CD41* cDNA. The predicted signal peptide is remarked in light green, the transmembrane domain in dark green, and the cytoplasmic region, containing the GFFKR sequence, in purple. The putative polyadenylation sequence is remarked in a black box. Potential N-glycosylation sites are indicated in red. Cysteine residues are marked as C in yellow. Putative cleavage sites are shown as ▲. Ca++ binding domains are remarked in pink. Primers used for cloning are marked with arrowheads.

Molecular Cloning, Characterization, Expression Analysis and Chromosomal Localization of the Gene Coding for
the Porcine αIIb Subunit of the αIIbβ3 Integrin Platelet Receptor
101

```
                    →                           W1
LNLDPVHLTIYTGPNGSH FG FSLDFYKKSHGSVSIVV GAP RTLGRNLEET
                          1                        2

GGVFLCPWKAKSVQCVALSFNLDDETRNVGAQTFQTFKARQG LG ASVLTW
      3                4                                  1
                                            W2
RDNVVA CAP WQHWNVLEKNEEAEKTPVGGCFVAQLQNSGRAEFSPCRANL
        2                              3

       ←   →                           W3
LSLVYVESK FN DRRYCEVGFSSAVTQAGELVL GAP GGYYFLSLMARAPIA
             1                      2               3

          ←   →                           W4
DIISSYRPGTLLWHVPTQKLTFDSDLPEYYESYLGYSVAVGE FD RNPNTT
                                              1

                                      ←   →        W5
EYIL GGP TWSMTLGAVEIFTSKHQRLHLLQGEQVASY FG HSVAVTDVNGD
   2        3              4                    1

GRHDLLV GAP LYMESRADHKLAEVGRVYLFLQSRGHHSLGTPSLLLTGTQ
      2          3                               4
   →                          W6
LYGR FG SAIAPLGDLNRDGYNDVAV AAP YGGPTGQGQVSVFLGQSEGLNS
                            1                2

          ←   →                    W7
QPSQVLHSPFAAGSA FG FSLRGATDIDDNGYPDLLV GAY GADKVVVYRAQ
      3           1                    2            3

       ←
PVVTATVQLMVQESL
      4
```

Fig. 3. Secondary structure of the porcine CD41 molecule. Sequences in β antiparallel sheets
are shown in green.

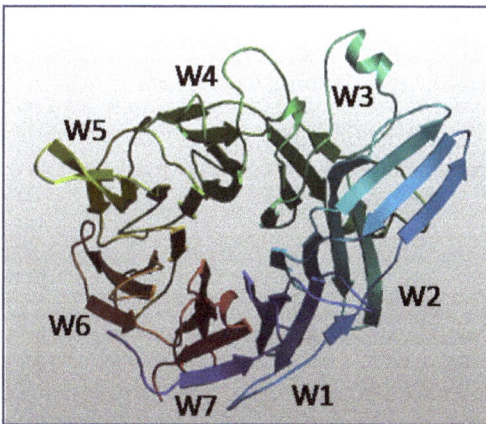

Fig. 4. Three-dimensional structure of the porcine CD41 molecule. The seven FG-GAP
tandem repeats are shown as W, each one composed by four β antiparallel chains.

3.2 Comparative analysis

The deduced protein sequence of the porcine CD41 was compared to their orthologous proteins from six different species: humans, cattle, horses, dogs, rats, rabbits, mice, zebrafish and xenopus (Figure5 and Table2).

```
PIG          MARALCLLHALWLLEWVQLLLGPGAAPPTWALNLDPVHLTIYTGPNGSHFGFSLDFYKKS  60
HUMAN        MARALCPLQALWLLEWVLLLLGPCAAPPAWALNLDPVQLTFYAGPNGSQFGFSLDFHKDS  60
HORSE        MARALRPLHALWLLEWMQLLLGPGTAPQAWALNLDPVRLTFYTGPNGSHFGFSLDFYKDS  60
RABBIT       MARALGPLPAFWFLEWALLLLGPGAGPPAWALNLDPVQLTIYTGLNSHFGFSLDFYKDS  60
DOG          MARAVCPLNALWLLEWVQLFLGPGAIPLGWALNLDPVQLTFYTGPNGSHFGFSLDFYKDN  60
COW          -------------------------PTWALNLDSVQFTVYTGPNGSHFGFSLDFYKNS  33
RAT          MARASCAWNTLWLLQWTPLFLGPSAAPPAWALNLDPVKFSVYTGPNGSHFGFSVDFHKDS  60
MOUSE        MARASCAWHSLWLLQWTPLFLGPSAVPPVWALNLDSEKFSVYAGPNGSHFGFSVDFHKDK  60
ZEBRA FISH   ----MDKKLEFSLLFLSILIFT----NHIRGFNLDLNQYTVFSGPEDSYFGFSVDFYQSS  52
XENOPUS      -----------MVPWLLLLLP----AFIQNLNLDK-KPQTLSGPPGSHFGFSMDFYNTA  43

PIG          HGSVSIVVGAPRTLG--RNLEETGGVFLCPWKAKSVQCVALSFNLD-DETRNVGAQTFQT  117
HUMAN        HGRVAIVVGAPRTLG--PSQEETGGVFLCPWRAEGGQCPSLLFDLR-DETRNVGSQTLQT  117
HORSE        RGSVSIVVGAPRTLG--RSQEEMGGAFLCPWKAEGGQCTSLSFDLN-DETRNTSSQIFQT  117
RABBIT       HGSVAIVVGAPRTLG--LGGQKETGGVFLCPWKAEGSPCSLLSFNLS-DEYRKTSSQLFQT  117
PERRO        HGRVAFVVGAPRTLG--RSQEETGGVFLCPWRAEGGQCTSLPFDLN-DETRHIGSHTFQT  117
COW          NGSVYVVVGAPRTLG--HSEEETGGVFLCPWKAEGGQCISLPFDLY-DETRSIGTQTFQT  090
RAT          HGSVSIVVGAPRALN--ANQEETGGVFLCPWKANNGTCTSLLFDLR-DETRKLSFQTFQT  117
MOUSE        HGSVSIVVGAPRALN--ASQEETGAVFLCPWKANGGKCNPLLFDLR-DETRNLGFQIFQT  117
ZEBRA FISH   SKSVSVVVGAPRANTNQSGVSHGGSVFMCPWATRGQSCQTLNFDQKGDENITFGNMLLMA  112
XENOPUS      DQGMSIVVGAPRMQTSQRNVTMGGGVFLCPWKPKGSSCVNIKFDSTGDRSIPFAGYTMKI  103

PIG          FKARQGLGASVLTWRDNVVACAPWQHWNVLEKNEE-AEKTPVGGCFVAQLQNSGRAEFSP  176
HUMAN        FKARQGLGASVVSWSDVIVACAPWQHWNVLEKTEE-AEKTPVGSCFLAQPESGRRAEYSP  176
HORSE        FKAQQGLGASVVSWSDYVVACAPWQHWNALEKTDE-AEKTPVGGCFVAQLENGRRAEYSP  176
RABBIT       FRARQGLGASVVSWNDIIVACAPWQQWNVLEKAAE-AEKTPVGGCFVAHLPSGRRAEYSP  176
DOG          FKSRQGLGASVVSWNDNIVACAPWQHWNVLEKTEE-AEKTPVGGCFVAQLRNGHRAEYSP  176
COW          FKAGQGLGASVVSWRDSIVACAPWQHWNVLDRNEEEAQKTPVGGCFVAHLQNGDRTEYSP  150
RAT          FKTGQGLGASVLSWNDVIVACAPWQHWNVLEKYDE-AEKTPVGGCFVAELQSGGRAEYSP  176
MOUSE        FKTGQGLGASVVSWNDVIVACAPWQHWNVLEKRDE-AEKTPVGGCFLAQLQSGGRAEYSP  176
ZEBRA FISH   HKSNQWLGASVRTYNNYILACAPLFHWNVLVDQEE-AMNTPVGNCQLLNMKTGELANYAP  171
XENOPUS      FKSNQWFGATVRTWNTAIVACAPFQQWNVMKLGSE-SGTTPTGTCYITNN-LEDIYEFAP  161

PIG          CRANLLSLVYVESKFND-RRYCEVGFSSAVTQAGELVLGAPGGYYFLSLMARAPIADIIS  235
HUMAN        CRGNTLSRIYVENDFSWDKRYCEAGFSSVVTQAGELVLGAPGGYYFLGLLAQAPVADIFS  236
HORSE        CRDNIMSHVYSKTYLGD-KRYCEAGFSSAVTQAGELVLGAPGGYYFLGLLARAPIANIVS  235
RABBIT       CRGNTMSHVYEKMYLRD-LRSCEAGFSSVITQEGELVLGAPGGYYYLGFLVRAPIANIIS  235
DOG          CRANTMSSVYVKNRFNQDKRYCEAGFSSAAVTQAGVLVLGAPGGYFFLGLLVRTPIDNIIS  236
COW          CRDNKMSQFYERNHFRDDRRYCEAGFSSVVTXAGELVLGAPGGYYFVSLLARAPIADIIS  210
RAT          CRSNTMSSVYSQGFSGD-KRYCEAGFSLAVTQAGELVLGAPGGYYFLGLLVRVPIENIIS  235
MOUSE        CRANTMSSVYAESFRGD-KRYCEAGFSLAVTQAGELVLGAPGGYYFLGLLARVPIENIIS  235
ZEBRA FISH   CREEYVYAITYTRG--YPDRRYCEAGFTTDITKNGRVVLGAPGGYYFQGQIITASLVNIMS  229
XENOPUS      CRESKMERHYEA-----DRRFCELGFSTDINKDGTLLAGAPWG-YFQGLYVTAGLPNILA  215

PIG          SYRPGTLLWHVPTQKLTFDSDLPEYYESYLGYSVAVGEFDRNPNTTEYILGGPTWSMTLG  295
HUMAN        SYRPGILLWHVSSQSLSFDSSNPEYFDGYWGYSVAVGEFDGDLNTTEYVVGAPTWSWTLQ  296
HORSE        SYRPGTLLWSVPTQRFTFDSMKPEYFDGYRGYSVAVGEFDEDLSTTEYVFGAPTWSWTLG  295
RABBIT       SYSPGVLLWTVPNQNFTFDYSNRKYFDGYRGYSVAVGEFDGDLSTTEYVLGAPTWSWTMG  295
DOG          SYRPGTLLWHVSSQSFTYDYSKPEYYDGYRGYSVAVGEFDGNLNTTEYVLGAPTWSCTLG  296
COW          SYRPSTLLWHVPTQ-FTYDQSHLQYYDGYRGYSVAVGNFDGNPNTTEYVFGAPTWSWTLG  269
RAT          TYRPGTLLWHVSNQRFSYDSSNPVYFHGYRGYSVAVGEFDGDLSTTEYIFGAPTWSWTLG  295
MOUSE        TYRPGTLLWHVSNQRFTYDNSNPVFFDGYRGYSVSVGEFDGDPSTTEYVSGAPTWSWTLG  295
ZEBRA FISH   SGSSFTPKHSMNGETKTPQRRD--YYDLYLGYSVAAGKFNND-NIPDYVVGVPNDLHTAG  286
XENOPUS      RPASSSLLQSYPGQQISPYIGS--SFDSYKGFSVAYGEFTGD-NTPEIVVGSP-KYQDRG  271
```

```
PIG         AVEIFT---SKHQRLHLLQGEQVASYFGHSVAVTDVNGDGRHDLLVGAPLYMESRADHKL 352
HUMAN       AVEILD---SYYQRLHRLRAEQMASYFGHSVAVTDVNGDGRHDLLVGAPLYMESRADRKL 353
HORSE       AVEILD---SNFQMLHRLHGEQMASYFGHSVAVTDVNGDRRHDLLVGAPLYMERRADRKL 352
RABBIT      AVEILD---SYFYRLHRLQGEQMASYFGHSVAVTDVNGDGRHDLLVGAPLFMASQADHKL 352
DOG         AVEILN---EYHQTLHRLHGEQMASYFGHSIAVTDVNGDGRHDLLVGAPLFMESRADRKL 353
COW         AVEILD---SYHQMLHRLHGEQMASYFGHSVAVTDVNGDGRHDLLVGAPLYMESRADRKL 326
RAT         AVEILD---SYYQTLHRLHGEQMASYFGHSVAVTDVNGDGRHDLLVGAPLYMESRVDRKL 352
MOUSE       AVEILD---SYYQPLHRLHGEQMASYFGHSVAVTDVNGDGRHDLLVGAPLYMESRADRKL 352
ZEBRA FISH  SVKIINGATVPLQIMKAISGTQIASYFGHSVAVTDINRDGWDDILIGAPLFMEQLSTQKF 346
XENOPUS     LVDIYT-VSNPWKTFISFLGKQVASYFGHSVAVTDVNNDGRDDVLVGAPLFMERRTRGKL 330

PIG         AEVGRVYLFLQSRGHHSLGTPSLLLTGTQLYGRFGSAIAPLGDLNRDGYNDVAVAAPYGG 412
HUMAN       AEVGRVYLFLQPRGPHALGAPSLLLTGTQLYGRFGSAIAPLGDLDRDGYNDIAVAAPYGG 413
HORSE       AEVGRVYLFLQPRSPQPLGPASLLLTGTRIYGRFGSAIAPLGDLNRDGYNDVAVAAPYGG 412
RABBIT      AEVGRVYLFLQPGPHLLGAPSLLLTGTQLYGRFGSAIAPLGDLNRDGYNDVAVAAPYGG 412
DOG         AEVGRVYLFLQPRGHQALGAPSLLLTGTQLYGRFGSAIASLGDLDRDGYNDVAVAAPYGG 413
COW         AEVGRVYLFLQTRGARMLGAPNLLLTGTQLYGRFGSAIAPLGDLNRDGYNDVAVAAPCGG 386
RAT         AEVGRVYLFLQPKGLQALSSPTLVLTGTQVYGRFGSAIAPLGDLNRDGYNDVAVAAPYGG 412
MOUSE       AEVGRVYLFLQPKGPQALSTPTLLLTGTQLYGRFGSAIAPLGDLNRDGYNDVAVAAPYGG 412
ZEBRA FISH  REVGQVYVYLQRNDFSFASRPNQILAGTYAYGRFGSAIAPLGDLDHDGFNDVAVGAP--G 404
XENOPUS     QEFGQVYVYLQRENRKFSFN-HPVLTGSQVYGRFGSSIAPLGDIDQDGFNDVAVGAPFGG 389

PIG         PTGQGQVSVFLGQSEGLNSQPSQVLHSPFA---AGSAFGFSLRGATDIDDNGYPDLLVGA 469
HUMAN       PSGRGQVLVFLGQSEGLRSRPSQVLDSPFP---TGSAFGFSLRGAVDIDDNGYPDLIVGA 470
HORSE       PDGRGQVLVFLGQSEGLSSHPSQVLDSPFS---TGSAFGFSLRGATDIDDNGYPDLLVGA 469
RABBIT      PSGRGQVLVYLGQSEGLNPHPSQVLDSPFP---AGSAFGFCLRGATDIDDNGYPDLLVGA 469
DOG         PSSLGQVLVYLGQSEGSSRPSQILDSPFP---AGSGFGFSLRGATDIDDNGYPDLLVGA 470
COW         PNGQGQVLVYLGQSEGLNPSPSQVLDSPFP---TGSGFGFSLRGATDIDDNGYPDLLVGA 443
RAT         PSGQGQVLIFLGQSEGLSPRPSQVLDSPFP---TGSGFGFSLRGSVDIDDNGYPDLIVGA 469
MOUSE       PSGQGQVLIFLGQSEGLSPRPSQVLDSPFP---TGSGFGFSLRGAVDIDDNGYPDLIVGA 469
ZEBRA FISH  SVDGGKVFIYLGKSGGLSTQYVQVIESPFRSLIDPPMFGFSIRGGTDIDDNGYPDLIIGA 464
XENOPUS     ESGGGCVFIYRGSPAGLSPQPSQILESPLP---PPAQFGFALRGGMDIDNNGYPDLLVGA 446

PIG         YGADKVVVYRAQPVVTATVQLMVQ-ESLNPAVKNCVQPQTKTPVSCFTIQMCVGATGHNI 528
HUMAN       YGANQVAVYRAQPVVKASVQLLVQ-DSLNPAVKSCVLPQTKTPVSCFNIQMCVGATGHNI 529
HORSE       YGANKVAVYRAQPVVMVSVQLLVN-DSLNPAVKNCVLPQKKTSVSCFDIQMCVGVTGHNI 528
RABBIT      YGADKVVVYRAQPVVMADVQLLVQ-DSLNPAVKNCVLHQTNTPVSCFNIQMCVGVTGHNI 528
DOG         YGASKVAVYRAQPVVVANVQLLVQ-DSLNPAVKNCILPQTKTPVSCFNIQMCVGATGHNI 529
COW         YGASKVAVYRAQPVVMVTVQLMVQ-DSLNPAVKTCVLSQTKTPVSCFNIQMCVGATGHNI 502
RAT         YGASKVAVYRAQPVVMATVQLMVQ-DSLNPTLKNCVLEQTKTPVSCFNVQMCVGATGHNI 528
MOUSE       YWASKVAVYRAQPGVMATVQLMVQ-DSLNPTLKNCVLDQTKTPVSCFNIQMCVGATGHNI 528
ZEBRA FISH  WGASKVVTYRAQAVVRTQARLSFFPDLLNPEDKFCQLQQSGTYITCFTIMACIRVSGHRI 524
XENOPUS     FHADKVFIFRTQPVVVLQASLFFNPEALNPDEKLCNFPQSGPAVSCFTIRVCAQASGRSL 506

PIG         PEKLRLNAELQLDRQKPHQSRRVLLLASQQASTVLDVDLTWRQSPTCHNTTAFLRDEADF 588
HUMAN       PQKLSLNAELQLDRQKPRQGRRVLLLGSQQAGTTLNLDLGGKHSPICHTTMAFLRDEADF 589
HORSE       PEKLRLNAELQLDRQKPRQGRRVLLLSSQQAGTTLHLDLGGRTSPNCRTIEAFLRDEADF 588
RABBIT      PQGLYLQAELQLDRQKPRQGRRVLLLGSQQASTTLSMDLGGRSRLCHNTTAFLRDEADF 588
DOG         PQQLSLNAELQLDRQKPRQGRRVLLLNSQLASSTLHLDLGGRHSPICHTTAFLRDEADF 589
COW         PEKLHLNAELQLDRQKPRQGRRVLLLDSQQAGTILNLDLRGRHNPNCSTATAFLRDEADF 562
RAT         PQKLHLKAELQLDLQKPRQARRVLLLASRQASLTLSLDLGGRNKPICHTIKAFLRDEADF 588
MOUSE       PQKLHLKAELQLDLQKPRQGRRVLLLASQQASLTLSLDLGGRDKPICHTTGAFLRDEADF 588
ZEBRA FISH  PQQIVFNTELQLDRMKQSMARRTLLLDSNQPYTNFQISVDRNSRDVCRNFTAYLLP--EF 582
XENOPUS     PKKISLSAELQLDRLKSRFARRTFFLDSSQPSKTIDMELQSNSAQLCQNLTPYLRGESEF 566

PIG         RDKLSPIVLSFNVSLQPEKNGDALTFMLHGDTHVQEQTRIILDCGEDQVCVPKLQLSANT 648
HUMAN       RDKLSPIVLSLNVSLPPTEAGMAPAVVLHGDTHVQEQTRIVLDSGEDDVCVPQLQLTASV 649
HORSE       RDKLSPIVLSLNVSLQPEKDGIAPALVLHGDTHVQEQTRIILDCGEDDVCVPQLHLTANV 648
RABBIT      RDKLSPIVLSFNVSLQPKEAGVAPAVVLHGNTHVQEQTRIILECGEDDVCVPQLHLTASL 648
DOG         RDKLSPIVLSLNVSLQPEKDGGAPALVLHGDTHVQEQTRIILDCGEDDLCVPQLQLTAIV 649
COW         RDKLSPIVLSFSVSLPPEKDGGAPALVLHGNTHVQEQ--------------------- 599
RAT         RDKLSPIVLSLNVSLPPEETGVAPAVVLHGVTHVQEQTRIILDCGEDNLCVPQLQLTATA 648
MOUSE       RDKLSPIVLSLNVSLPPEETGAPAVVLHGETHVQEQTRIILDCGEDDLCVPQLRLTATA 648
ZEBRA FISH  KDKLSPIFISVNYSLADSQ-----NAVLHGQSVAVGQTRIILNCGPDNVCIPDLQLKAVT 637
XENOPUS     KDKLSPIAMSVNFSLVRAQSMDTVQPTLHGTTFLQEQTNILLDCGDDNVCIPNLHLTANW 626
```

```
PIG        TGSPLLVGADNVLELHVVAANEGEGAHEAELVVHLPPGAHYMQALSNTKSFERLICTQKK  708
HUMAN      TGSPLLVGADNVLELQMDAANEGEGAYEAELAVHLPQGAHYMRALSNVEGFERLICNQKK  709
HORSE      TGSPLLIGADNVLKLQMDATNEGEGAYEAELAVQLPPGAHYMQALSNIEGFERLICDQKK  708
RABBIT     KGSPLLIGADNVLELQMVAANDGEGAYEAELVVHLPLGAHYMRAVSTMEGLERLICNQRK  708
DOG        MGSPLLIGADNVLELQMDAANEGEGAYEAELAVHLPPGAHYMRAISNIEGFERLICNQKK  709
COW        ------------------------------------------GFERLICNQKK  610
RAT        GDSPLLIGADNVLELKVNASNDGEGAYEAELAVHLPPGAHYIRAFSNVEGFERLVCTQKK  708
MOUSE      GDSPLLIGADNVLELKIEAANDGEGAYEAELAVHLPPGAHYMRALSNIEGFERLVCTQKK  708
ZEBRA FISH STEPILIGDENPALLIIEAENQGEGAYETELYISPPANTHYQGVLSNHEDFSALVCGQKK  697
XENOPUS    SADPLLIGIDNLVHVQFNAANLGEGAYEAELYVWLPNGAHYMQVLG--EAEEKILCSPKK  684

PIG        ENETKVVLCELGNPMKGDTQIEITMLVSVGNLEEAGEHVSFRLQIRSKNSQNPNSETVVL  768
HUMAN      ENETRVVLCELGNPMKKNAQIGIAMLVSVGNLEEAGESVSFQLQIRSKNSQNPNSKIVLL  769
HORSE      ENETKVVLCELGNPMKRNAQIEITMLVSVENLEEAGETVSLQLQIRSKNSKNPNSETLRL  768
RABBIT     ENQTKAVLCELGNPMK-QARIGITMLVSVGNLEDAGESVSFQLQIRSKNSQNPNSEAVLL  767
DOG        ENETKIVLCELGNPMKRNARIGITMLVSVENLEEAGEHVSFWLQIRSKNSQNPNSEAVLL  769
COW        ENETKVVLCELGNPMKSNAQIEVMMWVSVEKLEEAGEQVSFLLQIRSKNSQNPNSEMVEL  670
RAT        ENESRLALCELGNPMKKDTRIGITMLVSVEILEEAGDSVSFQLQIRSKNSQNPNSEAVLL  768
MOUSE      ENESRVALCELGNPMKKDTRIGITMLVSVENLEEAGESVSFQLQVRSKNSQNPNSKVVML  768
ZEBRA FISH ENGSVIVVCDLGNPLEAGQQLKAGLYFSMGDLEQVENHITFQMQIRSKNSQNSDSNLVQL  757
XENOPUS    GNESEIVVCELGNPMKNGAEIHADLQLSFSNLEDSGSTVTFQMQIKSRNTVNSASSLFLV  744

PIG        DVQVRAEAHLELRGNSFPASLLMAA-EGDWE---NSSDNLGPKVEHTYELHNNGPSTVSG  824
HUMAN      DVPVRAEAQVELRGNSFPASLVVAAEEGEREQ--NSLDSWGPKVEHTYELHNNGPGTVNG  827
HORSE      HVPVRAEARVELRGNSLPASLVVAAEEDDRK---NSSDSWGPKVEHTYELHNNGPGAVRG  825
RABBIT     AVPVRAAAQVELRGNSFPASLVLAEEGDQEQ------NSLDLKVEHTYELHNNGPGTVRG  821
DOG        DVPVRAEAHVKLRGNSFPASLVVVAEEDNRE---NSSESWGPKVEHTYELHNNGPGTVSG  826
COW        DVPVRAVAHVELRGNSFPASLVVAAEEGNGQ---NSSDSWGPKVEHTYELHNNGPGAVRG  727
RAT        PVAVRAEAAVELRGNSFPASLVVAAEEVDKEQ--DGLDSWVSRVEHTYELHNNGPGTVNG  826
MOUSE      PVAIQAEATVELRGNSFPASLVVAAEEGDREQ--EDLDSWVSRLEHTYELHNIGPGTVNG  826
ZEBRA FISH QVNVTAVASLEMRGVSSPVDCVLPISKWESKDYPEDLDEVGPLIEHVYELRNRGPSPVN-  816
XENOPUS    TMAVKVTASLELRGSSHPAEVILPLPNWEPREEWRKAQDYGEEVTHVYELHNSGPGSVH-  803

PIG        LHLRLHLPTGQSQTSDLLYIVDIQTQGGLQCSPQPSNPRQLDWG--LHSPTPSPVYPAH  882
HUMAN      LHLSIHLPG-QSQPSDLLYILDIQPQGGLQCFPQPPVNPLKVDWG--LPIPSPSPIHPAH  884
HORSE      LRLSLHLPS-QSQPSDLLYILDIQPQGGLQCSPQPSNPLKLDWG--LPTPSPSPVYHPR  882
RABBIT     LHLTIHLPG-QSQPSDLLYILGIEPQGGLQCSPQPSNPLKINWR--LPTPSPSPMHPGY  878
DOG        LHLHLCFPG-ESQPSDLLYILDIQPEGGLQCSPQPSINPFKLDWR--QPTPSPSPTSPGY  883
COW        LRLNLYLPS-QSQPSDLLYILDIHPQGGLQCASQPSNPLQLEWR--LPTPSPS---PAH  781
RAT        LSLIIHLPG-QSQPSDLLYILDVQPKGGLLCSTQPPPKLLKVDRS--LTTPSPSSIRRIH  883
MOUSE      LRLLIHIPG-QSQPSDLLYILDVQPQGGLLCSTQPSP---KVDWK--LSTPSPSSIRPVH  880
ZEBRA FISH VKLTLEFPV-SQNESYLLYVFANASEELISCQTDYAN----IDPRRLVKQESTNITVAEV  871
XENOPUS    VQLLLQSPE-MYHGDYFLYPLRLEVDDGMTCDNQSALNPLKLDILTSTEEPANYSSRSGD  862

PIG        HKRNRRQAVL              PGQKQPLSLQDRILLSCD-SARCTVVHCDLREMARG  927
HUMAN      HQRERREAFL--------------PEPEQPSRLQDPVLVSCD-SAPCTVVQCDLQEMARG  929
HORSE      HQRERREAFL--------------PGPMQPSRLQDVLNCD-SAPCTVVQCELEEMARG  927
RABBIT     RRERRHADLLE------------PQPSSAAGPRDPVLVSCD-SAPCTVVQCELEEMARG  924
DOG        HKRERRQASL--------------PGSSQPSGLQDPVLLSCK-SGPHTVVQCELQEMARG  928
COW        HKRDRRQAVL--------------PEEKQPSRLQDPILVSCD-SAPCTVVQCELQEMARG  826
RAT        HDRDRREASP-------------QGSKQTEQQDPVLVCNGSAPCTVVECELQEMVRG  928
MOUSE      HQRERRQAFL--------------QGPK-PGQQDPVLVSCDGSASCTVVECELREMVRG  924
ZEBRA FISH HHFNKRDLS--------------QKTENEQQWQHTVHVNCSSSEQCVVFDCVAAGLQRD  916
XENOPUS    HRLERRDLRRWGADEGMQEDGVNITKKDEKPPRNHTVLLNCSSFP-CWEVQCSVQNLERG  921

PIG        QRAMVTAQAFLWLPSLRQRPLDQFVLRSHAWFNVSSLPYAVPALSLPSGEAQVQTQVLRV  987
HUMAN      QRAMVTVLAFLWLPSLYQRPLDQFVLQSHAWFNVSSLPYAVPPLSLPRGEAQVWTQLLRA  989
HORSE      QRAMVTVRAFVWLPSLRQKLLDQFVLQSRAWFNVSSLPYAVPTLSLPSGEALVQTQLLRV  987
RABBIT     QRAMVTVLALLGLSSLRERPLDQFVLQSQAWFNVSSLPYAVPALSLPSGEALVQTQLLRV  984
DOG        QRAMVKVLAFLQLPSLQQRPLDQFVLESQAWFNVSSLPYAVPSLSLPSGETLVQTHLLRA  988
COW        QRVMVTVLALLSRSILQERPLDQFVLQSHAWFNVSSFPYSVPALSLPSGEAQVQTQLLRV  886
RAT        QRAMVTVQATLGLSILRQRPQEQFVLQSHAWFNVSSLPYSVPVVSLPSGKALVQTHLLRA  988
MOUSE      QRAMVTVQAMLGLSSLRQRPQEQFVLQSHAWFNVSSLPYSVPVVSLPSGQARVQTQLLRA  984
ZEBRA FISH ERAIVRVMSRLWVQTFLKRPYVNYVLHSTAHYEVMNVPSKIQPDVLPTGKAETHTKIIWR  976
XENOPUS    GRATVKLHSILWVPSFLKRQQQQFVLLSQGSFWVTSVPYKIQPAVLLYGNATANTTVLWV  981
```

```
PIG          RVLEDREVPLWWMLVGVLGGLLLLTLLVLAMWKCGFFKRNRPPLEESDEEEE------- 1039
HUMAN        L--EERAIPIWWVLVGVLGGLLLLTILVLAMWKVGFFKRNRPPLEEDDEEGE------- 1039
HORSE        L--EDRAIPIWWVLVGVLGGLLLLTLLVVAMWKVGFFKRNRPPLEEDEEDE------- 1036
RABBIT       L--EEKAVPIWWVLVGALGGLLLLILLVLAMWKVGFFKRNRLPLEEEDEDEE------ 1034
DOG          L--EERDIPIWWVLVGVLGGLLLLMLLVLAMWKGGFFKRNRPPLEEEEEE-------- 1036
COW          S--EEREIPMWWVLVGVLGGLLLLTFLILAMWKVGFFKRNRPPLEEDDEEE------0935
RAT          L--EERDIPVWWVLVGVLGGLLLLTLLVLAMWKAGFFKRNRPPLEEEEEEE------- 1037
MOUSE        L--EERAIPVWWVLVGVLGGLLLLTLLVLAMWKAGFFKRNRPPLEEDEEE-------- 1033
ZEBRA FISH   SPDGQEEVPLWWIVVSIVAGLLLLAALSTIFWKMGFFKRNRPPSDNDDDDDDDVTQQLN 1036
XENOPUS      SPDGQKEIPLWWIIVGALGGLLLLLALFVFVMWKLGFFRRTRPPSDDQEDLTSD----- 1034
```

Fig. 5. Comparison of the porcine CD41 amino acid sequence to other homologous molecules. The sequences were derived from GenBank entries with accession numbers shown in materials and methods. Signal peptide is in green. Heavy and light chains are shown by orange and black lines, respectively. Ca⁺⁺ binding domains are remarked in pink boxes. Potential N-glycosylation sites (N) and cysteine residues (C) are respectively marked in red and yellow in the respective sequences. Amino acids residues conserved in all the sequences are shown in light grey.

As shown in Table2, the longest porcine CD41 protein shares a 78% amino acid residue identity with those of humans, cattle and horses, 77% with dogs, 75% with rabbits, 73% with rats, 71% with mice, 42% with *Xenopus laevis* and 40% with zebrafish. Table2 also shows the percentages of amino acid residue identities of the different regions of the CD41 molecule. In general, both transmembrane and cytoplasmic domains are more preserved compared to the extracellular one. The phylogenetic tree of CD41 proteins shows that the counterpart closet to porcine CD41 was that of cows (Figure6).

ESPECIES	PROTEIN	EXTRACELLULAR	TARNSMEMBRANE	CYTOPLASMIC
HUMAN	78	77	80	85
COW	78	77	80	89
HORSE	78	77	80	78
DOG	77	77	80	83
RAT	71	71	84	84
MOUSE	73	72	84	84
RABBIT	75	74	80	80
ZEBRA FISH	40	41	53	55
XENOPUS	42	42	69	40

Table 2. Percentages of amino acids identities between the porcine (Po) and its constitutive blocks with those from humans (Hu), cow (Ca), horse (Ho), dog (Do), rat (Ra), mice (Mi), rabbit (Rb), zebrafish (Zf) and *Xenopus laevis* (Xe).

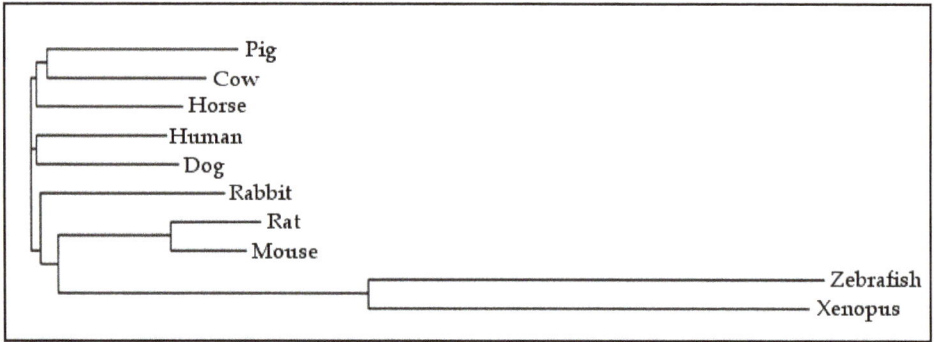

Fig. 6. Phylogenetic tree of the CD41 protein family.

3.3 Chromosome localization of porcine *CD41* gene

Chromosomal localization was carried out screening a pig-rodent somatic hybrid cell panel by PCR, using specific porcine CD41 primers (Table1). A specific amplification was observed in 9 (16, 20, 21, 22, 23, 24, 25, 26 and 27) of the 27 hybrid cells (Figure7), which enabled us to localize the porcine *CD41* gene in region p11-2/3p13 of chromosome 12 (*Sscr* 12) with a probability of 0.90 and an error margin lower than 0.1% (Chevalet et al., 1997, www.toulouse.inra.fr/lqc/pcr.htm).

The chromosome localization of the gene was confirmed by screening the INRA Minnesota porcine Radiation Hybrid (IMpRH) panel. The IMpRH mapping tool (Milan et al., 2000; www.imprh.toulouse.inra.fr) revealed that porcine *CD41* gene is closely linked to the SW957 marker (47cM; LOD=9) on the *Sscr* 12, p11-p13 region (Figure8).

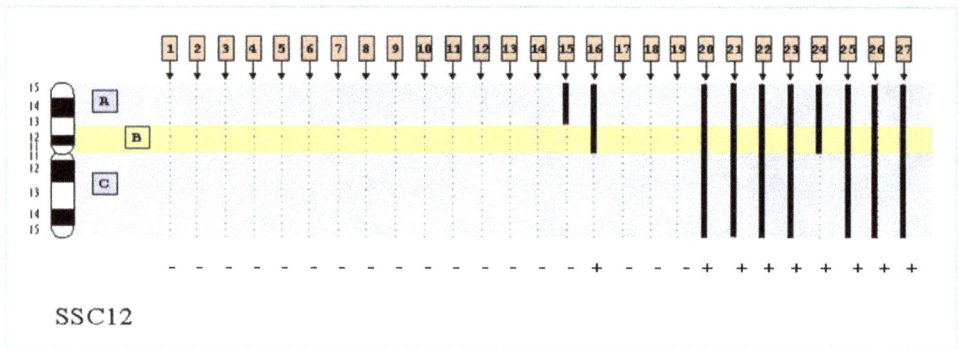

Fig. 7. Diagram showing results for the presence/absence of the *CD41* gene in the INRA somatic hybrid cell panel.

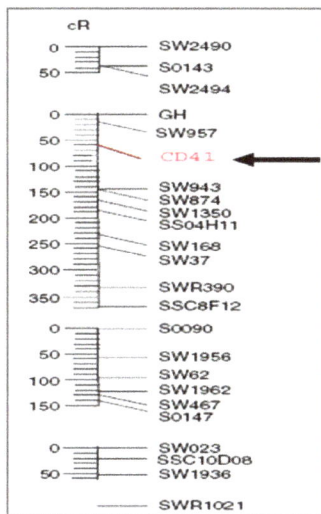

Fig. 8. Diagram showing the chromosomal localization of the *CD41* gene using the INRA
Minnesota porcine Radiation Hybrid (IMpRH) panel.

3.4 Cell and tissue expression of porcine *CD41* transcripts

To investigate the pattern of the porcine *CD41* mRNA expression, RT-PCR analysis was
conducted with a variety of pig adult tissues and cell types using VARP1/VARP2 gene-
specific primers (Table1). The highest level of CD41 transcripts was detected in platelets,
although a moderate level was detected in bone marrow and a low level in ganglions and
lungs. No *CD41* transcripts were detected in the rest of tissues and cells analyzed (Figure9).

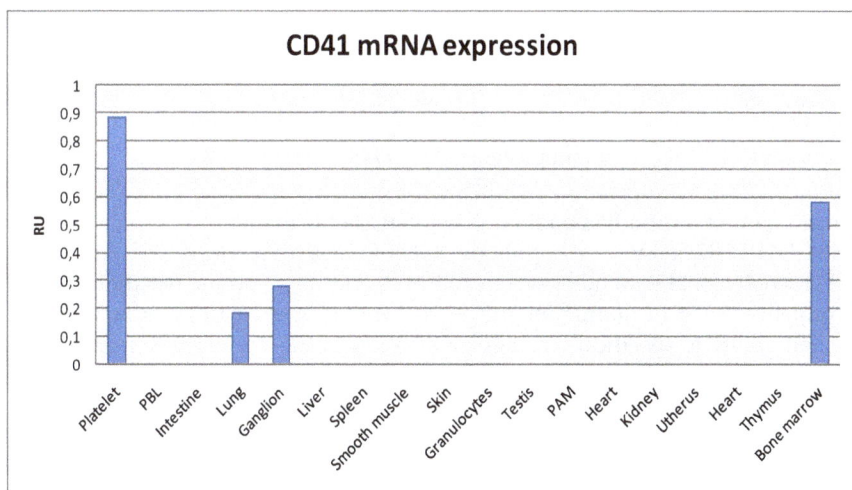

Fig. 9. RT-PCR expression patterns of *CD51* transcripts in different pig cells and tissues. RU:
Relative units. 18S RNA amplification was used as control.

3.5 Cell and tissue expression of porcine CD41 proteins

The precise localization of the *CD41* protein was studied by immunohistochemistry and by flow cytometry with antibodies developed against two different porcine CD41 recombinant proteins.

3.5.1 Expression and purification of porcine recombinant CD41 proteins

Two different cDNA fragments belonging to the functional region of the porcine CD41 protein were amplified and subcloned in the *pET-28b* expression vector. One, 996 bp long and amplified with primers CD41-F2 and CD41-R2 (Table1), contained the coding sequence for amino acids 32 to 363, and the other one, 677 bp long and amplified with primers CD41-F1 and CD41-R1, contained the coding sequence for amino acids 684 to 909, a highly antigenic region selected by the Jameson-Wolf method (Jameson & Wolf, 1988). The recombinant constructions, named respectively *pET-F2R2* and *pET-F1R1*, were transfected and expressed in *E. coli* (DE3). Two different recombinant CD41 proteins were purified: rCD41-F1R1 (about 26 kDa) and rCD41-F2R2 (about 46 kDa) (Figure10).

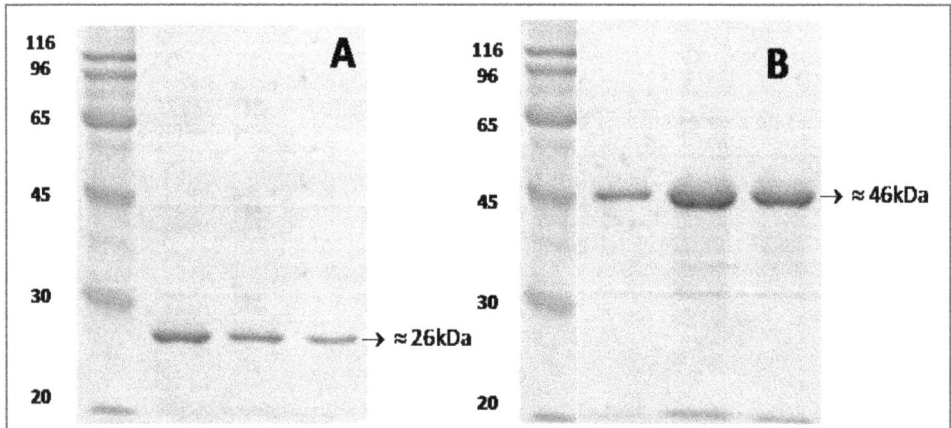

Fig. 10. Purified CD51 recombinant proteins. A: rCD41-F1R1. B: rCD4-F2R2.

3.5.2 Production of antibodies against porcine rCD41 proteins

An anti-rCD41-F2R2 monoclonal antibody (GE2B6), and two anti-rCD41-F2R2 and anti-rCD41-F1R1 polyclonal antibodies were produced, and their specific reactivity against the rCD41 proteins tested in immunoblottings. Before being used in immunohistochemical assays, their ability to specifically recognize the platelet CD41 molecules were carried out through immuno precipitations of platelets lysates in non-reduced conditions. An anti-porcine CD61 (JM2E5), previously produced by us (Pérez de la Lastra et al., 1997), was used as a positive control. Results are shown in Figure11 in which both polyclonal antibodies, the same as JM2E5, identified two proteins of 110 and 90 kDa, corresponding to the α and β chains of the receptor $\alpha_{IIb}\beta_3$. The antibodies α_{IIb} specific recognition was demonstrated through an immunoblotting of platelet lysates in non-reducing conditions (Figure12).

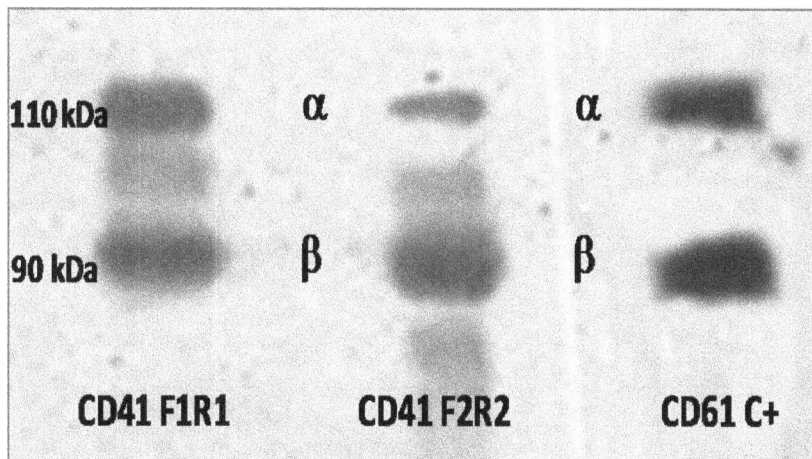

Fig. 11. Immunoprecipitation results of platelet lysates with anti-rCD41-F1R1 and anti-rCD4-F2R2 in non reducing conditions. JM2E5 anti-CD61 antibody was used as a control (C+).

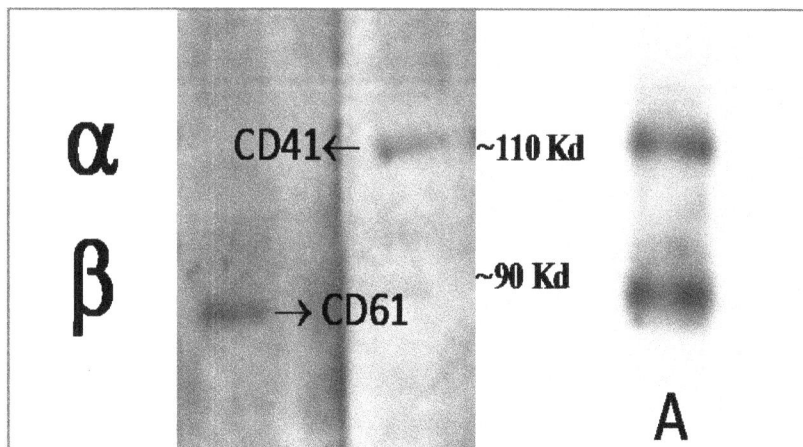

Fig. 12. Immunoblotting results of anti-rCD41-F2R2 and JM2E5 against a porcine platelet lysate.

3.5.3 Immunohistochemical detection of CD41 proteins

The reactivity of the anti-CD41 monoclonal and polyclonal antibodies was tested by immnohistochemestry on a variety of porcine tissues and cells types. Results are shown in Figures 13 and 14. Immunoreactivity was only detected in the membranes of megakaryocytes from bone marrow. No reactivity was detected in any of the tissues checked, including ganglion, in which a weak CD51 transcription was detected by RT-PCR.

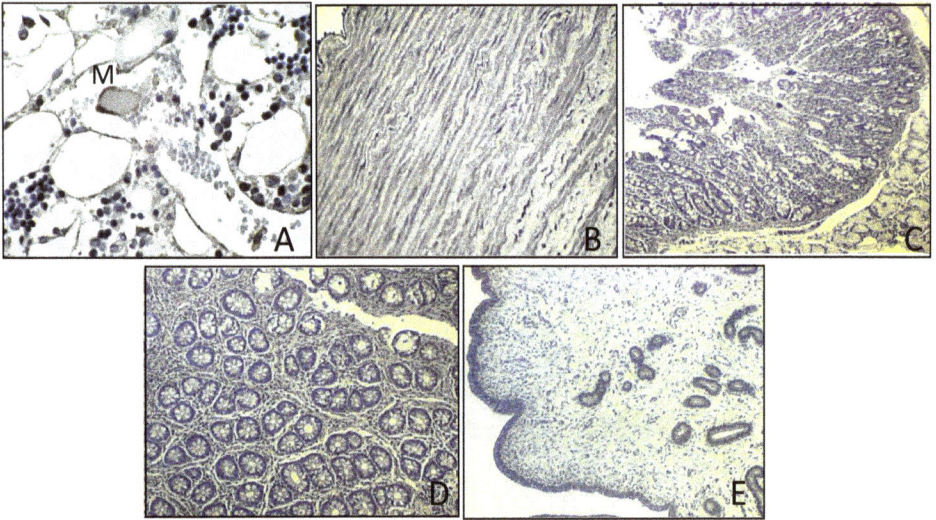

Fig. 13. Immunohistochemistry results with anti-rCD41-F2R2. A: Bone marrow (40X), MK, megakaryocyte. B: Aorta (20X). C: Small intestine (10X). D: Large intestine (10X). E: Uterus (10X).

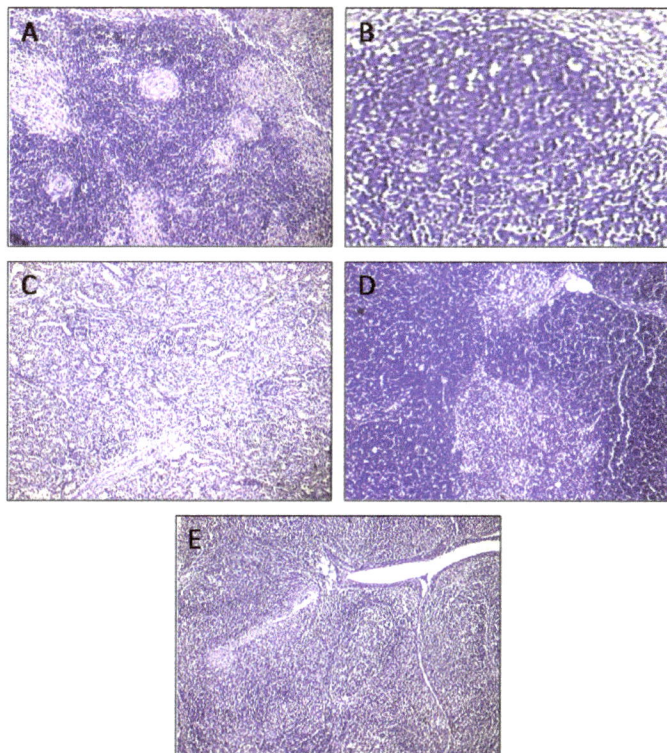

Fig. 14. Immunohistochemistry results with anti-rCD41-F2R2. A: Spleen (2X). B: Ganglyon
(20X). C: Kidney (10X). D: Thymus (10X). E: Tonsil (10X).

3.5.4 Detection of CD41 proteins by flow cytometry

In order to identify the possible PBL cells that express CD41 proteins we carried out a flow cytometry analysis by using both anti-CD41 polyclonal antibodies. Figure15 shows the results in platelets, lymphocytes, granulocytes and erythrocytes with anti-CD41-F2R2. CD41 proteins were only detected in platelets by both polyclonal antibodies.

Furthermore, to test the platelet porcine specificity of the antibodies produced in this study we test their reactivity with platelets from pigs, humans, dogs, horses, goats, chats, sheep and cows by flow cytometry. Both anti-CD41-F2R2 and anti-CD41-F1R1only reacted with porcine platelets (Figure 16), confirming the porcine CD41 specificity of both antibodies.

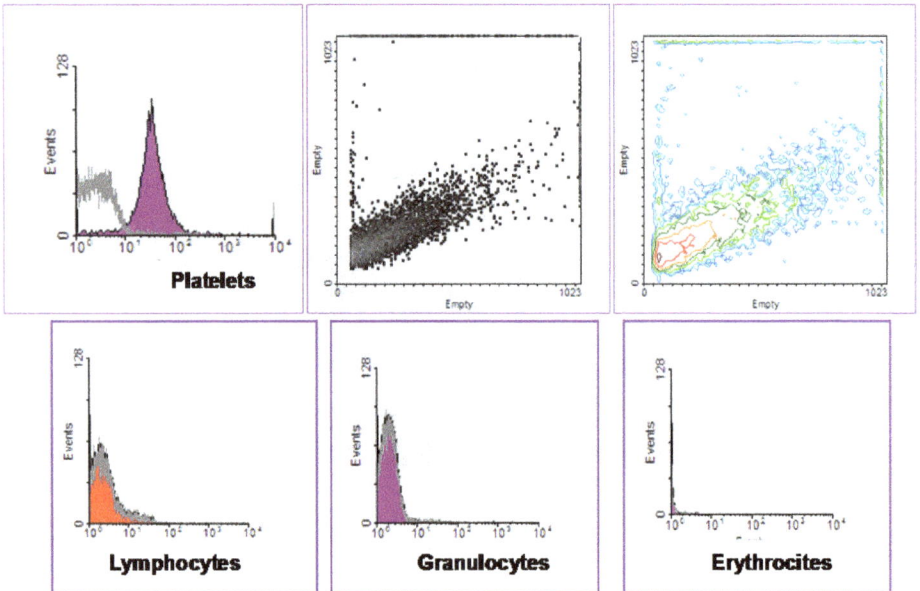

Fig. 15. Flow cytometry with anti-rCD41-F2R2 detecting expression in blood cells.

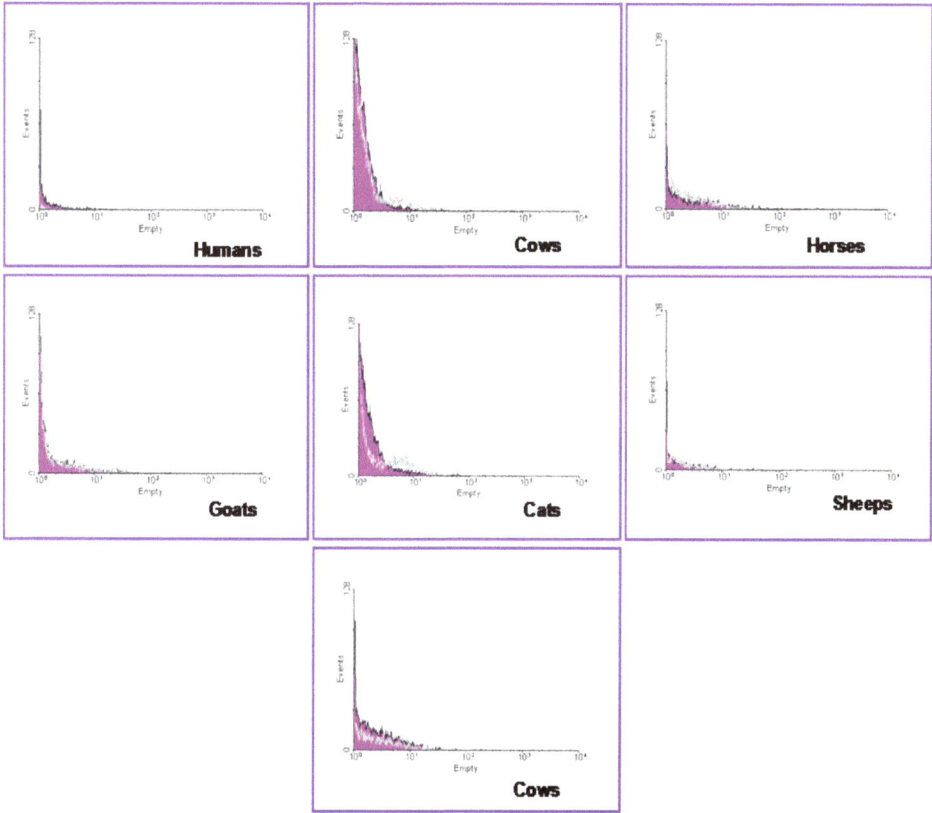

Fig. 16. Flow cytometry with anti-rCD41-F2R2 showing no CD41 expression in platelets
from different mammals.

3.6 Effect of the platelet activation on the expression of porcine CD41

Previous results obtained in our lab using a two dimension differential in gel electrophoresis
(2D-DIGE) technique had shown that the proteome of thrombin activated porcine platelets
showed a reduced number of proteins affected in their expression level, among which CD41
was not found. Although CD41 is strongly expressed in platelets, the membrane proteins are
usually poorly represented in the gels as a consequence of their high hydrophobicity. As we
had produced specific anti-CD41 polyclonal antibodies, we used the anti-rCD41-F2R2 to
check, using immunoblotting, if CD41 was or not present in a similar gel than that used in
our previous study. Results are shown in Figure 17 in which CD41 integrin was clearly
detected.

In order to test if the *CD41* transcripts level was or not modified in the platelets after
activation by thrombin, we carried out a real time quantitative PCR (rt-q-PCR) with RNAs
from unstimulated and stimulated platelets. Results are shown in Figure18 in which a
higher but not significant change in the *CD41* transcripts level was detected after the
activation by thrombin (the significant value is 1.5). Three replicates were assayed with very
similar results.

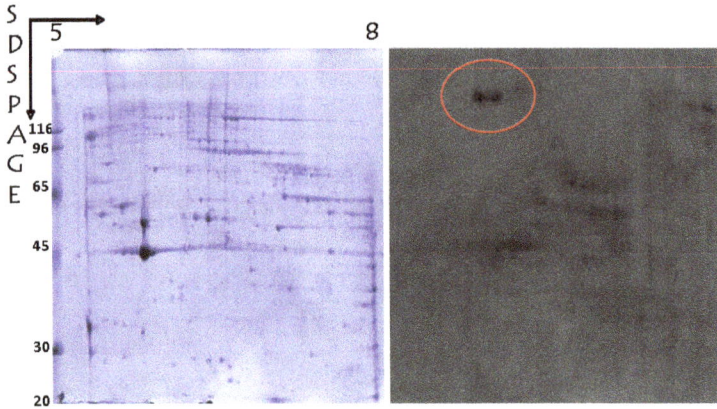

Fig. 17. A: Two dimension gel electrophoresis showing the platelet proteome stained with Comassie blue. B: Immunoblotting of platelet proteome with anti-rCD41-F2R2. Red circle shows detection of CD41 protein.

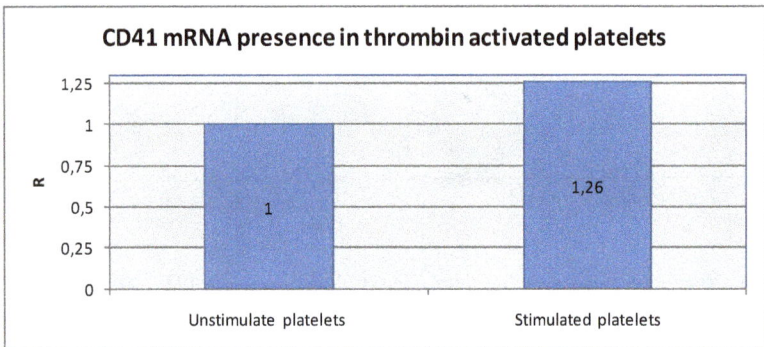

Fig. 18. Real time quantification of CD51 transcripts in unstimulated and thrombin stimulated platelets. R: Ratio (relative abundance to *β-actin* mRNA). Results represent the average of three replicates.

4. Discussion

In general, the study of genes expressed in platelets is difficult since platelets are enucleated cells that show a reduced level of protein synthesis, and megakaryocytes, the platelets precursors, represent only 0.1% of bone marrow cells (Bray et al., 1987). Nevertheless, in the present study we describe for first time the cloning and characterization of the full-length cDNA for the porcine CD41 (α_{IIb}) integrin chain.

The porcine CD41 proteins share common structural elements, including cytoplasmic, transmembrane and extracellular domains and the position of the proteolytic cleavage sites with the CD41 protein of other species. The porcine CD41 integrin showed an average of 75% amino acid identity with their mammal orthologous molecules, being the conservation in the transmembrane and cytoplasmic regions higher than in the extracellular one in all the species compared. The phylogenetic tree of CD41 family of proteins showed that the closest to porcine CD41 were those of cows and horses, and that the clusters of domestic mammals showed the less divergence in evolution. However, compared with other α mammal integrins, like α_v which show 90% of identity (Yubero et al., 2011), α_{IIb} integrins show lower level of conservation, which could be associated with the number of β chains with which they can form receptors: only one (β_3) for α_{IIb}, and at least five for α_v.

Porcine CD41 conserves all the main structural characteristics that define their functions in other species. The extracellular domain shows that porcine CD41 belongs to α integrins lacking I domain, a domain present in the NH_2 extreme of some integrins, like α_1, α_2 or β_2, which contains the functional sites to bind to ligands (Dickeson & Santoro, 1998; Humphries, 2000). Porcine and human α_{IIb}, the same as α_v, α_5 and α_8 ones, spreads the ligand binding sites among the first 334 NH_2 amino acid residues (Loftus et al., 1996). One characteristic of the α integrins lacking I domain is the presence of the seven FG-GAP tandem repeat sequences (W_1 to W_7; see Figures 3 and 4). Each FG-GAP sequence determines four antiparallel β chains, and the folding of all the seven FG-GAP sequences establishes the globular structure of the integrin, which contains the ligand binding sites. Fibrinogen is the main ligand for $\alpha_{IIb}\beta_3$ complex. In humans, the binding of fibrinogen to $\alpha_{IIb}\beta_3$ receptor requires the α_{IIb} chain Ala_{294} to Met_{314} residues, which are located in the third FG-GAP repeat (D'Souza et al., 1990). Other experiments, including molecular characterization of the Glanzmann thrombasthenia and mutagenesis analysis, have shown that residues Ala_{145}, Asp_{163}, Leu_{183}, Glu_{184}, Tyr_{189}, Tyr_{190}, Phe_{191} and Asp_{224} of the α_{IIb} chain, are also critics for the fibrinogen binding (Grimaldi et al., 1998; Honda et al., 1998; Tozer et al., 1999). The comparison between human and porcine α_{IIb} integrin sequences showed that all these critic residues are conserved in the porcine molecule.

On the other hand, the fibrinogen only binds to the activated $\alpha_{IIb}\beta_3$ integrins, this activation being mediated by Ca^{++} (Bennett & Vilaire, 1979). The molecular characterization of α_{IIb} carried out in this study showed that all the four Ca^{++} binding domains (consensus sequence DX[D/N]XDGXXD) were also highly conserved in the porcine α_{IIb} integrin when compared to that in humans.

The transmembrane region of the porcine α_{IIb} integrin is also highly conserved when compared to their homologous mammalian (80-84% of identity). The sequence GXXXG in this region is essential for a high affinity association of the transmembrane helixes (Senes et al., 2000). Changes as AXXXG or SXXXG in this sequence reduce significantly the affinity between them (Mendrola et al., 2002; Schneider & Engelman, 2004). Our results showed that the same GVLGG sequence was conserved in the α_{IIb} integrins from all the mammalian species compared, including that of the pig. It has been suggested and supported a "push-pull" mechanism for $\alpha_{IIb}\beta_3$ regulation in which the destabilization of the hetrodimeric α_{IIb} and β_3 transmembrane interactions push $\alpha_{IIb}\beta_3$ to its activated state, whereas processes that favor their homomeric association pull $\alpha_{IIb}\beta_3$ toward its active conformation (Li et al., 2005; Yin et al., 2006). This is in concordance with the high conservation of the GVLGG sequence in the transmembrane region of the porcine (and other mammal) α_{IIb} chains, since fibrinogen binding to $\alpha_{IIb}\beta_3$ is a prerequisite for platelets aggregation (Bennett, 2005). It is worthy to note that the porcine α_v integrin, also present in platelets membranes, contains an AVLGG sequence in the transmembrane region, as well as in all their mammalian homologous with which it was compared (Yubero et al, 2011).

The cytoplasmic region of the porcine α_{IIb} integrin is also highly conserved (80-89% of identity when compared to their homologous mammalians). A short GFFKR motif, which was involved in the activation of the integrin receptors, is present in the cytoplasmic region of all the human α integrins. In humans, mutations in the GFFKR motif of the $\alpha_{IIb}\beta_2$ integrin receptor induce a permanent activation of the integrins. As expected, in all the species compared in this study, the porcine α_{IIb} integrin contains this motif near to the transmembrane region. The porcine α_{IIb} integrin also contains in the cytoplasmic region the PPLEE motif, present in all the mammalian α_{IIb} integrin compared in this study, whose modifications determine changes that interfere with the specific recognizing of the ligands (Filardo & Cheresh, 1994).

Once the porcine α_{IIb} integrin was characterized, we used a porcine radiation hybrid panel and a somatic cell hybrid panel to map the pig CD41 (α_{IIb}) gene into swine chromosome 12 (Sscr 12), region p11(2/3)-p13. This chromosomal localization is in total concordance with heterologous painting data that demonstrate the correspondence between the swine Sscr 12 and the human Hsap 17 chromosomes (Rettenberger et al., 1995), where CD41 (α_{IIb}) gene maps in the human Hsap 17 q21 region (Bray et al., 1987), homologous to the porcine Sscr 12 p11-p13 one (Figure 19).

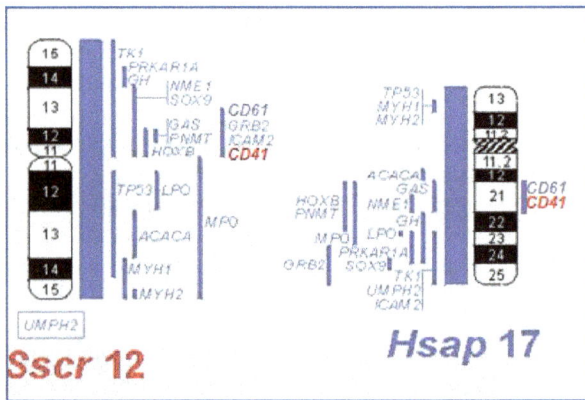

Fig. 19. Chromosomal localization of porcine and human CD41 genes showing the correspondence between porcine Sscr 12 and human Hsap 17 chromosomes.

It is interesting to note that the swine CD41 (α_{IIb}) and CD61 (β_3) genes are closely located - which confirms our previous results (Morera et al., 2002)-, the same as in humans, where both genes map together in chromosome 17, q21 region (Thornton & Poncz, 1999). This is exceptional for genes coding for α and β integrins belonging to the same receptor, and it must have a functional significance, as both genes are simultaneously expressed in human megakaryocytes (Bennett et al, 1983). Therefore, the chromosomal assignment of pig CD41(α_{IIb}) gene provides additional evidence of the conserved linkage homology in these chromosome regions among pigs and humans.

We also checked in this study the porcine α_{IIb} expression profile in different cells and tissues. When we used RT-PCR to detect the α_{IIb} transcripts level, we observed, as expected, a strong expression in platelets and in bone marrow. However, we also detected a lower expression in lymphatic ganglion and lung, which we explain by the probable presence of platelets or blood cells in them. When we used immnunohistochemisty and flow cytometry to locate accurately the CD41 (α_{IIb}) protein expression with specific antibodies produced by us, we confirmed that the presence of CD41 proteins was restricted to platelets and megakaryocytic membranes. The same restricted expression pattern of α_{IIb} proteins have been detected in other species, like humans and mice, although some studies suggest that α_{IIb} could be a differentiation marker expressed in early stages of the cellular hematopoietic differentiation (Mitjavila-García et al., 2002) or to be over expressed in tumor cells (Raso et al., 2004). In fact, the α_{IIb}/β_3 integrin expression in tumor cells has been controversial as $\alpha_{IIβ}$ and α_v integrin have similar structures, and although the role of the CD51 integrin in tumor metastasis and angiogenesis is well documented (Chen, 1992, 1997; Mitjans et al., 2000), these studies have been carried out using antibodies that could cross react with the α_v/β_3 receptor (Chen et al., 1992; Chen 2006). However, some studies have revealed that the $\alpha_{IIβ}/\beta_2$ receptor mediates interactions between platelets and tumor cells, detecting an over expression in the filopodia emitted by the platelets in the focal adhesion plates, with the filopodia being the first contact sites between tumor cells and platelets (Chopra et al., 1992).

In this sense, it is worthy to note that in our study we have produced the two first specific anti-porcine α_{IIb} antibodies, whose specificities we have demonstrated by flow cytometry in cross reactions against platelets from humans, dogs, horses, goats, cats, sheep and cows.

Finally, as α_{IIb}/β_3 is involved in adhesion and aggregation of platelets after their activation, we checked if the platelet activation was or not associated with changes in the α_{IIb} transcripts level. Changes in the proteome of platelet activated by thrombin, the strongest platelet activator, was previously studied in our laboratory, detecting some differential modification in only a small number of proteins, among which the CD41integrin was not included, even though a very sensitive two dimension differential in gel electrophoresis (2D-DIGE) technique was used (Esteso et al., 2008). As CD41 is strongly represented in platelets and it plays an essential role in their activation, we took advantage of the specific anti-porcine α_{IIb} antibodies produced for the studies presented in this chapter to check if CD41 integrin was or not present in the gels used to carry out those studies. Immunoblotting results clearly showed that CD41protein was detected in the platelet proteome, which confirmed our previous results that showed that platelets did not modify their CD41 protein level after thrombin activation. Moreover, although platelets are enucleated cells that lack their nucleus during the megakaryocytic cells cytoplasm fragmentation, it is well established that they conserve ribosomes, mRNAs, as well as the post-translationally modifying protein mechanisms (Dittrich et al., 2005). For this, we used a real time PCR to check if some change was produced in the α_{IIb} transcripts level as a consequence of the platelet activation by thrombin. Results showed that although a small increase was detected, this was not statistically significant. It is well established that most changes produced after platelets activation involve post-translational modifications that affect the interactions between transmembrane and cytoplasmic domains of α and β chains (Russ & Engelman, 1999). So, our results support that the changes produced after thrombin platelet activation, which seems to disrupt the helical interface between the integrin a and β subunit transmembrane domains, favoring homomeric α_{IIb} (and β_3) transmembrane domain interactions in the $\alpha_{IIb}\beta_3$ receptor (Luo et al., 2004; Li et al., 2005; Partridge et al., 2005; Yin et al., 2006), must be produced by post-translational regulation, without affecting neither the transcript nor the protein level in the α_{IIb}.

5. Conclusion

Integrins are a family of heterodimeric transmembrane glycoproteins consiting of varying combinations of noncovalently bound α and β chains that generate several receptors with different expression patterns and ligand binding profiles. $\alpha_{IIb}\beta_3$ (CD41/CD61) integrin is the most abundant platelet receptor being responsible for the platelet aggregation. Most of the studies with $\alpha_{IIb}\beta_3$ integrin have been carried out in humans and mice but little is known about the expression of $\alpha_{IIb}\beta_3$ integrin in porcine tissues, although pig is generally accepted as an optimal experimental model for different areas, as cardiovascular diseases, because of its similarity to humans. We have previously cloned and characterized the porcine gene coding for the β_3 (CD61) chain of the $\alpha_{IIb}\beta_3$ integrin; however, the one coding for α_{IIb} (CD41) chain -the only α subunit for the β_3 one- remained to be characterized.

We describe in this chapter the molecular cloning, the structural and comparative analysis, and the expression patterns of the porcine gene coding for the α_{IIb} integrin chain. Additionally, we also describe the chromosomal localization of the gene.

We used a combined strategy of PCR and RACE reactions to obtain a full porcine α_{IIb} cDNA sequence from platelet RNA. The pig α_{IIb} cDNA was 3336-pb long and contained an ORF 3111 b long that encodes a pre-α_{IIb} protein composed by 1036 amino acid residues, from which, 961, 26 and 10 belong to the NH2-extracellular, the transmembrane and the cytoplasmic-COOH domains, respectively. The porcine α_{IIb} shares with α_{IIb} from other species: identical structure, a high % amino acid identity, common domains (α-I, Ca++ binding, MIDAS), N-glycosylation sites, and the seven FG-GAP tandem repeats. However, in relation to other mammalian α chains, the porcine α_{IIb} shares lower identities with those homologous in mammals (78% with humans, horses and cows, 78% with dogs, 75% with rabbits, 73% with mice and 71% with rats). A phylogenetic tree identifies cows CD41 as the closest to pigs.

By using both somatic cell hybrid and irradiated cell hybrid panels, we localized the gene coding for the porcine α_{IIb} integrin in chromosome *Sscr* 12 region p11-(2/3 p13), in the same region where we previously localized the porcine β_3 integrin gene, region that corresponds to the human homologous *Hsap* 17(q21) in chromosome 17.

As expected, the porcine α_{IIb} mRNAs were predominantly detected in platelets, but they were also detected in bone marrow and ganglion, in which platelets or megakaryocytes –the platelets precursors- were probably presents. To locate accurately the pattern expression of the α_{IIb} protein, immunohistochemical, immunocytochemical and flow cytometry analysis were carried out. For this, monoclonal and polyclonal antibodies against porcine recombinant α_{IIb} integrins (rα_{IIb}) were previously produced. Citometry flow analysis determined the antibodies specificity for porcine platelets, being the first antibodies described with this characteristic. Immunohistochemical assays confirmed that the α_{IIb} expression is restricted to the membranes of megacariocytes present in bone morrow. Flow cytometry analysis of PBC confirmed the α_{IIb} expression in platelet but not in lymphocytes, erythrocytes or granulocytes.

Finally, we checked by RT-Q-PCR if any change was produced in the level of αIIb transcripts in thrombin activated platelets, no detecting significant ones. This result, together to previous ones obtained by us, support that no change were produced in neither the transcript nor the protein level of αIIb, supporting α_{IIb} post-translational changes in the $\alpha_{IIb}\beta_3$ platelet receptor after thrombin activation.

In conclusion, our results are of particular interest because the pig is an animal model system for a variety of immunological, developmental and pathological studies, and because α_{IIb} integrin plays an essential role in phenomenons so significant as thrombosis, homeostasis, tumors progression and invasion, and differentiation of cells from the myeloid lineage in the bone marrow.

6. References

Alam N, Goel HL, Zarif MJ, Butterfield JE, Perkins HM, Sansoucy BG, Sawyer TK & Languino LR, (2007). Theintegrin-growth factor receptor duet. *Journal of Cellular Physiology,* Vol.213, pp. 649–653.

Arce, C., Moreno, A., Millán, Y., Martín de las Mulas, J. & Llanes, D. (2002). Production and characterization of monoclonal antobodies against dog immunoglobulin isotypes. *Vet. Immunol. Immunopathol.,* Vol.6, No.88(1-2), pp. 31-41.

Bennett, J.S. & Vilaire, G. (1979). Exposure of platelet fibrinogen receptors by ADP and epinephrine. *J. Clin. Invest.,* Vol.64, pp. 1393-1401.

Bennett, J. S., Hoxie, J. A., Leitman, S. F., Vilaire, G. & Cines, D. B. (1983). Inhibition of fibrinogen binding to stimulated human platelets by a monoclonal antibody. *Proc. Natl. Acad. Sci. USA,* Vol.80, pp. 2417-2421.

Bennett, J.S. (2005). Structure and function of the platelet integrin alfaIIb/beta3. *J. Clin. Invest.,* Vol.115, pp. 3363-3369.

Bray, P.F., Rosa, J.P., Johnson, G.I., Shiu, D.T., Cook, R.G., Lau, C., Kan, Y.W., McEver, R.P. & Shuman, M.A. (1987). Platelets glycoprotein IIb. Chromosomal localization and tissue expression. *J. Clin. Invest.,* Vol.80, pp. 1812-1817.

Chen, Y. P., Djaffar, I., Pidard, D., Steiner, B., Cieutat, A. M., Caen, J. P. & Rosa, J. P. (1992). Ser-752-->Pro mutation in the cytoplasmic domain of integrin beta 3 subunit and defective activation of platelet integrin alpha IIb beta 3 (glycoprotein IIb-IIIa) in a variant of Glanzmann thrombasthenia. *Proc. Natl. Acad. Sci. USA,* Vol.89, pp. 10169-10173.

Chen, Y. Q., Trikha, M., Gao, X., Bazaz, R., Porter, A. T., Timar, J. & Honn, K. V. (1997). Ectopic expression of platelet integrin alphaIIb beta3 in tumor cells from various species and histological origin. *Int. J. Cancer,* Vol.72, pp. 642-648.

Chen, X. (2006). Multimodality imaging of tumor integrin alphavbeta3 expression. *Mini Rev Med Chem* Vol.6, pp. 227-234.

Chevalet, C., Gouzy, J. & SanCristobal-Gaudy, M. (1997). Regional assignment of genetic markers using a somatic cell hybrid panel: a WWW interactive program available for the pig genome. *Comput. Appl. Biosci.,* Vol.13, pp. 69-73.

Chopra, H., Timar, J., Rong, X., Grossi, I. M., Hatfield, J. S., Fligiel, S. E., Finch, C. A., Taylor, J. D. & Honn, K. V. (1992). Is there a role for the tumor cell integrin alpha IIb beta 3 and cytoskeleton in tumor cell-platelet interaction? *Clin. Exp. Metastasis,* Vol.10, pp. 125-137.

Clark. E. A. &Brugge J. S. (1995). Integrins and signal transduclion pathways: the road taken. *Science,* Vol. 268, pp.233-239.

Corbel, C. & Salaun, J. (2002). AlphaIIb integrin expression during development of the murine hemopoietic system. *Dev. Biol.,* Vol.243, pp. 301-311.

Dickeson, S. K., & Santoro, S. A. (1998). Ligand recognition by the I domain-containing integrins. *Cell Mol. Life Sci.,* Vol.54, pp. 556-566.

Dittrich, M., Birschmann I, StuhlfelderC, SickmannA,Herterich S, Nieswandt B, Walter U. & Dandekar T. (2005). Understanding platelets. *Thromb. Haemost.,* Vol.94, pp. 916–25.

D'Souza, S. E., Ginsberg, M. H., Burke, T. A. & Plow, E. F. (1990). The ligand binding site of the platelet integrin receptor GPIIb-IIIa is proximal to the second calcium binding domain of its alpha subunit. *J. Biol. Chem.,* Vol.265, pp. 3440-3446.

Edgar, R.C.(2004). MUSCLE: a multiple sequence alignment method with reduced time and and space complexity. *BMC Bioinformatics,* Vol.19, pp. 5, 113.

Esteso, G., Mora, M.I., Garrido; J.J., Corrales, F. & Moreno, A. (2008). Proteomic analysis of the porcine platelet proteome and alterations induced by thrombin activation. *Journal of Proteomics,* Vol.71, pp. 547-560.

Filardo, E.J. & Cheresh, D.A. (1994). A β turn in the cytoplasmic tail of the integrin αv subunit influences conformation and ligand binding of αvβ3. *J. Biol Chem.,* Vol.269, pp. 4641-4647.

García A, Prabhakar S, Brock C, Pearce A, Dwek R, Watson S., Hebestreit H, & Zitzmann N. (2004). Extensive analysis of the human platelet proteome by two-dimensional gel electrophoresis and mass spectrometry. *Proteomics,* Vol.4, pp. 656–68.

Grimaldi, C.M., Chen, F., Wu, C., Weis, H.J., Coller, B.S. & French, D.L. (1998). Glycoprotein IIb Leu214Pro mutation produces glanzmann thrombasthenia with both quantitative and qualitative abnormalities in GPIIb/IIIb. *Blood,* Vol.91, pp. 1562-1571.

Hawken, R. J., Murtaugh, J., Flickinger, G. H., Yerle, M., Robic, A., Milan, D., Gellin, J., Beattie, C. W., Schook, L. B. & Alexander, L. J. (1999). A first-generation porcine whole-genome radiation hybrid map. *Mamm. Genome,* Vol.10, pp. 824-830.

Hemler, M.E., Weitzman, J.B., Pasqualini, R., Kawaguchi, S., Kassner, P.D., & Berditchevsky, F.B. (1994). Structure, biochemical properties, and biological functions of integrin cytoplasmic domains. In Integrins: The Biological Problems. Y. Takada, editor. CRC Press, Boca Raton, Fl, pp. 1-35.

Honda, S., Tomiyama, Y., Shiraga, M., Tadokoro, S., Takamatsu, J., Saito, H., Kurata, Y. & Matsuzawa, Y. (1998). A two-amino acid insertion in the Cys146- Cys167 loop of the alphaIIb subunit is associated with a variant of Glanzmann thrombasthenia. Critical role of Asp163 in ligand binding. *J. Clin. Invest.,* Vol.102, pp. 1183-1192.

Humphries, M. J. (2000). Integrin structure. *Biochem. Soc. Trans.,* Vol.28, pp. 311-339.

Hynes, R.O., 1987. Integrins: a family of cell surface receptors. *Cell,* Vol.48, pp. 549-554.

Hynes, R. O. (1992). Integrins: versatility, modulation, and signaling in cell adhesion. *Cell,* Vol.69, pp. 11-25.

Hynes, R. O. (2002). Integrins: bidirectional, allosteric signaling machines. *Cell,* Vol.110, pp. 673-687.

Hynes, R.O., Lively, J.C., McCarty, J.H., Taverna, D., Xiao, Q. & Hodivala-Dilke, K. (2002). The diverse roles of integrins and their ligands in angiogenesis. Cold Spring Harb. Symp. *Quant. Biol.,* Vol.67, pp. 143-153.

Isenberg, G. (1991). Actin binding proteins - lipid interactions. *Journal of Muscle Research and Cell Motility.* Vol.12, pp. 136-144.

Jameson, B. A. & Wolf, H. (1988). The antigenic index: a novel algorithm for predicting antigenic determinants. *Comput. Appl. Biosc.,i* Vol.4, pp. 181-186.

Jiménez-Marín, A., Garrido, J.J, De Andrés, D.F., Morera, L., LLanes, D. & Barbancho, M. (2000). Molecular cloning and characterization of the pig homologue to human CD29, the integrin β1 subunit. *Transplantation,* Vol.70, No.4, pp.649-655.

Jiménez-Marin, A., Yubero, N., Esteso, G., Moreno, A., de Las Mulas, J. M., Morera, L., Llanes, D., Barbancho, M. & Garrido, J. J. (2008). Molecular characterization and expression analysis of the gene coding for the porcine beta(3) integrin subunit (CD61). *Gene,* Vol.408, pp. 9-17.

Li, W., Metcalf, D. G., Gorelik, R., Li, R. H., Mitra, N., Nanda, V., Law, P. B., Lear, J. D., DeGrado, W. F., and Bennett, J. S. (2005). A Push-Pul mechanism for integrin function. *Proc. Natl. Acad. Sci. U SA*, Vol.102, pp. 1424–1429.

Loftus, J. C., Halloran, C. E., Ginsberg, M. H., Feigen, L. P., Zablocki, J. A. & Smith, J. W. (1996). The amino-terminal one-third of alpha IIb defines the ligand recognition specificity of integrin alpha IIb beta 3. *J. Biol. Chem.*, Vol.271, 2033-2039.

Lunney, J.K., 2007. Advances in swine biomedical model genomics. *Int. J. Biol. Sci.*, Vol.3, No.3, pp. 179-184.

Luo, B.H., Springer, T.A. & Takagi, J. (2004) A specific interface between integrin transmembrane helices and affinity for ligand. *PLoS Biol.*, Vol.2, e153.

Mendrola, J.M., Beger, M.B., King, M.C. & Lemmon (2002). The single transmembrane domains of ErbB receptors self-associate in cell membranes. *J. Biol. Chem.*, Vol.277, pp. 4704-4712.

Mitjans, F., Meyer, T., Fittschen, C., Goodman, S., Jonczyk, A., Marshall, J.F., Reyes, G. & Piulats, J. (2000). In vivo therapy of malignant melanoma by means of antagonists of alfav integrins. *Int. J. Cancer* Vol.87, pp. 716-723.

Misdorp, W. (2003). Congenital and hereditary tumors in domestic animals, 2. Pigs: a review. *Vet. Q.*, Vol.25, pp. 17-30.

Mitjavila-Garcia, M. T., Cailleret, M., Godin, I., Nogueira, M. M., Cohen-Solal, K., Schiavon, V., Lecluse, Y., Le Pesteur, F., Lagrue, A. H. & Vainchenker, W. (2002). Expression of CD41 on hematopoietic progenitors derived from embryonic hematopoietic cells. *Development*, Vol.129, pp. 2003-2013.

Morera, L., Jiménez-Marín, A., Yerle, M., Llanes, D., Barbancho, M. & Garrido, J.J. (2002). A polymorphic microsatelite located on pig chromosome band 12p11-2/3p13, within the 3'-UTR of the ITGB3 gene. *Animal Genetics*, Vol.33, pp. 239-240.

Ody, C., Vaigot, P., Quere, P., Imhof, B. A. & Corbel, C. (1999). Glycoprotein IIb-IIIa is expressed on avian multilineage hematopoietic progenitor cells. *Blood*, Vol.93, pp. 2898-2906.

Partridge, A.W., Liu, S., Kim, S., Bowie, J.U. & Ginsberg, M.H. (2005). Transmembrane domain helix packing stabilizes integrin αIIbβ3 in the low affinity state. *J. Biol. Chem.*, Vol.280, pp. 7294–7300.

Pérez de la Lastra, J.M., Moreno, A., Pérez, J., Llanes, D. (1997). Characterization of the porcine homologe to human platelet glycoprotein IIb-IIIa (CD41/CD61) by a monoclonal antibody. *Tissue Antigens*, Vol.49, pp. 588-594.

Raso, E., Dome, B., Somlai, B., Zacharek, A., Hagman, W., Honn, K.V. & Timar, J. (2004). Molecular identification, localization and function of platelet-type 12 lipoxygenase in human melanoma progression, under experimental and clinical conditions. *Melanoma Res.*, Vol.14, pp. 245-250.

Rettenberger, G., Klett, C., Zechner, U., Kunz, J., Vogel, W. & Hameister, H. (1995). Visualization of the conservation of synteny between humans and pigs by heterologous chromosomal painting. *Genomics*, Vol.26, pp. 372-378.

Rojiani, M.V., Finlay, B.B., Gray, V. & Dedhar. S. (1991). In vitro interaction of a polypeptide homologous to the human Ro/SS-A antigen (calreticulin) with a highly conserved amino acid sequence in the cytoplasmic domain of integrin alpha subunits. *Biochemistry*, Vol.30, pp. 9859-9866.

Russ, W.P. & Engelman, D.M. (1999). TOXCAT: a measure of transmembrane helix association in a biological membrane. *Proc. Natl. Acad. Sci. USA*, Vol.96, pp. 863-868.

Schneider, D. & Engelman, D. M. (2004). Involvement of transmembrane domain interactions in signal transduction by alpha/beta integrins. J. Biol. Chem., Vol.279, pp. 9840-9846.

Senes, A., Gerstein, M. & Engelman, D. M. (2000). Statistical analysis of amino acid patterns in transmembrane helices: the GxxxG motif occurs frequently and in association with beta-branched residues at neighboring positions. *J. Mol. Biol.,* Vol.296, pp. 921-936.

Springer, T. A. (1997). Folding of the N-terminal, ligand-binding region of integrin alpha-subunits into a beta-propeller domain. *Proc. Natl. Acad. Sci. USA,* Vol.94, pp. 65-72.

Schwartz. M. A., Schaller. M. D. & Ginsberg. M. H. (1995). Integrins. Emerging paradigms of signal transduction. *Annu. Rev. Cell Dev. Biol.,* Vol.11, pp. 549-599.

Takada, Y., Elices, M. J., Crouse, C. and Hemler, M. E. (1989). The primary structure of the alpha 4 subunit of VLA-4: homology to other integrins and a possible cell-cell adhesion function. *Embo J.,* Vol.8, pp. 1361-1368.

Thornton, M. A. & Poncz, M. (1999). Characterization of the murine platelet alphaIIb gene and encoded cDNA. *Blood,* Vol.94, pp. 3947-3950.

Tozer, E.C., Baker, E.K., Ginsberg, M.H. & Loftus, J.C. (1999). A mutation in the alfa subunit of the platelet integrin alphaIIbeta3 identifies a novel region important for ligand binding. *Blood,* Vol.93, pp. 918-924.

Wall, C. D., Conley, P. B., Armendariz-Borunda, J., Sudarshan, C., Wagner, J. E., Raghow, R. & Jennings, L. K. (1997). Expression of alpha IIb beta 3 integrin (GPIIb-IIIa) in myeloid cell lines and normal CD34+/CD33+ bone marrow cells. *Blood Cells Mol. Dis.,* Vol.23, pp. 361-376.

Weiss, E. J., Bray, P. F., Tayback, M., Schulman, S. P., Kickler, T. S., Becker, L. C., Weiss, J. L., Gerstenblith, G. & Goldschmidt-Clermont, P. J. (1996). A polymorphism of a platelet glycoprotein receptor as an inherited risk factor for coronary thrombosis. *N. Engl. J. Med.,* Vol.334, pp. 1090-1094.

Xiong, J.P., Stehle, T., Diefenbach, B., Zhang, R., Dunker, R., Scott, D.L., Joachimiak, A., Goodman, S.L., Amin, M. & Arnaout, M.A. (2001). Crystal structure of the extracellular segment of integrin alphaV beta3. *Science,* Vol.294, pp. 339-345.

Yamada, K.M., 1991. Adhesive recognition sequences. *J. Biol. Chem.,* Vol.266, pp. 12809-12812.

Yerle, M., Echard, G., Robic, A., Mairal, A., Dubut-Fontana, C., Riquet, J., Pinton, P., Milan, D., Lahbib-Mansais, Y. & Gellin, J. (1996). A somatic cell hybrid panel for pig regional gene mapping characterized by molecular cytogenetics. *Cytogenet. Cell Genet.,* Vol.73, pp. 194-202.

Yerle, M., Pinton, P., Robic, A., Alfonso, A., Palvadeau, Y., Delcros, C., Hawken, R., Alexander, L., Beattie, C., Schook, L., *et al.* (1998). Construction of a whole-genome radiation hybrid panel for high-resolution gene mapping in pigs. *Cytogenet Cell Genet.,* Vol.82, pp. 182-188.

Yin, H., Litvinov, R.I., Vilaire, G., Zhu, H., Li, W., Caputo, G.A., Moore, D.T., Lear, J.D., Weisel, J.W., DeGrado, W.F. & Bennett, J.S. (2006). Activation of platelet aIIbb3 by an exogenous peptide corresponding to the transmembrane domain of aIIb. *J. Mol. Chem.,* Vol.281, No.48, pp. 36732-36741.

Yubero, N., Jiménez-Marín, Á., Barbancho, M., & Garrido, J.J. (2011). Two cDNAs coding for the porcine CD51 (α_v) integrin subunit: cloning, expression analysis, adhesion assays and chromosomal localization. *Gene,* Vol.481, pp. 29-40.

8

Identification of Molecules Involved in the Vulture Immune Sensing of Pathogens by Molecular Cloning

Crespo, Elena[1], de la Fuente, José[1,2] and Pérez de la Lastra, José M.[1]
[1]Instituto de Investigación en Recursos Cinegéticos (UCLM-CSIC-JCCLM), Ronda Toledo
[2]Department of Veterinary Pathobiology, Center for Veterinary Health Sciences,
Oklahoma State University, Stillwater
[1]Spain
[2]USA

1. Introduction

Vultures may have one of the strongest immune systems of all vertebrates (Apanius et al., 1983; Ohishi et al., 1979). Vultures are unique vertebrates able to efficiently utilize carcass from other animals as a food resource. These carrion birds are in permanent contact with numerous pathogens and toxins found in its food. In addition, vultures tend to feed in large groups, because carcasses are patchy in space and time, and feeding often incurs fighting and wounding, exposing vultures to the penetration of microorganisms present in the carrion (Houston & Cooper, 1975). Therefore, vultures were predicted to have evolved immune mechanisms to cope with a high risk of infection with virulent parasites.

Despite the potential interest in carrion bird immune system, little is known about the molecular mechanisms involved in the regulation of this process in vultures. The aim of this chapter is to describe the molecular cloning and characterization of two key molecules involved in the immune sensing of pathogens: a griffon vulture (Gyps fulvus) orthologue of TLR1 (CD281) and an orthologue of the alpha inhibitor of NF-κB (IκBα).

The toll-like receptor (TLR) family is an ancient pattern recognition receptor family, conserved from insects to mammals. Members of the TLR family are vital to immune function through the sensing of pathogenic agents and initiation of an appropriate immune response (Takeda & Akira, 2005). The rapid identification of Toll orthologues in invertebrates and mammals suggests that these genes must be present in other vertebrates (Takeda, 2005). During the recent years, members of the multigene family of Toll-like receptors (TLRs) have been recognised as key players in the recognition of microbes during host defence (Hopkinsn & Sriskandan, 2005). Recognition of pathogens by immune receptors leads to activation of macrophages, dendritic cells, and lymphocytes. Signals are then communicated to enhance expression of target molecules such as cytokines and adhesion molecules, depending on activation of various inducible transcription factors, among which the family NF-kappaB transcription factors plays a critical role. The involvement of NF-κB in the expression of numerous cytokines and adhesion molecules has supported its role as an evolutionarily conserved coordinating element in organism's response to situations of infection, stress, and injury. In many species, pathogen recognition, whether mediated via the Toll-like receptors or via the antigen-specific T- and B-cell receptors, initiates the activation of distinct signal transduction pathways that activate nuclear factor-kappa B (NF-κB) (Ghosh et al., 1998). TLR-mediated NF-κB activation is also an evolutionarily conserved event that occurs in phylogenetically distinct species ranging from insects to mammals.

The identification of orthologues of TLRs in other species, particularly in those showing a strong immune system, together with the elucidation of their TLR-mediated signal transduction pathways, would contribute to our understanding of how these receptors have evolved and the importance of different orthologues to resistance to different pathogens.

2. Strategy for cloning of vulture TLR1 and IκBα

In order to identify key components of the vulture system for sensing of pathogens, we constructed and screened a cDNA library from vulture peripheral blood monuclear cells (PBMC) using specific probes for TLR1 and IκBα. Since the majority of toll-like receptors are expressed in leukocytes and lymphoid tissues in human and other vertebrates, we decided to use vulture PBMC as the source of RNA to obtain a specific probes for TLR1 and IκBα and to construct a cDNA library. Using this strategy we cloned cDNAs encoding for griffon vulture (Gyps fulvus) orthologues of mammalian TLR1 (CD281) and for the alpha inhibitor of NF-κB (IκBα). The tissue and cell expression pattern of vulture TLR1 and IκBα were analyzed by real-time RT-PCR and correlated with the ability to respond to various pathogenic challenges.

2.1 Design of specific probes for vulture TLR1 and IκBα

To obtain specific probes for vulture TLR1 and IκBα, total RNA was isolated from vulture PBMC and from cells and tissues using the Ultraspec isolation reagent (Biotecx Laboratories, Houston TX, USA). Ten micrograms of total RNA was heated at 65 °C for 5 min, quenched on ice for 5 min and subjected to first strand cDNA synthesis. The RNA was reverse transcribed using an oligo dT12 primer by incubation with 200 U RNase H- reverse transcriptase (Invitrogen, Barcelona, Spain) at 25°C for 10 min, then at 42°C for 90 min in the presence of 50 mM Tris-HCl, 75 mM KCl, 3 mM MgCl2, 10 mM DTT, 30 U RNase-inhibitor and 1mM dNTPs, in a total volume of 20 μl.

For the vulture TLR probe, a partial fragment of 567 bp showing sequence similarity to human TLR-1 was amplified by PCR from vulture PBMC cDNA using two oligonucleotide primers TLR1/2Fw (5'-GAT TTC TTC CAG AGC TG–3') and TLR1/3Rv (5'-CAA AGA TGG ACT TGT AAC TCT TCT CAA TG -3'), which were designed based on regions of high homology among the sequences of human and mouse TLR1 (GenBank, accession numbers NM_003263 and NM_030682, respectively). Cycling conditions were 94°C for 30 s, 52°C for 30 s and 72°C for 1.5 min, for 30 cycles.

For the vulture IκBα probe, a partial fragment of 336 bp showing sequence similarity to human and chicken IκBα was amplified by PCR from vulture PBMC cDNA using two oligonucleotide primers IκBα-Fw (5'-CCT GAA CTT CCA GAA CAA C-3') and IκBα-Rv (5'-GAT GTA AAT GCT CAG GAG CCA TG-3'), which were designed based on regions of high homology among the sequences of human and chicken IκBα (GenBank, accession numbers

M69043 and S55765, respectively). Cycling conditions were 94°C for 30 s, 52°C for 30 s and 72°C for 1.5 min, for 30 cycles.

The obtained PCR products were cloned into pGEM-T easy vector using a TA cloning kit (Promega, Barcelona, Spain) and sequenced bidirectionally to confirm their respective specificities. These fragments were DIG-labelled following the recommendation of the manufacturer (Roche, Barcelona, Spain) and used as probes to screen 500 000 plaque colonies of the vulture-PBMC cDNA library.

2.2 cDNA library construction and screening

Total RNA (500 µg) was extracted from PBMC (pooled from 6 birds) using the Ultraspec isolation reagent (Biotecx). mRNA (20 µg) was extracted by Dynabeads (Dynal biotech-Invitrogen, Barcelona, Spain) and used in the construction of a cDNA library in Lambda ZAP vector (Stratagene, La Jolla, CA, USA) by directional cloning into EcoRI and XhoI sites. The cDNA library was plated by standard protocols at 50 000 plaque forming units (pfu) per plate and grown on a lawn of XL1-Blue E. coli for 6-8 h. Screening of the library was performed with DIG labelled probes. Plaques were transferred onto Hybond-N+ membranes (Amersham, Barcelona, Spain) denatured in 1.5 M NaCl/0.5M NaOH, neutralised in 1.5 M NaCl/0.5 Tris (pH 8.0) and fixed using a cross-linker oven (Stratagene). The filters were then pre-incubated with hybridisation buffer (5XSSC [1XSSC is 150 mM NaCl, 15 mM trisodium citrate, pH 7.7], 0.1% N-laurylsarcosine, 0.02% SDS and 1% blocking reagent (Roche)) at 65 °C for 1 h and then hybridised with hybridisation buffer containing the DIG-labelled probe, overnight at 65 °C. The membranes were washed at high stringency (2XSSC, 0.1% SDS; 2x5 min at ambient temperature followed by 0.5XSSC, 0.1% SDS; 2x15 min at 65 °C). DiG-labelled probes were detected using phosphatase-labelled anti-digoxygenin antibodies (Roche) according to the manufacturer's instructions. Positive plaques on membranes were identified, isolated in agar plugs, eluted in 1 ml of SM buffer (0.1M NaCl, 10 mM MgSO4, 0.01% gelatin, 50 mM Tris-Hcl, pH 7.5) for 24 h at 4°C and replated. The above screening protocol was then repeated. Individual positive plaques from the secondary screening were isolated in agar plugs and eluted in SM buffer. The cDNA inserts were recovered using the Exassist/SOLR system (Stratagene). Individual bacterial colonies containing phagemid were grown up in LB broth (1% NaCl, 1% trytone, 0.5% yeast extract, pH 7.0) containing 50 µg/ml ampicillin. Phagemid DNA was purified using a Bio-Rad plasmid mini-prep kit and sequenced.

3. Structural analysis of vulture TLR1 and IκBα sequences

Sequences were analyzed using the analysis software LaserGene (DNAstar, London, UK) and the analysis tools provided at the expasy web site (http://www.expasy.org). PEST regions are sequences rich in Pro, Glu, Asp, Ser and Thr, which have been proposed to constitute protein instability determinants. The analysis of the PEST region for the putative protein was made using the webtool PESTfind at http://www.at.embnet.org/toolbox/pestfind. The potential phosphorylation sites were calculated using the NetPhos 2.0 prediction server at http://www.cbs.dtu.dk/services/NetPhos. The prediction of the potential attachment of small ubiquitin-related modifier (SUMO) was made using the webtool SUMOplot™.

The alignment of vulture TIR domain sequences with TLR-1 from other species and of the vulture IκBα sequences with IκBα from other species was done using the program ClustalW v1.83 with Blosum62 as the scoring matrix and gap opening penalty of 1.53. Griffon vulture TLR-1 and IκBα sequences were deposited in the Genbank under accession numbers DQ480086 and EU161944, respectively.

3.1 Vulture TLR1

The screening of the vulture PBMC cDNA library for TLR1 yielded seven clones with identical open reading frame (ORF) sequences. The fact that the screening of 500,000 vulture cDNA clones resulted in 7 identical sequences suggested that this TLR receptor is broadly represented in PBMC, possibly illustrating its important role in pathogen recognition during vulture innate immune response. This result was consistent with the real time RT-PCR analysis of TLR1 transcripts in vulture cells.

The largest clone (2,355 bp) contained an ORF that encoded a 650 amino acid putative vulture orthologue to TLR1, flanked by 319 bp 5'UTR and a 83 bp 3'UTR that contained a potential polyadenylation signal, AATAAA, 21 bp upstream of the poly (A) tail (Fig. 1). The predicted molecular weight of the putative vulture TLR1 was of 74.6 KDa. The predicted protein sequence had a signal peptide, an extracellular portion, a short transmembrane region and a cytoplasmic segment (Fig. 1). In assigning names to the vulture TLR, we looked at the closest orthologue in chicken and followed the nomenclature that was proposed for this species (Yilmaz et al., 2005). Therefore, the discovered sequence was identified as vulture TLR1.

3.1.1 Amino acid sequence comparison of vulture TLR1 with other species

The comparison of the deduced amino acid sequence of vulture TLR1 with the sequence of chicken, pig, cattle, human and mouse TLR1 indicated that the deduced protein had a higher degree of similarity to chicken (64% of amino acid similarity) than to pig (51%), cattle (51%), human (51%) and mouse (48%) sequences (Fig. 2). Protein sequence similarity was different on different TLR domains (Fig. 2).

Amino acid sequence of vulture TLR1 was aligned with the orthologous sequence of chicken (Gallus gallus), pig (Sus scrofa), cattle (Bos taurus), human (Homo sapiens) and mouse (Mus musculus) based on amino acid identity and structural similarity. Identical amino acid residues to vulture TLR1 from the aligned sequences are shaded. Gaps were introduced for optimal alignment of the sequences and are indicated by dashes (-). GenBank or Swiss protein accession numbers are: DQ480086, Q5WA51, Q59HI9, Q706D2, Q5FWG5 and Q6A0E8, respectively.

For the TLRs, it is assumed that the structure of the ectodomain has evolved more quickly than the structure of the TIR (Johnson et al., 2003). Similarly to other TLR receptors, the degree of homology of vulture TLR1 was higher in the transmembrane and cytoplasmic domains than in the extracellular domain.

The vulture TLR1 with 650 amino acids is probably the TLR with the shortest length and the smallest predicted MW (74.6 kDa). Recently, a chicken isoform of TLR1 (Ch-TLR1 type 2) was identified *in silico* and predicted to have a similar number of residues than vulture TLR1 (Yilmaz et al., 2005). However, this receptor also contains an additional transmembrane region in its N-terminal end, and the pattern of expression in tissues is also different from that ChTLR1 type 1 (Yilmaz et al., 2005).

```
            cccagttctcagaagcatgcttcacaaatacggatcatactatgtgacttacacgcttatc      61
            aggcaaaagtctctgaagtttcccataaaggatattctgaagaaagtttgaaggtactca      121
            taaataatttgactgaatgccaggatataggaaggagaaagaaaattaagcacatgtgga      181
            agaattgtatccttctttcacctagtccctggatattgatgaaattttgtcctaagaaga      241
            aataacgacttgaaggattagaacaaaggtggacagataagagaagtattgagcatctcc      301
            aaggaaacagaaaccagtatgacagaaaatatgagatctctcagaaactttttttctttac      361
                                 M  T  E  N  M  R  S  L  R  N  F  F  L  Y   14
            aagtgtctgtttgcattaactttttggaattgtgtcagcctgtct↓tggaaaatgaactc      421
             K  C  L  F  A  L  T  F  W  N  C  V  S  L  S  V  E  N  E  L   34
            ttcacatctgtttctaacgaagatggttctgacaaaaaaatcaagagcctgccactcctc      481
             F  T  S  V  S  N  E  D  G  S  D  K  K  I  K  S  L  P  L  L   54
            tatacaaatagtcatcagtccaaagctaattttgactgggttgtgatacaaaatactaca      541
             Y  T  N  S  H  Q  S  K  A  N  F  D  W  V  V  I  Q (N) T  T   74
            gaaagcctatcgttgtcagaaatcacaaatgacaatgtaaaaaaattagtagcattatta      601
             E  S  L  S  L  S  E  I  T  N  D  N  V  K  K  L  V  A  L  L   94
            tctaatttcagacaaggctccaggttacaaaatctgacactgacaaatgtgtcagttgac      661
             S  N  F  R  Q  G  S  R  L  Q (N) L  T  L  T (N) V  S  V  D   114
            tggaatgctcttattgaaacttttcagactgtatggcactcacccattgaatacttcagt      721
             W  N  A  L  I  E  T  F  Q  T  V  W  H  S  P  I  E  Y  F  S   134
            gttaacggtgtaacacaattgtcggacatcgaaagctatgactttgactattcaggtacg      781
             V  N  G  V  T  Q  L  S  D  I  E  S  Y  D  F  D  Y  S  G  T   154
            tctatgaaagcggtcacaatgaagaaagttttaatcacagatctgtacttctcacagaat      841
             S  M  K  A  V  T  M  K  K  V  L  I  T  D  L  Y  F  S  Q  N   174
            gacctatacaaaatatttgcagacatgaatattgcagccttgacaatagctgaatcagag      901
             D  L  Y  K  I  F  A  D  M  N  I  A  A  L  T  I  A  E  S  E   194
            atgatacatatgctgtgtccttcgtctgcagagtcccctttagatacttaaattttttaaag      961
             M  I  H  M  L  C  P  S  S  S  D  S  P  F  R  Y  L  N  F  L  K   214
            aacgatttaacagatctgcttttttcaaaaatgtgacaaattaattcaactggagacatta      1021
             N  D  L  T  D  L  L  F  Q  K  C  D  K  L  I  Q  L  E  T  L   234
            atcttgccgaagaataaatttgagagccttttccaaggtaagcttcatgactagccgtatg      1081
             I  L  P  K  N  K  F  E  S  L  S  K  V  S  F  M  T  S  R  M   254
            aaatcactgaaatacctggacatcagcagcaacttgctgagtcacgtggagctgatgtg      1141
             K  S  L  K  Y  L  D  I  S  S  N  L  L  S  H  D  G  A  D  V   274
            caatgccaatgggctgagtctctgacagagttggacctgtcctcaaatcagttgacggat      1201
             Q  C  Q  W  A  E  S  L  T  E  L  D  L  S  S  N  Q  L  T  D   294
            gccgtgtttgagtgcttgccagtcaacatcagaaaactcaacctccaaaacaatcacatc      1261
             A  V  F  E  C  L  P  V  N  I  R  K  L  N  L  Q  N  N  H  I   314
            accagtgtccccaagggaatggctgagctgaaatccttgaaagagctgaacctggcatcg      1321
             T  S  V  P  K  G  M  A  E  L  K  S  L  K  E  L  N  L  A  S   334
            aacaggctggctgacctgccggggtgcagtggctttacgtcgctggagttcctgaacgta      1381
             N  R  L  A  D  L  P  G  C  S  G  F  T  S  L  E  F  L  N  V   354
            gagatgaattcgatcctcaccccatctgccgacttcttccagagctgcccacaggtcagg      1441
             E  M  N  S  I  L  T  P  S  A  D  F  F  Q  S  C  P  Q  V  R   374
            gagctgcaagccgggcacaaacccattcaagtgttcctgtgaaactgcaagactttatccgt      1501
             E  L  Q  A  G  H  N  P  F  K  C  S  C  E  L  Q  D  F  I  R   394
            ctggccgaggcagtctgggggaagctgtttggctggccagcggcgtatgtgtgcgagtac      1561
             L  A  R  Q  S  G  G  K  L  F  G  W  P  A  A  Y  V  C  E  Y   414
            ccggaagacttgcaaggaacgcagctgaaggacttccacctgactgaactggcttgcaac      1621
             P  E  D  L  Q  G  T  Q  L  K  D  F  H  L  T  E  L  A  C  N   434
            acggtgctcttgctggtgacagctctgctgctgacgctggtggtggctgtcgtggcc      1681
             T  V  L  L̲  L̲  V̲  T̲  A̲  L̲  L̲  L̲  T̲  L̲  V̲  L̲  V̲  A̲  V̲  V̲  A̲   454
            tttctgtgcatctacttggatgtgccgtggtacgtgcggatgacgtggcagtggacgcag      1741
             F̲  L̲  C̲  I̲  Y̲  L̲  D  V  P  W  Y  V  R  M  T  W  Q  W  T  Q   474
            acaaagcggagggcttggcacagccaccccgaagagcaggagaccattctgcagtttcac      1801
             T  K  R  R  A  W  H  S  H  P  E  E  Q  E  T  I  L [Q  F  H]  494
            gcgttcatttcctacagcgagcgcgattcgttcgttgtggggtgaagaacgagctgatcccgaac      1861
            [A  F  I  S  Y  S  E  R  D  S  L  W  V  K  N  E  L  I  P  N]  514
            ctggagaagggggagggctgtgtacaactgtgccagcacgagaggaactttatccccggc      1921
            [L  E  K  G  E  G  C  V  Q  L  C  Q  H  E  R  N  F  I  P  G]  534
            aagagcattgtggagaacatcattaactgcattggagaagagctacaggtcgatctttgtg      1981
            [K  S  I  V  E  N  I  I  N  C  I  E  K  S  Y  R  S  I  F  V]  554
            ttgtctcccaactttgtgcagagcgagtggtgtcactatgagctgtactttgcccatcac      2041
            [L  S  P  N  F  V  Q  S  E  W  C  H  Y  E  L  Y  F  A  H  H]  574
            aaattattcagtgagaattccaacagcttaatcctcattttactggagccgatccctccg      2101
            [K  L  F  S  E  N  S  N  S  L  I  L  I  L  L  E  P  I  P  P]  594
            tacattatccctgccaggtatcacaagctgaaggctctcatgcaaagcgaacctacctg      2161
            [Y  I  I  P  A  R  Y  H  K  L  K  A  L  M  A  K  R  T  Y  L]  614
            gagtggccaaaggagagaggagcaagcatcccctttctggctaacctgagggcagctatt      2221
            [E  W  P  K  E  R  S  K  H  P  L  F  W  A  N  L  R  A  A  I]  634
            agcattaacctgctaatggctgatggaaagaggtgtgggggaaacagattaagaatctttc      2281
            [S  I  N] L  L  M  A  D  G  K  R  C  G  E  T  D  *          650
            taatggagtttcttccatttttttcttggtgaagcaataaatgctttatgatttccaaaaa      2341
            aaaaaaaaaaaaaa
```

Fig. 1. Nucleotide and deduced amino acid sequence of vulture TLR1. Complete sequence of the full-length Vulture TLR obtained from the cDNA library (GenBank accession number:

DQ480086). Translated amino acid sequence is also shown under nucleotide sequence. Numbers to the right of each row refer to nucleotide or amino acid position. The cleavage site for the putative signal peptide is indicated by an arrow. LRRs domains are shaded. Potential N-glycosylation sites are circled. The predicted transmembrane segment is underlined. The initiation codon (atg) and the polyadenilation site are underlined. The translational stop site is indicated by an asterisk. The cysteines critical for the maintenance of the structure of LRR-CT are in bold.

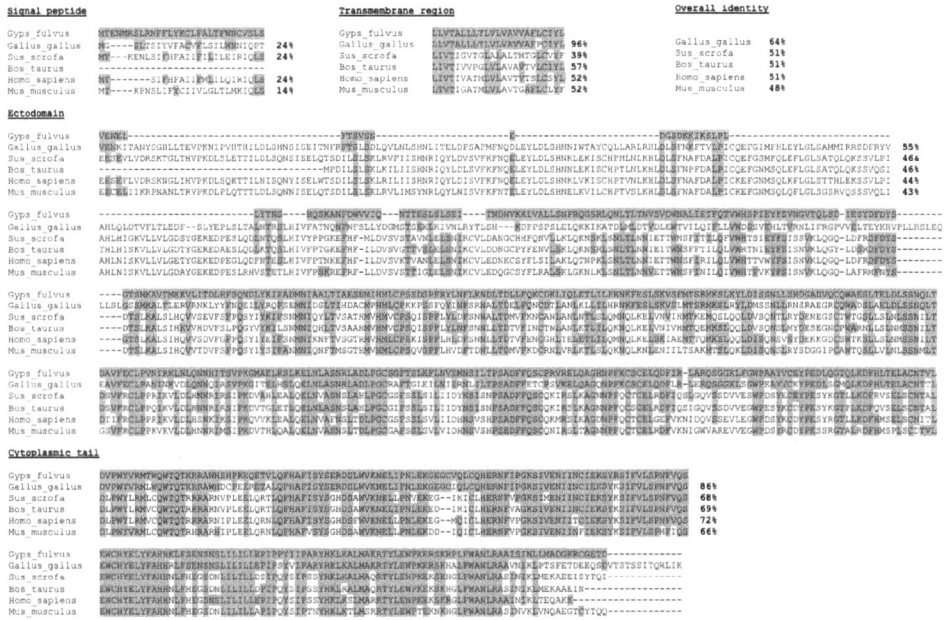

Fig. 2. Alignment of amino acid sequences of TLR1 from different species.

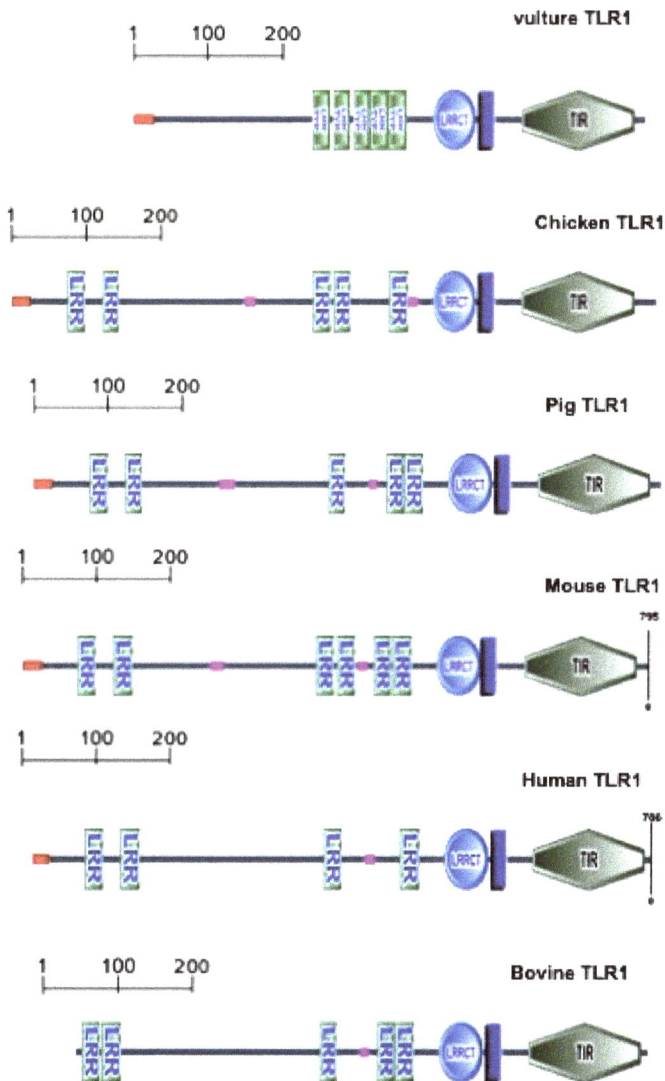

Fig. 3. Schematic structure of TLR1 from various species

Comparison of the structure obtained from the SMART analysis (at expasy web server) of the amino acid sequence from human, bovine, pig, mouse, chicken and vulture TLR1. Each diagram shows a typical structure of a member of the toll-like receptor family. Vulture TLR1 consists of an ectodomain containing five leucine rich repeats (LRRs) followed by an additional leucine rich repeat C terminal (LRR-CT) motif. The Vulture TLR has a transmembrane segment and a cytoplasmic tail which contains the TIR domain. Genbank or swiss accession number for proteins are DQ480086 (vulture), Q5WA51 (chicken), Q59HI9 (pig), Q706D2 (bovine), Q5FWG5 (human) and Q6A0E8 (mouse).

In general, the structure of vulture TLR1 shows similarity to chicken and mammalian TLR1 (Table 1). However, vulture TLR1 exhibits some structural features that could influence its functional role as pathogen receptor (Fig. 3). For example, it is possible that the smaller size of vulture TLR1, the lower number of N-glycosylation sites and the grouping of its LRRs in the proximal half of its ectodomain have functional implications.

Structural feature	G fulvus	G gallus	S scrofa	B taurus	H sapiens	M musculus
Amino acid residues	650	818	796	727	786	795
Number of LRRs	5	5	5	5	4	6
N-glycosylation sites	3	5	4	6	7	8
Predicted MW(KDa)	74.60	94.46	90.94	83.04	90.29	90.67
Length of ectodomain	409	569	560	521	560	558

Table 1. Structural features of TLR1 receptor from Griffon vulture (G fulvus), Chicken (G gallus), pig (S scrofa), cattle (B Taurus) human (H sapiens) and mouse (M musculus) amino acid sequences. The theoretical molecular weight, number of LRRs, and of glycosilation sites was calculated using the software available at the expasy web server (http://www.expasy.org). Genbank or Swiss accession number for proteins are DQ480086 (G fulvus), Q5WA51 (G gallus), Q59HI9 (S scrofa), Q706D2 (B Taurus), Q5FWG5 (H sapiens) and Q6A0E8 (M musculus).

The set of Toll proteins for humans and insects each contain widely divergent LRR regions, and this is viewed as providing the potential to discriminate between different ligands. Perhaps these features provide vulture TLR1 some advantages on pathogen recognition. TLR glycosylation is also likely to influence receptor surface representation, trafficking and pattern recognition (Weber et al., 2004).

3.2 Vulture IκBα

The screening of the vulture PBMC cDNA library yielded one clone that contained an ORF that encoded a 313 amino acid putative vulture orthologue to IκBα, flanked by 15 bp 5'UTR and a 596 bp 3'UTR (Fig. 4).

The predicted molecular weight of the putative vulture IκBα was of 35170 Da. Structurally, the vulture I kappa B alpha molecule could be divided into three sections: a 70-amino-acid N terminus with no known function, a 205-residue midsection composed of five ankyrin-like repeats, and a very acidic 42-amino-acid C terminus that resembles a PEST sequence. Examination of the Griffon vulture sequence revealed the features characteristic of an IκB molecule (Fig. 4) The putative vulture IκBα protein was composed of a N-terminal regulatory domain, a central ankyrin repeat domain (ARD), required for its interaction with NF-κB, and a putative PEST-like sequence in the C-terminus (Fig. 5), which is similar to IκBα proteins from other organisms (Jaffray et al., 1995). Together with the N-terminal regulatory domain and the central ARD domain, the presence of an acidic C-terminal PEST region rich in the amino acids proline (P), glutamic acid (E), serine (S) and threonine (T) is characteristic of IκBα inhibitors (Luque & Gelinas, 1998). PEST regions have been found in the C-terminus of avian IκBα (Krishnan et al., 1995) and mammalian IκBα and it was also present in the vulture IκBα sequence (Fig. 5). Particularly, the PEST sequence of IκBα seems to be critical for its calpain-dependent degradation (Shumway et al., 1999).

```
cggagccctgccgctatgatcagcgcccgccgcctcgtcgagccgccggttatggagggc    60
                M   I   S   A   R   R   L   V   E   P   P   V   M   E   G    15
tacgagcaagcgaagaaagagcgccagggcggcttcccgctcgacgaccgccacgacagc   120
  Y   E   Q   A   K   K   E   R   Q   G   G   F   P   L   D   D   R   H   D   S    35
ggcttggactccatgaaggaggaagagtaccggcagctggtgaaggagctggaggacata   180
  G   L   D   S   M   K   E   E   E   Y   R   Q   L   V   K   E   L   E   D   I    55
cgcctgcagccccgcgagccgcccgcctgggcgcagcagctgacggaggacggagacact   240
  R   L   Q   P   R   E   P   P   A   W   A   Q   Q   L   T   E   D   G   D   T    75
tttctccacttggcgattattcacgaggaaaaagccctgagcctggaggtgatccggcag   300
  F   L   H   L   A   I   I   H   E   E   K   A   L   S   L   E   V   I   R   Q    95
gcggccggggaccgtgctttcctgaacttccagaacaacctcagccagactcctcttcac   360
  A   A   G   D   R   A   F   L   N   F   Q   N   N   L   S   Q   T   P   L   H    115
ctggcagtgatcaccgatcagcctgaaattgccgagcatcttctgaaggccggatgcgac   420
  L   A   V   I   T   D   Q   P   E   I   A   E   H   L   L   K   A   G   C   D    135
ctggaactcagggacttccgaggaaacacccccctgcatattgcctgccagcagggctcc   480
  L   E   L   R   D   F   R   G   N   T   P   L   H   I   A   C   Q   Q   G   S    155
ctcaggagcgtcagcgtcctcacgcagtactgccagccgcaccacctcctcgctgtcctg   540
  L   R   S   V   S   V   L   T   Q   Y   C   Q   P   H   H   L   L   A   V   L    175
caggcaaccaactacaacgggcatacatgtctccatttggcatctattcaaggatacctg   600
  Q   A   T   N   Y   N   G   H   T   C   L   H   L   A   S   I   Q   G   Y   L    195
cctattgtcgaatacttgctgtccttgggagcagatgtaaatgctcaggagccatgcaat   660
  A   I   V   E   Y   L   L   S   L   G   A   D   V   N   A   Q   E   P   C   N    215
ggcagaacggcactacatttggctgtcgacctgcagaattcagacctggtgtcgcttctg   720
  G   R   T   A   L   H   L   A   V   D   L   Q   N   S   D   L   V   S   L   L    235
gtgaaacatggggcggacgtgaacaaagtgacctaccaaggctattccccctatcagctc   780
  V   K   H   G   A   D   V   N   K   V   T   Y   Q   G   Y   S   P   Y   Q   L    255
acatggggaagagacaactccagcatacaggaacagctgaagcagctgaccacagccgac   840
  T   W   G   R   D   N   S   S   I   Q   E   Q   L   K   Q   L   T   T   A   D    275
ctgcagatgttgccagaaagtgaggacgaggagagcagtgaatcggagcctgaattcaca   900
  L   Q   M   L   P   E   S   E   D   E   E   S   S   E   S   E   P   E   F   T    295
gaggatgaacttatatacgatgactgccttattggaggacgacagctggcattttaaagc   960
  E   D   E   L   I   Y   D   D   C   L   I   G   G   R   Q   L   A   F   *    313
agagctatctgtgaaaagaagtgactgtgtacatatgtatagaaaaaggactgacttc**at**   1020
**tta**aaaagaaagtcgcaatgcaaagggaaaaaccaggagggaaatactacactgcccagc   1080
aaggagcacataattgtaacaggttctggcctgtgtttaaatacaggagtgggatgtgta   1140
acatcagtagggatctgtgattattcacaccacctgataaagagccacatagccaatctt   1200
ctcagccctacaaaggtaacagactacacatccaacctgctggttacagagagctatctt   1260
gtggtgttaagtaccacgaggaatgcgtgtcgcctcgtggcaaggcaggctcataccaac   1320
cccccatcttctcggagactgcgtgttaatctgcgttgggctggtggtgctccctggcc   1380
ttactgaccggcctcagctgctcttggtggggtgtcccaggtggaggagtcaaaccaagg   1440
gactggtgacctcctgactgttagaagaaagtagcaataatgttaactgtgggcattgga   1500
aactgtgtgtttcacaccatgtgtgtcataattgctacacttttttagcaattg   1553
```

Fig. 4. Nucleotide and deduced amino acid sequence of vulture IκBα. Complete sequence of the full-length vulture IκBα obtained from the cDNA library (GenBank accession number: EU161944). Translated amino acid sequence is also shown under nucleotide sequence. Numbers to the right of each row refer to nucleotide or amino acid position. Ankyrin domains are shaded. The PEST region is underlined. The ATTTA domain is in bold. Phosphorylation sites Ser-35 and Ser-39 are circled. The translational stop site is indicated by an asterisk.

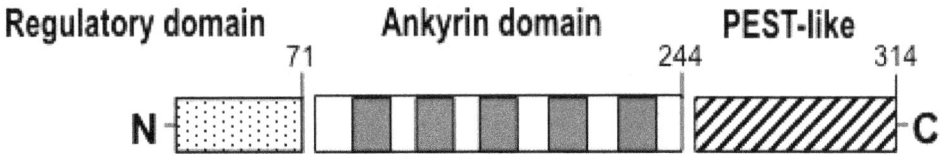

Fig. 5. Schematic structure of vulture IκBα.

Structure obtained from the SMART analysis (at expasy web server) of the amino acid sequence from vulture IκBα. Each box shows a typical structure of a member of the IκBα inhibitor. Vulture IκBα consists of an N-terminus regulatory domain, a central ankyrin domain containing five ankyrin repeats followed by an additional PEST-like motif. Number shows the amino acid flanking the relevant domains.

Classical activation of NF-kappaB involves phosphorylation, polyubiquitination and subsequent degradation of IκB. Several residues are known to be important in the N-terminal regulatory domain (Luque & Gelinas, 1998, Luque et al., 2000). In nonstimulated cells, NF-kappa B dimers are maintained in the cytoplasm through interaction with inhibitory proteins, the IκBs. In response to cell stimulation, mainly by proinflammatory cytokines, a multisubunit protein kinase, the I kappa B kinase (IKK), is rapidly activated and phosphorylates two critical serines in the N-terminal regulatory domain of the I kappa Bs. Phosphorylated IκBs are recognized by a specific E3 ubiquitin ligase complex on neighboring lysine residues, which targets them for rapid degradation by the 26S proteasome, which frees NF-kappaB and leads to its translocation to the nucleus, where it regulates gene transcription (Karin & Ben-Neriah, 2000). It has been demonstrated that phosphorylation of the N-terminus residues Ser-32 and Ser-36 is the signal that leads to inducer-mediated degradation of IκBα in mammals (Brown et al., 1997; Good et al., 1996). As can be observed in the alignment of Figure 6, the griffon vulture equivalent residues seem to be Ser-35 and Ser-39, which are part of the conserved sequence DSGLDS (Luque et al., 2000; Pons et al., 2007). This observation suggests that the phosphorylation of these serine residues could trigger the IκBα inducer-mediated degradation in vulture in a similar manner to that in mammals. Unlike ubiquitin modification, which requires phosphorylation of S32 and S36, the small ubiquitin-like modifier (SUMO) modification of IκBα is inhibited by phosphorylation. Thus, while ubiquitination targets proteins for rapid degradation, SUMO modification acts antagonistically to generate proteins resistant to degradation (Desterro et al., 1998; Mabb & Miyamoto, 2007). This SUMO modification occurs primarily on K21 (Mabb & Miyamoto, 2007). This residue was also conserved in the IκBα sequence from human, mouse, pig, rat and vulture, but not from chicken (Fig. 6).

Amino acid sequence of vulture IκBα was aligned with the orthologous sequence of chicken (Gallus gallus), pig (Sus scrofa), cattle (Bos taurus), human (Homo sapiens) and mouse (Mus musculus) based on amino acid identity and structural similarity. Identical amino acid residues to vulture IκBα from the aligned sequences are shaded. Gaps were introduced for optimal alignment of the sequences and are indicated by dashes (-). SUMOlation sites are squared and phosphorylation sites are circled. GenBank or Swiss protein accession numbers are: DQ480086, Q5WA51, Q59HI9, Q706D2, Q5FWG5 and Q6A0E8, respectively. Griffon vulture IκBα sequence was deposited in the Genbank under accession number EU161944.

```
Vulture   MISARRLVEPPVMEGYEQA-KKERQGGFPL-DDRHISGLDSKEEFYRQLVKELEDIRLQP
Chicken   MLSAHRPAEPPAVEGCEPP-RKERQGGLLPPDDRHISGLDSKEEFYRQLVRELEDIRLQP
Human     MFQAAERPQEWAMEGPRDGIKKER---LL--DDRHISGLDSKDEHYEQMVKELQEIRLEP
Mouse     MFQPAGHGQDWAMEGPRDGIKKER---LV--DDRHISGLDSKDEHYEQMVKELREIRLQP
Pig       MFQPAEPGQEWAMEGPRDAIKKER---LL--DDRHISGLDSNKDEHYEQMVKELREIRLEP
Rat       MFQPAGHGQDWAMEGPRDGIKKER---LV--DDRHISGLDSKDELYEQMVKELREIRLQP


Vulture   REPP----AWAQQLTEDGDTFLHLAIIHEEKAISLEVIRQAAGDRAFLNFQNNLSQTPLH
Chicken   REPPARPHAWAQQLTEDGDTFLHLAIIHEEKAISLEVIRQAAGDAAFLNFQNNLSQTPLH
Human     QEVPRGSEPWKQQLTEDGDSFLHLAIIHEEKALTMEVIRQVKGDLAFLNFQNNLQQTPLH
Mouse     QEAPLAAEPWKQQLTEDGDSFLHLAIIHEEKPLTMEVIGQVKGDLAFLNFQNNLQQTPLH
Pig       QEAPRGAEPWKQQLTEDGDSFLHLAIIHEEKALTMEVVRQVKGDLAFLNFQNNLQQTPLH
Rat       QEAPLAAEPWKQQLTEDGDSFLHLAIIHEEKTLTMEVIGQVKGDLAFLNFQNNLQQTPLH


Vulture   LAVITDQPEIAEHLIKAGCDLELRDFRGNTPLHIACQQGSLRSVSVLTQYCQPHHLLAVL
Chicken   LAVITDQAEIAEHLIKAGCDLDVRDFRGNTPLHIACQQGSLRSVSVLTQHCQPHHLLAVL
Human     LAVITNQPEIAEALLGAGCDPELRDFRGNTPLHLACEQGCLASVGVITQSCTTPHLHSIL
Mouse     LAVITNQPGIAEALIKAGCDPELRDFRGNTPLHLACEQGCLASVAVLTQTCTPQHLHSVL
Pig       LAVITNQPEIAEALLEAGCDPELRDFRGNTPLHLACEQGCLASVGVLTQPRGTQHLHSIL
Rat       LAVITNQPGIAEALIKAGCDPELRDFRGNTPLHLACEQGCLASVAVLTQTCTPQHLHSVL


Vulture   QATNYNGHTCLHLASIQGYLAIVEYLLSLGADVNAQEPCNGRTALHLAVDLQNSDLVSLL
Chicken   QATNYNGHTCLHLASIQGYLAVVEYLLSLGADVNAQEPCNGRTALHLAVDLQNSDLVSLL
Human     KATNYNGHTCLHLASIHGYLGIVELLVSLGADVNAQEPCNGRTALHLAVDLQNPDLVSLL
Mouse     QATNYNGHTCLHLASIHGYLAIVEHLVTLGADVNAQEPCNGRTALHLAVDLQNPDLVSLL
Pig       QATNYNGHTCLHLASIHGYLGIVELLVSLGADVNAQEPCNGRTALHLAVDLQNPDLVSLL
Rat       QATNYNGHTCLHLASIHGYLGIVEHLVTLGADVNAQEPCNGRTALHLAVDLQNPDLVSLL


Vulture   VKHGADVNKVTYQGYSPYQLTWGRDNSSIQEQLKQLTTADLQMLPESEDEESSISIP---
Chicken   VKHGPDVNKVTYQGYSPYQLTWGRDNASIQEQLKLLTTADLQILPESEDEESSESP---
Human     LKCGADVNRVTYQGYSPYQLTWGRFSTRIQQQLGQLTLENLQMLPESEDEESYDTSEFT
Mouse     LKCGADVNRVTYQGYSPYQLTWGRFSTRIQQQLGQLTLENLQMLPESEDEESYDTTS--
Pig       LKCGADVNRVTYQGYSPYQLTWGRFSTRIQQQLGQLTLENLQMLPESEDEESYDTTS--
Rat       LKCGADVNRVTYQGYSPYQLTWGRFSTRIQQQLGQLTLENLQTLPESEDEESYDTTS--


Vulture   ERTEDELIYDDCLIGGRQLAF
Chicken   ERTEDELMYDDCCIGGRQLTF
Human     ERTEDELRYDDCVFGGQRLTL
Mouse     ERTEDELRYDDCVFGGQRLTL
Pig       ERTEDELRYDDCVLGGQRLTL
Rat       ERTEDELRYDDCVFGGQRLTL
```

Overall identity	
Chicken	91%
Human	73%
Mouse	74%
Pig	73%
Rat	73%

Fig. 6. Alignment of amino acid sequences of IκBα from different species.

A common characteristic of the IκB proteins is the presence of ankyrin repeats, which interact with the Rel-homology domain of NF-κB (Aoki et al., 1996; Luque & Gelinas, 1998). In the vulture sequence, five ankyrin repeats were detected using the Simple Modular Architecture Research Tool (SMART) at EMBL (Table 2). Five ankyrin repeats also exist in human and other vertebrates IκBα (Jaffray et al., 1995). It is possible that individual repeats have remained conserved because of their important structural and functional roles in regulating NF-κB.

Compared with other species, vulture IκBα exhibited the lowest number of predicted SUMOlation sites (Table 2).

Structural feature	G fulvus	G gallus	H sapiens	S scrofa	R norvegicus	M musculus
Amino acid residues	313	318	317	314	314	314
Number of ankyrin repeats	5	5	5	5	5	5
Phosphorylation sites	14	13	14	15	14	13
Predicted MW(KDa)	35,17	35,40	35,61	35,23	35,02	35,02
SUMOlation sites	2	3	4	4	5	5

Table 2. Structural features of IκBα from Griffon vulture (G fulvus), Chicken (G gallus), human (H sapiens), pig (S scrofa), rat (R norvegicus) and mouse (M musculus) amino acid sequences. The theoretical molecular weight, number of ankyrin repeats, SUMOlation and of phosphorylation sites was calculated using the software available at the expasy web server (http://www.expasy.org). Genbank or Swiss accession number for proteins are EU161944 (G fulvus), Q91974 (G gallus), P25963 (H sapiens), Q08353 (S scrofa), Q63746 (R norvegicus), and Q9Z1E3 (M musculus).

3.2.1 Amino acid sequence comparison of vulture IκBα with other species

The comparison of the deduced amino acid sequence of vulture IκBα with the sequence of chicken, human, mouse, pig, and rat IκBα indicated that the deduced protein had a higher degree of similarity to chicken (91% of amino acid similarity) than to human (73%), mouse (74%), pig (73%) and rat (73%) sequences (Fig. 6). The analysis of the vulture IκBα sequence using the software NetPhos 2.0 (http://www.cbs.dtu.dk/services/NetPhos) revealed 14 potential phosphorylation sites: 10 Ser (S35, S39, S89, S160, S251, S263, S282, S287, S288, and S290), 1 Thr (T295) and 3 Tyr (Y16, Y45, and Y301). Although many of these residues were conserved in the aligned sequences from chicken, human, mouse, pig and rat IκBα, two phosphorylation sites (Y16 and S160) were distinctive to the vulture sequence (Fig. 6).

4. Detection of vulture TLR1 and IκBα expression in tissues

In order to better understand the biological roles of TLR1 and IκBα, we analyzed their tissue expression pattern. The presence of transcripts encoding vulture TLR1 and IκBα in tissues was determined by real time RT-PCR. Biological samples were collected from vultures (about 8-10 months old) that were provisionally captive at the Centre for Wild Life Protection, "El Chaparrillo", Ciudad Real, Spain. Blood was obtained by puncture of the branquial vein, located in the internal face of the wing, and collected in 10 ml tubes with EDTA as anti-coagulant. Blood (10 ml) was diluted 1:1 (vol:vol) with PBS (Sigma) and the mononuclear fraction containing PBMC was obtained by density gradient centrifugation on Lymphoprep (Axis-Shield, Oslo, Norway). All vulture tissues used for cDNA preparation were obtained fresh from euthanised birds that were impossible to recover.

RT-PCR was performed on a SmartCycler® II thermal cycler (Cepheid, Sunnyvale, CA, USA) using the QuantiTect® SYBR® Green RT-PCR Kit (Quiagen, Valencia, CA, USA), following the recommendations of the manufacturer. We used primers GfTLR-Fw (5'-GCT TGC CAG TCA ACA TCA GA-3') and GfTLR-Rv (5'-GAA CTC CAG CGA CGT AAA GC-3'), which amplify a fragment of 158 bp of vulture TLR1 and primers IκBα -L (5'- CTG CAG GCA ACC AAC TAC AA -3') and IκBα –R (5'- TGA ATT CTG CAG GTC GAC AG-3'), which amplify a fragment of 165 b of vulture IκBα. Cycling conditions were: 94°C for 30 sec, 60°C for 30 sec, 72°C for 1 min, for 40 cycles. As an internal control, RT-PCR was performed on the same RNAs using the primers BA-Fw (5'-CTA TCC AGG CTG TGC TGT CC-3') and BA-Rv (5'-TGA GGT AGT CTG TCA GGT CAG G-3'), which amplify a fragment of 165 bp from the conserved housekeeping gene beta-actin. Control reactions were done using the same procedures, but without RT to control for DNA contamination in the RNA preparations, and without RNA added to control contamination of the PCR reaction. Amplification efficiencies were validated and normalized against vulture beta actin, (GenBank accession number DQ507221) using the comparative Ct method. Experiments were repeated for at least three times with similar results. Tissues used for the study were artery, liver, lung, bursa cloacalis, heart, small intestine, peripheral blood mononuclear cells (PBMC), large intestine and kidney.

The level of TLR1 mRNA was higher in kidney, small intestine and PBMC (Fig. 7).

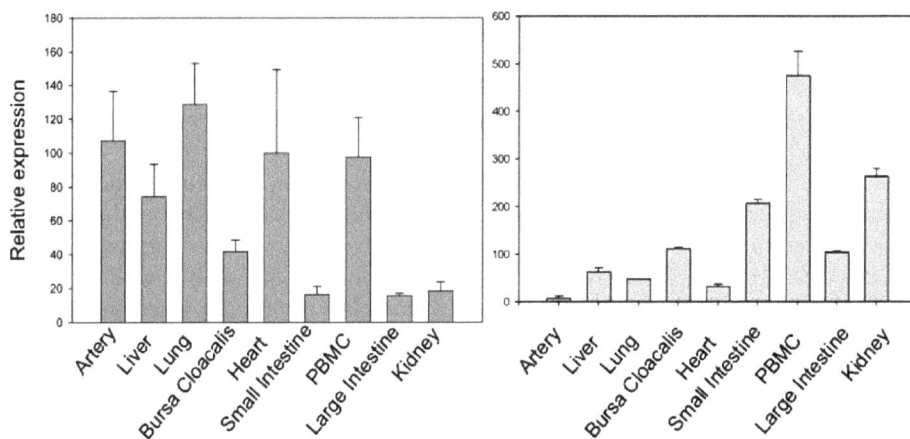

Fig. 7. Relative expression of TLR1 and IκBα mRNA transcripts in vulture cells and tissues.

Real time RT-PCR was used to examine the relative amount of TLR1 (right) and IκBα (left) transcripts in vulture cells and tissues. The data were normalised using the beta-actin gene and calculated by the delta Ct method.

Moderate vulture TLR1 mRNA levels were observed in Bursa cloacalis and large intestine, whereas the lowest TLR1 mRNA levels were found in liver, heart and artery (Fig. 6). It has been reported that the patterns of TLR tissue expression are variable, even among closely related species (Zarember & Godowski 2002). Likewise, the intensity and the anatomic location of the innate immune response may vary considerably among species (Rehli, 2002). Consistent with its role in pathogen recognition and host defense, the tissue and cell expression pattern of vulture TLR1, as revealed by real time RT-PCR, correlated with vulture ability to respond to various pathogenic challenges. The expression of vulture TLR1 was higher in cells such as circulating PBMC and intestinal epithelial cells that are immediately accessible to microorganisms upon infection.

The analysis of the relative expression of IκBα mRNA transcripts, using real-time RT-PCR, demonstrated that vulture IκBα mRNAs were higher in lung, artery, heart, and in PBMC cells (Fig. 7), which was consistent with its role in numerous physiological processes. Interestingly, the expression of vulture IκBα mRNA was observed in tissues at which the lowest expression of vulture Toll-like receptor was found. This is consistent with the role of IκBα as inhibitor of the TLR-signalling pathway.

5. Analysis of the evolutionary relationship of vulture TLR and IκBα

The dendrogram of sequences was calculated based on the distance matrix that was generated from the pairwise scores and the phylogenetic trees were constructed based on the multiple alignment of the sequences using the PHYLIP (Phylogeny Inference Package) available at the expasy.org web page. All ClustalW phylogenetic calculations were based around the neighbor-joining method of Saitou and Nei (Saitou & Nei, 1987).

For the analysis of the evolutionary relationship of vulture and other vertebrate TLR and IκBα, a phylogenetic tree was constructed with the TIR-domain sequences of human, macaque, bovine, pig, mouse, Japanese pufferfish and chicken TLR1. GenBank or swiss protein accessions numbers Q5WA51, Q706D2, Q6A0E8, Q59HI9, Q5H727 and Q5FWG5, respectively. The phylogenetic analysis of the TIR domain of vulture TLR1 revealed separate clustering of TLR1 from birds, fish, mouse and other mammals (Fig. 8B)

For the TLRs, it is assumed that the structure of the ectodomain has evolved more quickly than the structure of the TIR (Johnson, 2003). Similarly to other TLR receptors, the degree of homology of vulture TLR1 was higher in the transmembrane and cytoplasmic domains than in the extracellular domain. As expected, phylogenetic analysis of the TIR domains revealed separate clustering of TLR1 from birds, fish and mammals (Fig. 8B), suggesting independent evolution of the Toll family of proteins and of innate immunity (Beutler & Rehli, 2002; Roach et al., 2005).

The unrooted trees were constructed by neighbor-joining analysis of an alignment of the ankirin repeats of IκBα sequences from vulture and other species (A) or the alignment of the TIR domains of TLR1 from vulture and other species (B). The branch lengths are proportional to the number of amino acid differences. GenBank or swiss protein accessions numbers of chicken (Gallus gallus), human (Homo sapiens), mouse (Mus musculus), rat (Rattus norvegicus), African frog (Xenopus laevis), cattle (Bos taurus), zebrafish (Danio rerio), Mongolian gerbil (Meriones unguiculatus), Rainbow trout (Oncorhynchus mykiss) and pig (Sus scrofa) sequences used for the phylogenetic tree were Q91974, P25963, Q08353, Q63746, Q1ET75, Q6DCW3, Q8WNW7, Q6K196, Q1ET75, Q8QFQ0 and Q9Z1E3, respectively.

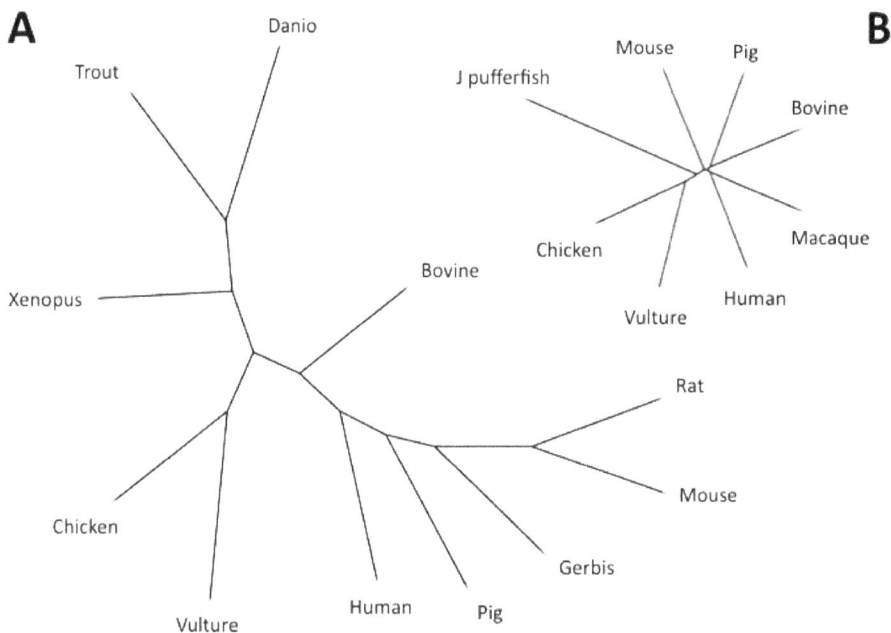

Fig. 8. Phylogenetic trees illustrating the relationship between TLR and IκBα sequences from vulture and other species.

For the analysis of the evolutionary relationship of vulture and other vertebrate IκBα, a phylogenetic tree was constructed with the sequences of chicken, human, mouse, rat, African frog, cattle, zebrafish, Mongolian gerbil, Rainbow trout and pig IκBα. The phylogenetic analysis of the ankyrin domain of vulture IκBα revealed separate clustering of IκBα from rodents, fish and other species and the sequence of vulture IκBα clustered together with that of chicken IκBα (Fig 8A). The IκB family includes IκBα, IκBβ, IκBγ, IκBε, IκBζ, Bcl-3, the precursors of NFκB1 (p105), and NF-κB2 (p100), and the Drosophila protein Cactus (Hayden et al., 2006; Karin & Ben-Neriah, 2000; Totzke et al., 2006; Gilmore, 2006). Why multiple IκB proteins now exist in vertebrates has been a subject of great interest, and much effort has been expended on establishing the roles of individual members of this protein family in the regulation of NF-κB. The recent identification of a novel member of IκB family (IκBζ) indicates that there might exist species-specific differences in the regulation of NF-κB (Totzke et al., 2006).

Evolutionarily, the IκB protein family is quite old, as members have been found in insects, birds and mammals (Ghosh & Kopp, 1998). However, the finding that individual ankyrin repeats within each IκB molecule are more similar to corresponding ankyrin repeats in other IκB family members, rather than to other ankyrin repeats within the same IκB, suggests that all IκB family members evolved from an ancestral IκB molecule (Huguet, et al., 1997).

Consistent with the hypothesis that all these factors evolved from a common ancestral RHD-ankyrin structure within a unique superfamily, explaining the specificities of interaction between the different Rel/NF-kappa B dimers and the various I kappa B inhibitors (Huguet, et al., 1997).

Recently, the presence of two IkappaB-like genes in Nematostella encoded by loci distinct from nf-kb suggested that a gene fusion event created the nfkb genes in insects and vertebrates (Sullivan et al., 2007). This is consistent with the hypothesis that interactions between transcription factors of the Rel members and members of the IκB gene family evolved to regulate genes mainly involved in immune inflammatory responses (Bonizzi & Karin, 2004).

NF-kappaB represents an ancient, generalized signaling system that has been co-opted for immune system roles independently in vertebrate and insect lineages (Friedman & Hughes, 2002). Therefore, while these proteins share a basic three-dimensional structure as predicted by their shared ankyrin repeat pattern and sequence, a possible evolutionary scenario based on this phylogenetic tree could be that subtle differences in the amino acid substitutions in the ankyrin repeats and flanking sequences occurred throughout evolution, which contributed to their specificity of interaction with various members of the Rel family.

6. Conclusions

In summary, the molecular cloning methods reported herein identified and characterized the vulture orthologues to TLR1 (CD281) and to IκBα, the first NF-κB pathway element from the griffon vulture G. fulvus. These results have implications for the understanding of the evolution of pathogen-host interactions. Particularly, these studies help to highlight a potentially important regulatory pathway for the study of the related functions in vulture immune system (Perez de la Lastra & de la Fuente, 2007; 2008). Despite the overall structure of vulture TLR1 and expression pattern was similar to that of chicken, pig, cattle, human and mouse TLR, vulture TLR1 had differences in the length of the ectodomain, number and position of LRRs and N-glycosylation sites that makes vulture TLR1 structurally unique with possible functional implications.

Strong selective pressure for recognition of and response to pathogen-associated molecular patterns (PAMPs) has probably maintained a largely unchanged TLR signalling pathways in all vertebrates. The IκBα gene reported here expands our understanding of the immune regulatory pathways present in carrion birds that are in permanent contact with pathogens. Current investigations should focus on the cloning and characterization of other members of NF-κB signalling cascade and genes controlled by this signalling pathway. At this point it is difficult to understand the implications of the structural differences between vulture TLR1, chicken TLR1 and TLR1 in different mammalian species. A greater understanding of the functional capacity of non-mammalian TLRs and, particularly in carrion birds that are in permanent contact with pathogens, has implications for the understanding of the evolutionary pressures that defined the TLR repertoires in present day animals.

7. Acknowledgements

This work was supported by the Junta de Comunidades de Castilla-La Mancha (JCCM), project PII1I09-0243-4350.

8. References

Aoki, T., et al., The ankyrin repeats but not the PEST-like sequences are required for signal-dependent degradation of IkappaBalpha. Oncogene, 1996. 12(5): p. 1159-64.

Apanius, V., S.A. Temple, and M. Bale, Serum proteins of wild turkey vultures (Cathartes aura). Comp Biochem Physiol B, 1983. 76(4): p. 907-13.

Beutler, B. and M. Rehli, Evolution of the TIR, tolls and TLRs: functional inferences from computational biology. Curr Top Microbiol Immunol, 2002. 270: p. 1-21.

Bonizzi, G. and M. Karin, The two NF-kappaB activation pathways and their role in innate and adaptive immunity. Trends Immunol, 2004. 25(6): p. 280-8.

Brown, K., et al., The signal response of IkappaB alpha is regulated by transferable N- and C-terminal domains. Mol Cell Biol, 1997. 17(6): p. 3021-7.

Desterro, J.M., M.S. Rodriguez, and R.T. Hay, SUMO-1 modification of IkappaBalpha inhibits NF-kappaB activation. Mol Cell, 1998. 2(2): p. 233-9.

Friedman, R. and A.L. Hughes, Molecular evolution of the NF-kappaB signaling system. Immunogenetics, 2002. 53(10-11): p. 964-74.

Ghosh, S., M.J. May, and E.B. Kopp, NF-kappa B and Rel proteins: evolutionarily conserved mediators of immune responses. Annu Rev Immunol, 1998. 16: p. 225-60.

Gilmore, T.D., Introduction to NF-kappaB: players, pathways, perspectives. Oncogene, 2006. 25(51): p. 6680-4.

Good, L.F., et al., Multiple structural domains within I kappa B alpha are required for its inducible degradation by both cytokines and phosphatase inhibitors. Biochem Biophys Res Commun, 1996. 223(1): p. 123-8.

Hayden, M.S., A.P. West, and S. Ghosh, NF-kappaB and the immune response. Oncogene, 2006. 25(51): p. 6758-80.

Hopkins, P.A. and S. Sriskandan, Mammalian Toll-like receptors: to immunity and beyond. Clin Exp Immunol, 2005. 140(3): p. 395-407.

Houston, D.C. and J.E. Cooper, The digestive tract of the whiteback griffon vulture and its role in disease transmission among wild ungulates. J Wildl Dis, 1975. 11(3): p. 306-13.

Huguet, C., P. Crepieux, and V. Laudet, Rel/NF-kappa B transcription factors and I kappa B inhibitors: evolution from a unique common ancestor. Oncogene, 1997. 15(24): p. 2965-74.

Jaffray, E., K.M. Wood, and R.T. Hay, Domain organization of I kappa B alpha and sites of interaction with NF-kappa B p65. Mol Cell Biol, 1995. 15(4): p. 2166-72.

Johnson, G.B., et al., Evolutionary clues to the functions of the Toll-like family as surveillance receptors. Trends Immunol, 2003. 24(1): p. 19-24.

Karin, M. and Y. Ben-Neriah, Phosphorylation meets ubiquitination: the control of NF-[kappa]B activity. Annu Rev Immunol, 2000. 18: p. 621-63.

Krishnan, V.A., et al., Structure and regulation of the gene encoding avian inhibitor of nuclear factor kappa B-alpha. Gene, 1995. 166(2): p. 261-6.

Luque, I. and C. Gelinas, Distinct domains of IkappaBalpha regulate c-Rel in the cytoplasm and in the nucleus. Mol Cell Biol, 1998. 18(3): p. 1213-24.

Luque, I., et al., N-terminal determinants of I kappa B alpha necessary for the cytoplasmic regulation of c-Rel. Oncogene, 2000. 19(9): p. 1239-44.

Mabb, A.M. and S. Miyamoto, SUMO and NF-kappaB ties. Cell Mol Life Sci, 2007. 64(15): p. 1979-96.

Ohishi, I., et al., Antibodies to Clostridium botulinum toxins in free-living birds and mammals. J Wildl Dis, 1979. 15(1): p. 3-9.

Perez de la Lastra, J.M. and J. de la Fuente, Molecular cloning and characterisation of the griffon vulture (Gyps fulvus) toll-like receptor 1. Dev Comp Immunol, 2007. 31(5): p. 511-9.

Perez de la Lastra, J.M. and J. de la Fuente, Molecular cloning and characterisation of a homologue of the alpha inhibitor of NF-kB in the griffon vulture (Gyps fulvus). Vet Immunol Immunopathol, 2008. 122: p. 318-25

Pons, J., et al., Structural studies on 24P-IkappaBalpha peptide derived from a human IkappaB-alpha protein related to the inhibition of the activity of the transcription factor NF-kappaB. Biochemistry, 2007. 46(11): p. 2958-72.

Rehli, M., Of mice and men: species variations of Toll-like receptor expression. Trends Immunol, 2002. 23(8): p. 375-8.

Roach, J.C., et al., The evolution of vertebrate Toll-like receptors. Proc Natl Acad Sci U S A, 2005. 102(27): p. 9577-82.

Saitou, N. and M. Nei, The neighbor-joining method: a new method for reconstructing phylogenetic trees. Mol Biol Evol, 1987. 4(4): p. 406-25.

Shumway, S.D., M. Maki, and S. Miyamoto, The PEST domain of IkappaBalpha is necessary and sufficient for in vitro degradation by mu-calpain. J Biol Chem, 1999. 274(43): p. 30874-81.

Sullivan, J.C., et al., Rel homology domain-containing transcription factors in the cnidarian Nematostella vectensis. Dev Genes Evol, 2007. 217(1): p. 63-72.

Takeda, K. and S. Akira, Toll-like receptors in innate immunity. Int Immunol, 2005. 17(1): p. 1-14.

Takeda, K., Evolution and integration of innate immune recognition systems: the Toll-like receptors. J Endotoxin Res, 2005. 11(1): p. 51-5.

Totzke, G., et al., A novel member of the IkappaB family, human IkappaB-zeta, inhibits transactivation of p65 and its DNA binding. J Biol Chem, 2006. 281(18): p. 12645-54.

Weber, A.N., M.A. Morse, and N.J. Gay, Four N-linked glycosylation sites in human toll-like receptor 2 cooperate to direct efficient biosynthesis and secretion. J Biol Chem, 2004. 279(33): p. 34589-94.

Yilmaz, A., et al., Identification and sequence analysis of chicken Toll-like receptors. Immunogenetics, 2005. 56(10): p. 743-53.

Zarember, K.A. and P.J. Godowski, Tissue expression of human Toll-like receptors and differential regulation of Toll-like receptor mRNAs in leukocytes in response to microbes, their products, and cytokines. J Immunol, 2002. 168(2): p. 554-61.

9

Molecular Cloning, Expression, Purification and Immunological Characterization of Proteins Encoded by Regions of Different Genes of *Mycobacterium tuberculosis*

Shumaila Nida Muhammad Hanif, Rajaa Al-Attiyah and
Abu Salim Mustafa
Department of Microbiology, Faculty of Medicine, Kuwait University
Kuwait

1. Introduction

Tuberculosis (TB) is an international infectious disease problem, but people in the developing countries are the major sufferers (World Health Organization, 2010). The disease in humans is primarily caused by a pathogenic mycobacterial species known as *Mycobacterium tuberculosis*. The current estimates suggest that about 1/3rd of the world population is latently infected with *M. tuberculosis*, 9 million people develop the active disease and 1.7 million die of TB each year (World Health Organization, 2009). Among infectious diseases, TB is the second top most killer of adults, after HIV/AIDS, and is among the overall top 10 causes of death in the world (World Health Organization, 2008). The global problem of TB is so serious that it has been declared "a global emergency" by the World Health Organization (WHO), the first declaration of its kind by WHO. Several factors have contributed in the deterioration of the global problem of TB, which include the increase in the incidence of drug-resistant TB (multi and extensive drug resistance), migration of people, due to political and economic reasons, from poor and highly endemic countries to countries with low endemicity, and co-infection with *M. tuberculosis* and HIV (Borgdorff et al., 2010; Chiang et al., 2010; Semba et al., 2010). The worldwide control of TB requires cost-effective, sensitive and specific methods to diagnose latent as well as active TB, and vaccines that can provide protection in all human populations irrespective of their immune status, geographical locations and environmental conditions (Mustafa, 2009, 2010).

Tuberculin skin test (TST) is considered the standard test for the diagnosis of infection with *M. tuberculosis*. The test involves intradermal injection of purified protein derivative (PPD), which is a poorly defined and crude mixture of secreted and somatic proteins present in the culture filtrate of *M. tuberculosis*. For TST, PPD is injected in the forearm and the ensuing delayed type hypersensitivity (DTH) skin reaction (induration) is read after 48 to 72 h. A DTH reaction with induration of 10 mm or greater is considered positive in immunocompetent subjects (Mahadevan et al., 2005; Moffitt & Wisinger, 1996). Although, PPD has been used for TB diagnosis and epidemiological studies for more than half a century, it cannot distinguish between active disease, prior sensitization by contact with *M.*

tuberculosis, BCG vaccination, or cross-sensitization by environmental mycobacteria (Hill et al., 2006; Leung et al., 2005). This is because PPD contains a large number of mycobacterial antigens (>200), some of which are shared among all the pathogenic mycobacteria belonging to the *M. tuberculosis* complex (*M. tuberculosis, M. bovis,* and *M. africanum*), environmental non-tuberculous mycobacteria (NTM), and the substrains of *M. bovis* bacille Calmette-Guerin (BCG) vaccines. Thus, although response to PPD is an important aid in the diagnosis of TB and can give an indication of exposure to mycobacteria, it is often impossible to distinguish BCG vaccination and exposure to NTM from *M. tuberculosis* infection. Thus, TST has poor diagnostic value, especially in geographic areas where BCG vaccine is routinely used or the environmental burden of non-tuberculous mycobacteria is high (Kurup et al., 2006). Therefore, there is a need to identify and produce *M. tuberculosis*-specific antigens, which can be useful as skin test reagents for specific detection of infection with *M. tuberculosis*.

Vaccination is considered the most important tool to provide protection against TB, and attempts have been made to vaccinate humans, against TB, since the discovery of the TB bacillus by Koch in 19th century. However, the best vaccine available till date is the live attenuated strain of *M. bovis*, i.e. BCG, developed by Calmette and Guerin in 1921. Since then BCG has been used as a vaccine to prevent TB. In spite of its widespread application for about 90 years and many advantages like being inexpensive, safe at birth, given as a single shot, and provision of some protection against childhood TB and severe manifestations of the disease (Trunz et al., 2006), BCG vaccine remains the most controversial vaccine in current use (Fine, 2001). This is because its protective efficacy has varied widely in different parts of the world (Haile & Kallenius, 2005). In particular, BCG has failed to provide protection in developing countries against pulmonary TB, the most common manifestation of TB, in adults (Crampin et al., 2009; Narayanan, 2006). Furthermore, the widespread use of BCG vaccination faces two additional problems. First is the interference with TST in the diagnosis of TB because BCG vaccination induces a DTH response that cannot be distinguished from infection with *M. tuberculosis*, and therefore it becomes difficult to use TST for diagnostic or epidemiological purposes (Mustafa & Al-Attiyah, 2003). Second, BCG, being a live vaccine, cannot be used in all groups of people because it may cause disease by itself in immunocompromised people (Hesseling et al., 2007). The WHO has recommended that children with symptoms of HIV or AIDS should receive all the vaccines except BCG (World Health Organization, 2007). An ideal vaccine against TB should be safe and induce protective immunity both in immunocompetent and immunocompromised individuals. The development of a better BCG vaccine or alternative vaccines requires the identification and production of *M. tuberculosis*-specific antigens recognized by protective immune responses.

2. Identification of *M. tuberculosis*-specific genomic regions and encoded antigens

The existence of three *M. tuberculosis / M. bovis*-specific genomic regions deleted in BCG, i.e. regions of difference (RD)1, RD2 and RD3, was first described by Mahairas et al. in 1996 by employing subtractive genomic hybridization to identify genetic differences between *M. tuberculosis, M. bovis* and BCG (Mahairas *et al.*, 1996; Table 1). Later, in 1999, studies using comparative genome analysis identified 16 RDs that were present in *M. tuberculosis* H37Rv (RD1 to RD16) but deleted/absent in some or all strains of *M. bovis* and/or BCG (Behr et al., 1999; Table 1). Among these, 11 regions (RD1, RD4 to RD7, RD9 to RD13 and RD15)

Molecular Cloning, Expression, Purification and Immunological Characterization of Proteins Encoded by Regions
of Different Genes of Mycobacterium tuberculosis

143

covering 91 open reading frames (ORFs) of *M. tuberculosis* H$_{37}$Rv were deleted from all vaccine strains of BCG (Table 1). It was suggested that the identification of antigens encoded by the genes in *M. tuberculosis*-specific RDs may identify new *M. tuberculosis*-specific antigens with diagnostic and / or vaccine potential (Mustafa, 2001, 2002).

RD designation	No. of ORFs	ORFs designation	ORFs deleted / absent in	
			M. bovis strains	BCG strains
RD1	8 9 14	ORF1A-ORF1K Rv3871-Rv3879 ORF2-ORF15	Not deleted	All strains
RD2	11	Rv1978-Rv1988	Not deleted	Connaught , Danish, Frappier, Glaxo, Pasteur, Phipps, Prague, Tice
RD3	14	Rv1573-Rv1586c	3/8	All strains
RD4	3	Rv0221-Rv0223c	8/8	All strains
RD5	5	Rv3117-Rv3121	8/8	All strains
RD6	11	Rv1506c-Rv1516c	8/8	All strains
RD7	8	Rv2346c-Rv2353c	8/8	All strains
RD8	4	Rv0309-Rv0312	Not deleted	Connaught, Frappier
RD9	7	Rv3617-Rv3623	8/8	All strains
RD10	3	Rv1255c-Rv1257c	8/8	All strains
RD11	5	Rv3425-Rv3429	8/8	All strains
RD12	4	Rv2072c-Rv2075c	8/8	All strains
RD13	16	Rv2645-Rv2660c	4/8	All strains
RD14	8	Rv1766-Rv1773c	Not deleted	Pasteur
RD15	15	Rv1963c-Rv1977	8/8	All strains
RD16	6	Rv3400-Rv3405c	Not deleted	Moreau

Table 1. Designation of *M. tuberculosis* RDs deleted/absent in *M. bovis* and/or BCG, the number of ORFs predicted in each ORFs designations.

To analyze the *M. tuberculosis*-specific RDs for cell mediated immune (CMI) responses involving T helper (Th)1 cells that are associated with protection against TB, a novel approach of overlapping synthetic peptides was used (Al-Attiyah & Mustafa, 2008). Each peptide was 25 aa in length and overlapped with the neighboring peptides by 10 aa. Since the length of Th1 cell epitopes is usually between 8 to 10 aa (Mustafa, 2000, 2005a), the overlapping synthetic peptide strategy minimized the possibility of missing the potential Th1 cell epitopes. A total of 1648 peptides were synthesized and peptide pools corresponding to each RD were tested for reactivity with human PBMC in Th1-cell assays, i.e. antigen-induced proliferation and IFN-γ secretion. The results showed that RD1 was the most important region encoding proteins with Th1-cell reactivity (Al-Attiyah & Mustafa, 2008, 2010; Mustafa & Al-Attiyah, 2009; Mustafa et al., 2011). The analysis of individual peptide pools of RD1 ORFs identified three major (Rv3873, Rv3874 and Rv3875) and three moderate (Rv3871, Rv3872 and Rv3876) antigens stimulatory for Th1 cells in antigen-induced proliferation and IFN-γ assays (Hanif et al., 2008; Mustafa 2005b; Mustafa et al., 2008). However, Rv3873 and Rv3876 proteins were equally good stimulators of Th1 cells from TB patients and healthy donors, whereas Rv3872, Rv3874, Rv3875 proteins were better stimulators in TB patients and Rv3871 protein in healthy donors (Mustafa et al., 2008). When tested with PBMC in a cattle model of TB, Rv3872, Rv3873, Rv3874 and Rv3875 were also found to be major to moderate stimulators of Th1 cells present in the peripheral blood of M. tuberculosis-infected Cattle (Mustafa et al., 2002). The overall results showed that M. tuberculosis-specific proteins of RD1 Rv3872, Rv3873, Rv3874 and Rv3875 are major Th1-cell antigens and therefore could be useful as antigens in specific diagnosis and developing new vaccines against TB. However, for these applications, it was essential that the genes encoding these proteins are cloned and expressed in suitable vectors and the recombinant proteins are obtained in a purified form.

The predictions for all ORFs with Rv numbers are according to the predictions of Behr et al. (Behr et al., 1999), whereas RD1 ORF1A to ORF1K and ORF2 to ORF15 are according to the predictions of Mahairas et al. and Mustafa, respectively (Mahairas et al., 1996; Mustafa, 2005c). These RDs are expected to encode a total of 134 ORFs, of which 91 ORFs are predicted in RD1, RD4 to RD7, RD9 to RD13 and RD15, which are present in all strains of M. tuberculosis, but absent/deleted in all strains of BCG (Mustafa, 2001). RD3, although deleted in all BCG strains, is also deleted from some clinical strains of M. tuberculosis as well (Behr et al., 1999).

3. Molecular cloning, expression, purification and immunological characterization of proteins encoded by RD genes

In order to immunologically characterize the putative RD proteins encoded by M. tuberculosis-specific genes deleted/absent in BCG, previous studies attempted to clone and express six ORFs of RD1, i.e. ORF10 to ORF15 (Amoudy et al., 2006), as recombinant proteins in E. coli. However, these studies were successful in expressing five of the six targeted proteins as fusion proteins, fused with glutathione S-transferase (GST), by using the commercially available pGEX-4T vectors, but only two of them could be purified using glutathione-Sepharose affinity columns (Ahmed et al., 1999a, Amoudy et al., 2007; Amoudy & Mustafa, 2008). The problems included degradation of the mycobacterial proteins, non-binding of fusion proteins to affinity columns and the presence of contaminating E. coli proteins in purified preparations (Ahmed et al., 1999b, Amoudy & Mustafa, 2008). These studies have shown that even though high level expression of several mycobacterial proteins could be achieved by fusing them with GST, their purification was unsatisfactory and sometimes difficult due to improper folding of the fusion proteins and co-elution of E. coli proteins from the column along with the mycobacterial proteins. This is mainly due to the fact that the binding of GST containing fusion proteins on glutathione-Sepharose column is dependent on the proper folding of the GST tag and some *E. coli* proteins bind to the column non-specifically. However, binding of proteins fused with the 6x His tag to Ni-NTA agarose is not affected by the conformation of the expressed proteins and, consequently, proteins containing this tag can be purified even under denaturing conditions (Hendrickson et al., 2000). Therefore a new vector, pGES-TH-1, was developed using pGEX-4T-1 vector as the backbone (Ahmad et al., 2003). In addition to the features offered by the pGEX-4T vectors for high level expression of fusion proteins, the new vector was expected to allow relatively easy purification of recombinant proteins on the highly versatile Ni-NTA-agarose affinity matrix. The experimental utility of the new vector was demonstrated by expressing and purifying to homogeneity two proteins of RD1 (Rv3872 and Rv3873) and three proteins of RD15 (Mce3A, Mce3D and Mce3E) of *M. tuberculosis* (Ahmad et al., 2003, 2004, 2005). All these proteins were expressed in *E. coli*, in fusion with GST. The recombinant RD proteins were purified and isolated, free of GST, by affinity purification on glutathione-Sepharose and/or Ni-NTA-agarose affinity matrix and cleavage of the purified fusion proteins by thrombin protease. All recombinant proteins were more than 90% pure and were immunologically reactive with antibodies present in sera of TB patients and/or immunized animals (Ahmad et al., 2003, 2004, 2005; El-Shazly et al., 2007). These results demonstrated the utility of the newly constructed pGES-TH-1expression vector with two affinity tags for efficient expression and purification of recombinant RD proteins expressed in *E. coli*, which could be used for further diagnostic and immunological investigations both in animals as well as in humans.

In a recent study, we extended the work to achieve high level expression and purification of three low molecular weight proteins encoded by *M. tuberculosis*-specific genomic regions of RD1 and RD9, i.e. Rv3874, Rv3875 and Rv3619c, using the pGES-TH-1 expression vector (Hanif et al., 2010a). The complete cloning, expression and purification strategy for these proteins is shown in figure 1, by using Rv3875 as an example (Fig. 1). In brief, DNA corresponding to each gene was amplified using gene-specific forward (F) and reverse (R) primers and genomic DNA of *M. tuberculosis* by polymerase chain reaction (PCR). In addition to gene-specific sequences, in order to facilitate cloning in the pGEST-TH-1 vector, the F and R primers had *BamH* 1 and *Hind* III restriction sites at their 5′ ends, respectively. The PCR-amplified DNA were analyzed by agarose gel electrophoresis, which showed the amplified fragments corresponded to the size of the respective genes. The PCR-amplified DNA were ligated to pGEM-T Easy vector DNA, yielding recombinant plasmids pGEM-T/Rv3874, pGEM-T/Rv3875 and pGEM-T/Rv3619c. The identities of the cloned DNA fragments were determined based on size by digestion of the recombinant plasmids with *EcoR* I. Furthermore, The DNA fragments corresponding to *rv3874*, *rv3875* and *rv3619c* genes were excised from recombinant pGEM-T Easy using the restriction enzymes *BamH* 1 and *Hind* III and subcloned in the expression vector pGES-TH-1, which was predigested with the same restriction enzymes. The identity of each gene in recombinant pGES-TH-1 was confirmed by restriction digestion and DNA sequencing. *E. coli* cells were transformed with recombinant pGES-TH-1 and induced for the expression of fusion proteins, which were detected by sodium dodecyl sulphate–polyacrylamide gel electrophoresis (SDS–PAGE) using lysates of induced *E. coli* cells. The results showed the expression of proteins that corresponded to the size of GST/Rv3874, GST/Rv3875 and GST/Rv3619c. The identity of the expressed fusion proteins was established by Western immunoblotting with anti-GST and anti-penta His antibodies.

The SDS–PAGE analysis of cell-free extracts and pellets of sonicates of induced *E. coli* cells containing pGES-TH/Rv3874, pGES-TH/Rv3875 and pGES-TH/Rv3619c showed that recombinant GST-Rv3874 and GST-Rv3875 proteins were present in the soluble fraction, whereas the recombinant GST-Rv3619c protein was present in the pellet, which could be solublized best in 4 M urea. To purify the recombinant RD proteins, the soluble/solublized fractions were loaded on to glutathione-Sepharose affinity matrix and the GST-free RD proteins were released from the GST fusion partner bound to the column matrix by cleavage with the enzyme thrombin protease. The cleaved RD proteins were eluted and analyzed for purity by SDS–PAGE and staining with Coomassie Blue. The results showed that the recombinant Rv3874 and Rv3875 proteins were contaminated with another protein of nearly 70 kDa, whereas Rv3619c protein was nearly homogeneous (more than 95% pure). The partially purified Rv3874 and Rv3875 proteins were further purified on Ni-NTA agarose affinity matrix, and the analysis of eluted fractions showed the presence of a single sharp band in SDS–PAGE gels, which suggested that Rv3874 and Rv3875 preparations became free of the 70-kDa contaminant and were nearly homogeneous (more than 95% pure) (Hanif et al., 2010a). These results further strengthen the suggestion that pGES-TH-1 is useful for high level expression and efficient purification of recombinant mycobacterial proteins (Ahmed et al., 2003; 2004). The reason for Rv3619c requiring only one column (glutathione-Sepharose) for purification could be the presence of the fusion protein GST-Rv3619c in the pellet of induced *E. coli* cultures, which lacked the contaminating *E. coli* protein of 70 kDa. On the other hand, GST-Rv3874 and GST-Rv3875 proteins were present in the soluble fraction that also contained *E. coli* protein of 70 kDa, which was capable of binding to glutathione-Sepharose column nonspecifically, and was eluted from the column along with Rv3874 and Rv3875. However, the subsequent use of Ni-NTA matrix efficiently removed the contaminating *E. coli* protein and made the recombinant Rv3874 and Rv3875 proteins homogeneously pure (Hanif et al., 2010a).

The immunogenicity of all the three pure recombinant proteins (Rv3874, Rv3875 and Rv3619c) was evaluated in rabbits. The animals were immunized with the purified RD proteins and the sera were tested for antibody reactivity with the full-length recombinant proteins, pools of synthetic peptides covering the sequence of each protein and their individual peptides, as described previously (Hanif et al., 2010a). In brief, polyclonal antibodies were raised in rabbits against the purified and GST-free Rv3874, Rv3875 and Rv3619c recombinant proteins by emulsifying purified proteins (50 µg/ml) with an equal volume of incomplete Freund's adjuvant and injected intramuscularly in the right and left thigh. The rabbits were boosted twice with the same amount of protein at 2 weeks intervals. The animals were bled from the ear vein before the immunization and 2 weeks after the last immunization. The sera were tested for antigen-specific antibodies using Western immunoblotting. The results showed that pre-immunized rabbit sera did not have antibodies to any of these proteins, and the sera from immunized rabbits had antibodies reactive with the immunizing proteins only, thus confirming the specificity of the polyclonal antibodies to the immunizing antigens. In addition, the results suggested that the rabbits used were not exposed to *M. tuberculosis* and the antibody epitopes of a given protein were not crossreactive with other proteins. Moreover, ELISA was performed to detect antibodies in rabbit sera against full-length purified recombinant proteins and overlapping synthetic peptides corresponding to each protein. The ELISA results further showed specificity of the antibody reactivity with the immunizing proteins and positivity with synthetic peptides, which suggested that linear epitopes of each protein were recognized by B cells, after immunization with full-length proteins. Furthermore,

Molecular Cloning, Expression, Purification and Immunological Characterization of Proteins Encoded by Regions
of Different Genes of Mycobacterium tuberculosis

147

testing of sera with individual peptides of each protein demonstrated that rabbit antibodies recognized several linear epitopes that were scattered throughout the sequence of each protein (Hanif et al., 2010a). Interestingly, previous studies using pools of synthetic peptides have shown that all of these proteins (Rv3874, Rv3875 and Rv3619c) are major T cell antigens in humans, and the linear T cell epitopes were scattered throughout the sequence of these proteins (Hanif et al., 2008; Mustafa et al. 2000, 2003, 2008). A further analysis of the sequence of these proteins for B cell and T cell epitope prediction using appropriate prediction servers further suggested that B-cell and T-cell epitopes were scattered throughout the sequence of each protein (Hanif et al., 2010a; Mustafa et al., 2008; Mustafa and Shaban, 2006). Thus, both prediction and experimental results confirm the strong immunogenicity of Rv3874, Rv3875 and Rv3619c proteins for inducing antigen-specific immunological reactivity.

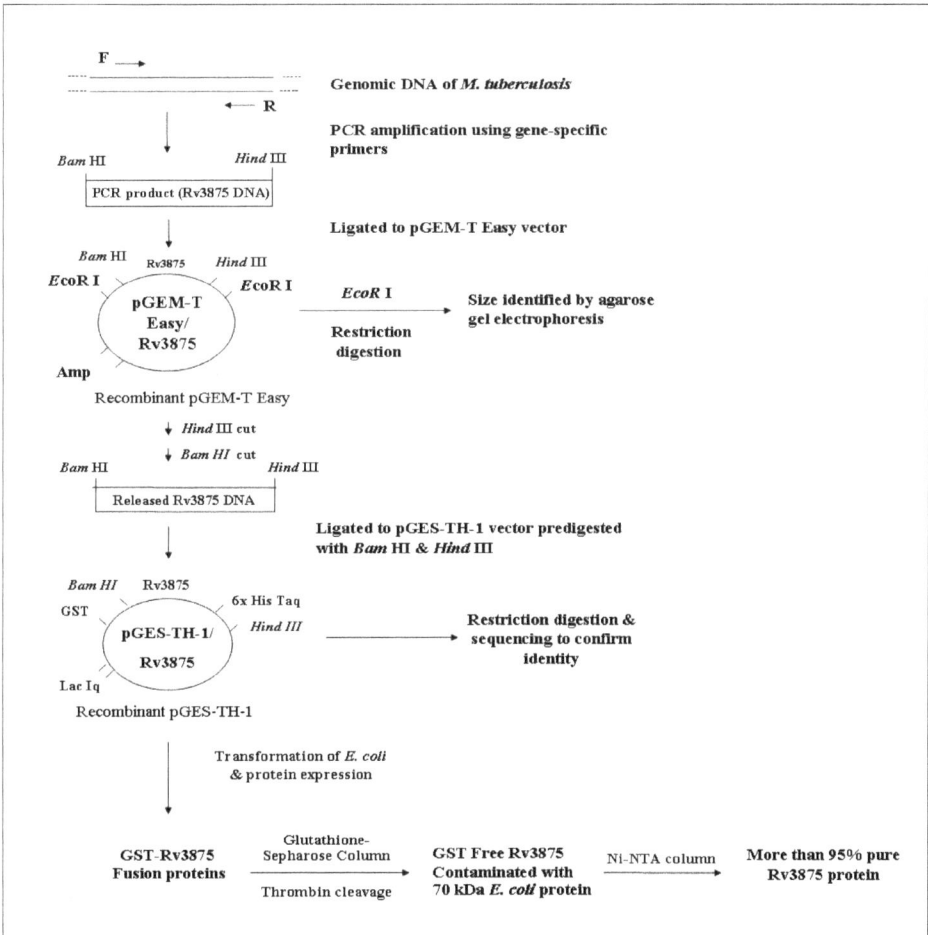

Fig. 1. Strategy and the schematic presentation of the steps involved in the cloning of *rv3875* gene in pGES-TH-1 vector, followed by expression of the fusion protein (GST-Rv3875) and purification of GST-free recombinant Rv3875 protein.

4. Potential of the purified recombinant RD proteins as *M. tuberculosis*-specific skin test antigens

Although, PPD in tuberculin skin test has been widely used as a diagnostic and epidemiological tool for tuberculosis monitoring in humans and animals for many years, there are many practical and theoretical problems related to the skin test, as stated above in the introduction of this chapter. To overcome the antigenic crossreactivity of PPD, we have tested four purified recombinant RD1 proteins (Rv3872, Rv3873, Rv3874, Rv3875) and one RD9 protein (Rv3619c) of *M. tuberculosis* for their ability to induce DTH skin responses in guinea pigs immunized with *M. tuberculosis* (Hanif et al., 2010b). Like RD1, RD9 is also a region that has been suggested to be deleted in all strains of BCG (Behr et al., 1999). However, using overlapping synthetic peptides, Rv3619c protein has been previously shown to induce moderate T-cell responses both in TB patients and healthy subjects (Mustafa and Al-Attiyah, 2004). Furthermore, Rv3619c belongs to the same family of proteins as Rv3874 and Rv3875 of RD1, i.e. ESAT6-family (Mustafa and Al-Attiyah, 2004). The evaluation of these proteins for induction of DTH responses as well as their specificity to *M. tuberculosis*-injected animals is important in the context of their potential usefulness as reagents for specific diagnosis to replace PPD, which has both *M. tuberculosis*-specific as well as crossreactive antigens.

In order to investigate the diagnostic potential of the purified recombinant proteins Rv3872, Rv3873, Rv3874, Rv3875, and Rv3619c, DTH skin responses were studied in guinea pigs injected with heat killed *M. tuberculosis* and live BCG, *M. avium* and *M. vaccae*. Two to four weeks later, the guinea pigs were challenged intradermally in the flank region with 1 µg of mycobacterial sonicates and purified recombinant proteins. The DTH responses were quantitated by measuring erythema at the sites of injections after 24 h. The results showed that all mycobacterial sonicates induced positive DTH responses in *M. tuberculosis*, BCG, *M. avium* and *M. vaccae* injected guinea pigs, which, like PPD, have crossreactive antigens. The purified recombinant proteins Rv3872, Rv3873, Rv3874 and Rv3875 elicited positive DTH responses in *M. tuberculosis* injected group, but not in BCG, *M. avium* and *M. vaccae* injected guinea pigs; whereas Rv3619c elicited positive DTH responses in *M. tuberculosis* and BCG injected animals but not in *M. avium* and *M. vaccae* injected guinea pigs (Hanif et al., 2010b). These results confirm the expression of all the proteins in *M. tuberculosis* and furthermore demonstrate their *in vivo* immunogenicity in guinea-pigs, which is the most sensitive animal model of TB. In addition, the induction of species-specific DTH responses suggests that the use of *M. tuberculosis*-specific recombinant RD1 antigens (Rv3872, Rv3873, Rv3874 and Rv3875) may lead to the development of DTH skin tests that can differentiate between tuberculous infection verses BCG vaccination and exposure to environmental mycobacterial species of *M. avium* and *M. vaccae*. In addition, the use of *M. tuberculosis* complex-specific Rv3619c protein in skin tests may differentiate between infection with *M. tuberculosis* and BCG vaccination verses exposure to environmental mycobacteria. However, these suggestions should be followed-up and studies should be extended to natural hosts of TB, including humans, for evaluation of the identified recombinant proteins as species-specific skin test antigens to replace PPD.

Molecular Cloning, Expression, Purification and Immunological Characterization of Proteins Encoded by Regions
of Different Genes of Mycobacterium tuberculosis
149

5. Cloning, expression and immunogenicity of RD genes in DNA vaccine vectors

The failure of BCG vaccine in humans has prompted the research to develop alternative vaccines against TB. Among the novel vaccine candidates, plasmid DNA-based TB vaccines have drawn close attention because of their unique features compared to conventional live or subunit vaccines, including induction of strong Th1 based CD4+ responses as well as cytotoxic T cell responses (Huygen, 2006). Therefore potency of plasmid DNAs expressing a variety of immunogenic *M. tuberculosis* antigens has been intensively evaluated in the past (Romano & Huygen., 2009). Unfortunately, their performance is generally not superior to BCG, especially in large animals. However, the licensure of a DNA vaccine in horses highlights the potential of DNA vaccine technology in the prevention of TB infection (Ulmer et al., 2006). Besides, it is generally believed that novel TB vaccines will be tested in the context of the widely used BCG and perhaps different kinds of vaccines are needed for the eradication of TB (Mitsuyama & McMurray, 2007; Sander & McShane, 2007). As a result, enhancement of TB DNA vaccine efficacy has become the active field of current research (Okada & Kita, 2010). In this context, Baldwin et al. have shown that inclusion of the secretion signal peptide from tissue plasminogen activator (tPA) into a DNA vaccine construct resulted in stronger immune responses to Ag85A, and provided sustained protection upon *M. tuberculosis* challenge in mice, as compared to the DNA vaccine construct based on the parent plasmid lacking tPA (Baldwin et al., 1999). Furthermore, a plasmid DNA vaccine expressing the heat shock protein 65 (HSP65-DNA vaccine) provided improved protective and therapeutic effects in mice when fused with human interleukin-2 (hIL-2) (Changhong et al., 2009).

To further study and compare the effect of tPA and hIL-2, we have used two plasmid vectors i.e. pUMVC6 and pUMVC7, to clone and express *M. tuberculosis*-specific RD proteins and studied antigen-specific cellular immune responses after immunization of mice with the recombinant plasmids (Hanif et al., 2010c). The plasmid vectors pUMVC6 and pUMVC7 are eukaryotic expression vectors, which have been prepared by University of Michigan Vector Core (UMVC) and made commercially available by Aldevron, USA. Both vectors have CMV promoter at 5′ end of the cloning site (Figs. 2, 3). pUMVC6 is characterized by having a secretion signal peptide from hIL-2 (hIL2 secretory peptide) as an immuno-stimulatory sequence (Fig. 2), whereas pUMVC7 has a signal peptide for targeting peptides to a secretory pathway by fusion to the tPA signal peptide (Fig. 3).

To prepare DNA vaccine constructs using the above plasmids, DNA fragments corresponding to five RD genes, i.e. *rv3872 (pe35)*, *rv3873 (ppe68)*, *rv3874 (esxA)*, *rv3875 (esxB)* and *rv3619c (esxV)*, were PCR amplified from genomic DNA of *M. tuberculosis* by using gene-specific forward (F) and reverse (R) primers. To facilitate cloning of PCR amplified DNA in pUMVC6 and pUMVC7 vectors, in addition to gene-specific sequences, the F and R primers had appropriate restriction sites at 5′ ends, i.e. *Bam*H I and *Bam*H I, and *Bam*H I and *Xba* I restriction sites for pUMVC6 and pUMVC7, respectively. The amplified DNA were first cloned into pGEM-T Easy vector, before subcloning into pUMVC6 and pUMVC7 (Figs. 2, 3 with Rv3875 DNA shown as an example). This was done to facilitate the subcloning of appropriate DNA into the eukaryotic expression vectors pUMVC6 and pUMVC7, as has been demonstrated previously for prokaryotic expression vectors pGEX-4T and pGES-TH-1 (Ahmed et al., 1999b, 2003; Hanif et al., 2010a). The cloning in pGEM-T Easy successfully yielded five recombinant plasmids, one for each gene, i.e. pGEM-T/Rv3872, pGEM-

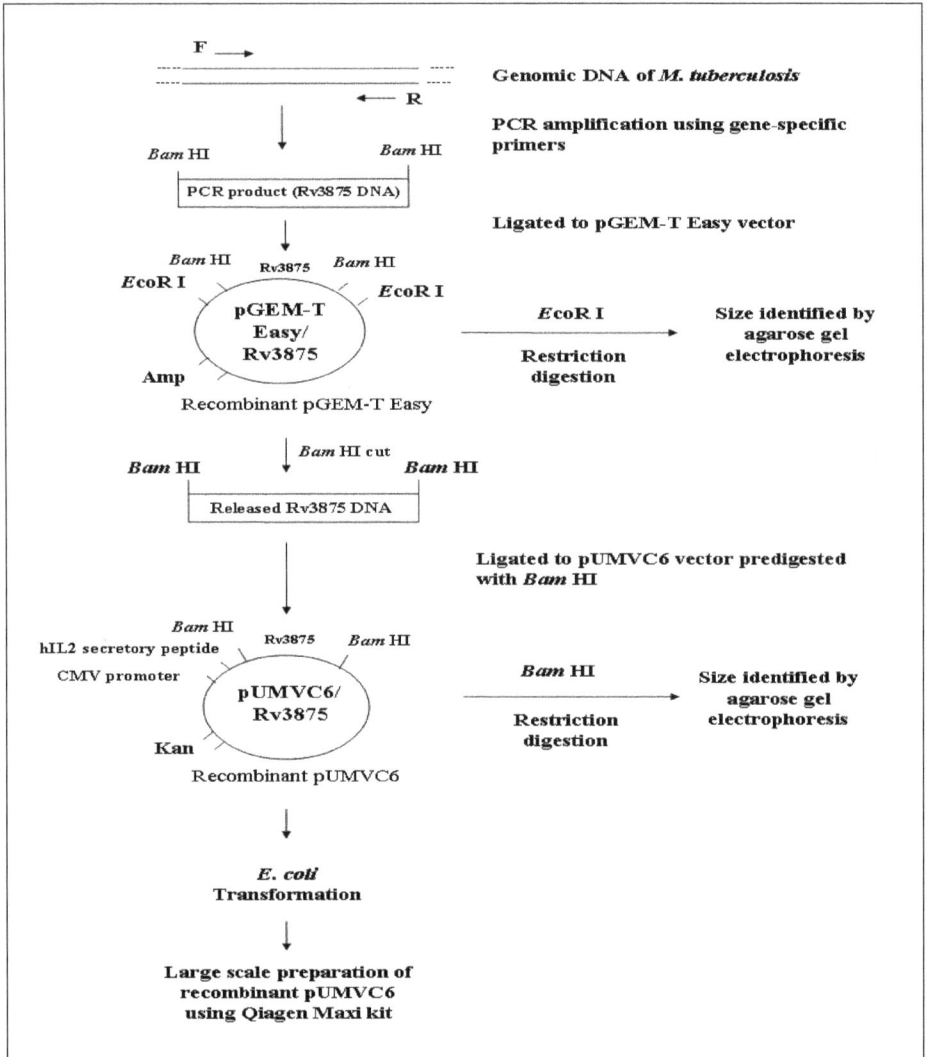

Fig. 2. The strategy and schematic presentation of the steps involved in the construction of recombinant plasmid pUMVC6/Rv3875, purification of recombinant plasmid DNA on a large scale.

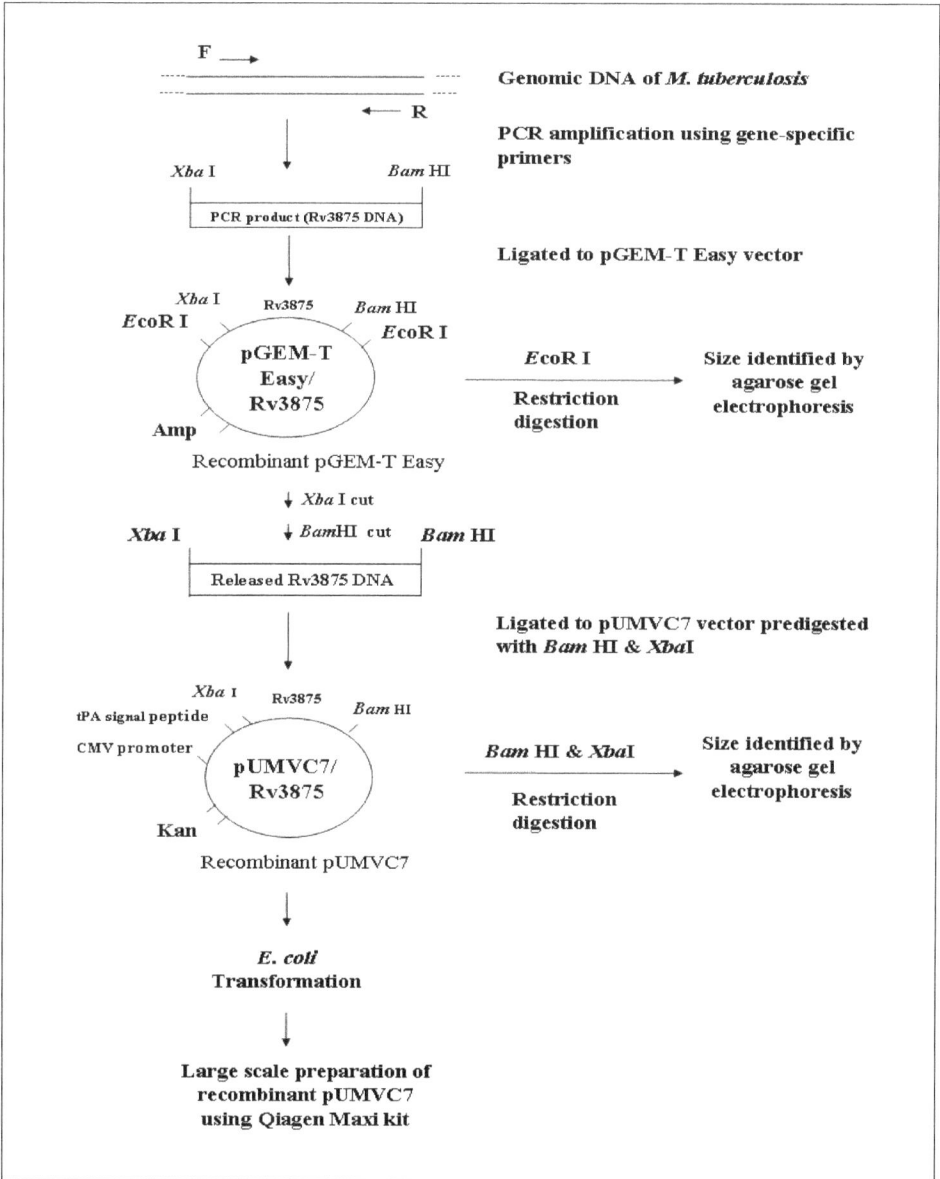

Fig. 3. The strategy and schematic presentation of the steps involved in the construction of recombinant plasmid pUMVC7/Rv3875, purification of recombinant plasmid DNA on a large scale.

T/Rv3873, pGEM-T/Rv3874, pGEM-T/Rv3875 and pGEM-T/Rv3619c. The analysis of DNA fragments released from the recombinant pGEM-T Easy plasmids after digestion with *Eco*R I showed that the cloned DNA corresponded to the expected molecular size of *rv3872, rv3873, rv3874, rv3875* and *rv3619c* genes. To prepare DNA vaccine constructs, the recombinant pGEM-Ts were single digested with *Bam*H I for subcloning into pUMVC6 and double digested with *Bam* HI and *Xba* I for subcloning into pUMVC7. These restriction digestions released the DNA fragments corresponding to *rv3872, rv3873, rv3874, rv3875* and *rv3619c* genes with *Bam*H I/*Bam*H I and *Bam*H I/*Xba* I cohesive ends. All the genes were subcloned into plasmid vectors by their ligation to pUMVC6 and pUMVC7 DNA predigested with *Bam*H I/*Bam*H I and *Bam*H I/*Xba* I, respectively. The identity of each cloned gene was determined by restriction digestion of recombinant plasmids with the restriction enzymes *Bam*H I for pUMVC6; and *Bam*H I and *Xba* I for pUMVC7, which released the cloned DNA corresponding to the size expected for each gene. *E. coli* cells were transformed with the parent and recombinant pUMVC6 and pUMVC7 plasmids, and large quantities of the plasmids were purified from the transformed *E. coli* cells grown in vitro by using Qiagen Endofree Mega kits (Qiagen, Valencia, CA, USA) according to the manufacturer's instructions. The overall strategies of gene amplification, cloning and large-scale purification of recombinant pUMVC6 and pUMVC7 plasmid DNA are shown in figures 2 and 3, respectively, using *rv3875* as an example (Fig. 2, 3). In this way, a total of 10 recombinant DNA plasmids (five for each vector) were constructed (Hanif et al., 2010c).

To study the expression and immunogenicity of the RD1 and RD9 proteins cloned in the recombinant vaccine constructs of pUMVC6 and pUMVC7, immunization studies were performed in mice. Groups of 6–8 week old BALB/c mice (five mice in each group) were immunized intramuscularly with three doses of purified parent or recombinant plasmid DNA, each dose containing 100 µg of DNA, and given 3 weeks apart. After 3 weeks of the third immunization, spleen cells were isolated from each immunized mouse for cellular immune responses using antigen-induced proliferation as an indicator (Hanif et al., 2010c). The results demonstrated that spleen cells of mice un-immunized or immunized with the parent plasmids did not show positive antigen-induced proliferation responses to any of the antigens corresponding to the cloned genes (Hanif et al., 2010c). The results further showed that both the recombinant vectors induced antigen-induced cellular proliferation to Rv3872, Rv3874, Rv3875 and Rv3619c proteins. However, antigen-induced proliferation responses were observed in response to a given protein only with spleen cells of mice immunized with the recombinant DNA vaccine construct expressing that protein, except for Rv3873, which failed to induce proliferation responses in animals immunized with pUMVC6/Rv3873 or pUMVC7/Rv3873 (Hanif et al., 2010c). Moreover, recombinant pUMCV6 induced relatively better responses than recombinant pUMCV7. The improved responses with recombinant pUMCV6 suggest that hIL2 secretory protein acted as a better adjuvant and enhanced cellular immune responses, assessed by antigen-induced proliferation of spleenocytes, to the fused mycobacterial proteins more effectively than the tPA signal peptide. The relevance of antigen-specific cellular proliferation induced by DNA vaccine constructs with protection against *M. tuberculosis* challenge has been demonstrated in the mouse model of TB (Fan et al., 2009).

Although, the results of this study, showed induction of antigen-specific cellular immune responses to the antigens encoded by genes present in DNA vaccine constructs, are interesting; the work should be extended to demonstrate their protective efficacy in challenge experiments, with *M. tuberculosis*, using various animal models of TB, e.g. mice, guinea-pigs, rabbits and monkeys etc. Such vaccines may be useful in both prophylactic and therapeutic applications in humans, If found effective in animals. Furthermore, DNA-based vaccines expressing *M. tuberculosis*-specific antigens may even be useful in BCG vaccinated subjects as preventive vaccines, because revaccination with BCG has not shown beneficial effects (Rodrigues et al., 2005, Roth et al., 2010), and may even be combined with BCG to improve its protective efficacy (Fan et al., 2007).

6. Conclusions

The modified pGES-TH-1 expression vector was extremely useful in obtaining highly purified recombinant preparations of Rv3872, Rv3873, Rv3874, Rv3875 and Rv3619c proteins of *M. tuberculosis*. All of these recombinant proteins were immunogenic in rabbits and the antibody epitopes were scattered throughout the sequence of each protein. These results suggest that pGES-TH-1 vector could be employed in obtaining pure recombinant proteins, predicted to be encoded by hypothetical genes present in *M. tuberculosis*-specific genomic regions, for their immunological characterization. Furthermore, the induction of species-specific DTH responses suggests that the use of *M. tuberculosis*-specific RD1 antigens may lead to the development of a DTH skin test that discriminates between tuberculous infection verses BCG vaccination and exposure to environmental mycobacteria (*M. avium* and *M. vaccae*). In addition, the use of *M. tuberculosis* complex-specific Rv3619c protein in skin tests may differentiate between tuberculosis infection and BCG vaccination verses exposure to environmental mycobacteria. Furthermore, DNA vaccine constructs encoding RD1 and RD9 genes induced antigen-specific cellular responses in immunized mice to the respective proteins, and thus could be useful as safer vaccines to immunize against TB. However, to confirm their usefulness as anti-TB vaccines, further studies with the DNA vaccine constructs of pUMVC6 and pUMVC7 may be performed to determine their protective efficacy in appropriate animal models of TB (mice, guinea-pigs, rabbits and monkeys etc.) after challenging the immunized animals with live *M. tuberculosis*.

7. Acknowledgements

This study was supported by Research Administration, Kuwait University, Kuwait grant YM01/03.

8. References

Ahmad, S.; Akbar P. K.; Wiker, H. G.; Harboe, M. & Mustafa, A. S. (1999b). Cloning, expression and immunological reactivity of two mammalian cell entry proteins encoded by the mce1 operon of *Mycobacterium tuberculosis*. *Scandinavian Journal of Immunology,* Vol. 50, No. 5, (November 1999), pp. 510-518.

Ahmad, S.; Ali, M. M. & Mustafa, A. S. (2003). Construction of a modified vector for efficient purification of recombinant *Mycobacterium tuberculosis* proteins expressed in *Escherichia coli. Protein Expression and Purification.* Vol. 29, No. 2, (June 2003), pp. 167-175.

Ahmad, S.; Amoudy, H. A.; Thole, J. E.; Young, D. B. & Mustafa, A. S. (1999a). Identification of a novel protein antigen encoded by a *Mycobacterium tuberculosis*-specific RD1 region gene. *Scandinavian Journal of Immunology*, Vol. 49, No. 5, (May 1999), pp. 515-522.

Ahmad, S.; El-Shazly, S.; Mustafa, A. S. & Al-Attiyah, R. (2004). Mammalian cell-entry proteins encoded by the mce3 operon of *Mycobacterium tuberculosis* are expressed during natural infection in humans. *Scandinavian Journal of Immunology*, Vol. 60, No. 4, (October 2004), pp. 382-391.

Ahmad, S.; El-Shazly, S.; Mustafa, A. S. & Al-Attiyah, R. (2005). The six mammalian cell entry proteins (Mce3A-F) encoded by the mce3 operon are expressed during in vitro growth of *Mycobacterium tuberculosis*. *Scandinavian Journal of Immunology*, Vol. 62, No. 1, (July 2005), pp. 16-24.

Al-Attiyah, R. & Mustafa, A. S. (2008). Characterization of human cellular immune responses to novel *Mycobacterium tuberculosis* antigens encoded by genomic regions absent in *Mycobacterium bovis* BCG. *Infection and Immunity*, Vol. 76, No. 9, (September 2008), pp. 4190-4198.

Al-Attiyah, R. & Mustafa, A. S. (2010). Characterization of human cellular immune responses to *Mycobacterium tuberculosis* proteins encoded by genes predicted in RD15 genomic region that is absent in *Mycobacterium bovis* BCG. *FEMS Immunology and Medical Microbiology*, Vol. 59, No. 2, (July 2010), pp. 177-187.

Amoudy, H. A.; Ahmad, S.; Thole, J. E. & Mustafa, A. S. (2007). Demonstration of *in vivo* expression of a hypothetical open reading frame (ORF-14) encoded by the RD1 region of *Mycobacterium tuberculosis*. *Scandinavian Journal of Immunology*, Vol. 66, No. 4, (October 2007), pp. 422-425.

Amoudy, H. A.; Al-Turab, M. B. & Mustafa, A. S. (2006). Identification of transcriptionally active open reading frames within the RD1 genomic segment of *Mycobacterium tuberculosis*. *Medical Principles and Practice*, Vol. 15, No. 2, (June 2006), pp. 137-144.

Amoudy, H. A. & Mustafa, A. S. (2008). Amplification of Six Putative RD1 Genes of *Mycobacterium tuberculosis* for Cloning and Expression in *Escherichia coli* and Purification of Expressed Proteins. *Medical Principles and Practice*, Vol. 17, No. 5, (August 2008), pp. 378-384.

Baldwin, S. L.; D'Souza, C. D.; Orme, I. M.; Liu, M. A.; Huygen, K., Denis, O.; Tang, A.; Zhu, L.; Montgomery, D. & Ulmer, J. B. (1999). Immunogenicity and protective efficacy of DNA vaccines encoding secreted and nonsecreted forms of *Mycobacterium tuberculosis* Ag85A. *Tubercle and Lung Diseases*, Vol. 79, No. 4, (August 1999), pp. 251-259.

Behr, M. A.; Wilson, M. A.; Gill , W. P.; Salamon, H.; Schoolnik, G. K.; Rane, S.& Small, P. M. (1999). Comparative genomics of BCG vaccines by whole-genome DNA microarray. *Science*, Vol. 284, No. 5419, (May 1999), pp. 1520-1523.

Borgdorff, M. W.; Van Den Hof, S.; Kremer, K.; Verhagen, L.; Kalisvaart, N.; Erkens, C. & van Soolingen, D. (2010). Progress towards tuberculosis elimination: secular trend, immigration and transmission. *European Respiratory Journal*, Vol. 36, No. 2, (August 2010), pp. 339-347.

Changhong, S.; Hai, Z.; Limei, W.; Jiaze, A.; Li, X.; Tingfen, Z.; Zhikai, X. & Yong, Z. (2009). Therapeutic efficacy of a tuberculosis DNA vaccine encoding heat shock protein 65 of *Mycobacterium tuberculosis* and the human interleukin 2 fusion gene. *Tuberculosis (Edinb)*, Vol. 89, No. 1, (January 2009), pp. 54-61.

Chiang, C. Y.; Centis, R.; Migliori, G. B. (2010). Drug-resistant tuberculosis: past, present, future. *Respirology*, Vol. 15, No. 3, (April 2010), pp. 413-432.

Crampin, A. C.; Glynn, J. R. & Fine, P. E. (2009). What has Karonga taught us? Tuberculosis studied over three decades. *The International Journal of Tuberculosis and Lung Diseases*, Vol. 13, No. 2, (February 2009), pp. 153-164.

El-Shazly, S.; Mustafa, A. S.; Ahmad, S. & Al-Attiyah R. (2007). Utility of three mammalian cell entry proteins of *Mycobacterium tuberculosis* in the serodiagnosis of tuberculosis. *The International Journal of Tuberculosis and Lung Diseases*, Vol. 11, No. 6, (June 2007), pp. 676-682.

Fan, X.; Gao, Q. & Fu, R. (2007). DNA vaccine encoding ESAT-6 enhances the protective efficacy of BCG against *Mycobacterium tuberculosis* infection in mice. *Scandinavian Journal of Immunology*, Vol. 66, No. 5, (November 2007), pp. 523-528.

Fan, X.; Gao, Q. & Fu, R. (2009). Differential immunogenicity and protective efficacy of DNA vaccines expressing proteins of *Mycobacterium tuberculosis* in a mouse model. *Microbiol Research*, Vol. 164, No. 4, (August 2009), pp. 374-382.

Fine, P. E. M. (2001). BCG: the challenge continues. *Scandinavian Journal of Infectious Diseases*, Vol. 33, No. 4, (November 2001), pp. 243-245.

Haile, M. & Kallenius, G. (2005). Recent developments in tuberculosis vaccines. *Current Opinion in Infectious Diseases*, Vol. 18, No. 3, (June 2005), pp. 211-215.

Hanif, S. N. M.; Al-Attiyah, R. & Mustafa, A. S. (2010a). Molecular cloning, expression, purification and immunological characterization of three low-molecular weight proteins encoded by genes in genomic regions of difference of *Mycobacterium tuberculosis*. *Scandinavian Journal of Immunology*, Vol. 71, No. 5, (May 2010), pp. 353-361.

Hanif, S. N. M.; Al-Attiyah, R. & Mustafa, A. S. (2010b). Species-specific antigenic *Mycobacterium tuberculosis* proteins as tested by delayed-type hypersensitivity response. *The International Journal of Tuberculosis and Lung Diseases*, Vol. 14, No. 4, (April 2010), pp. 489-494.

Hanif, S. N. M.; Al-Attiyah, R. & Mustafa, A. S. (2010c). DNA vaccine constructs expressing *Mycobacterium tuberculosis*-specific genes induce immune responses. *Scandavian Journal of Immunology*, Vol. 72, No. 5, (November 2010), pp. 408-415.

Hanif, S. N. M.; El-Shamy A. M.; Al-Attiyah, R. & Mustafa, A. S. (2008). Whole blood assays to identify Th1 cell antigens and peptides encoded by *Mycobacterium tuberculosis*-specific RD1 genes. *Medical Principles and Practice*, Vol. 17, No. 3, (April 2008), pp. 244-249.

Hendrickson, R. C.; Douglass, J. F.; Reynolds, L. D., McNeill, P. D.; Carter, D.; Reed, S. G. & Houghton, R. L. (2000). Mass spectrometric identification of Mtb81, a novel serological marker for tuberculosis. *Journal of Clinical Microbiology*, Vol. 38, No. 6, (June 2000), pp. 2354-2361.

Hesseling, A. C.; Marais, B. J.; Gie, R. P.; Schaaf, H. S.; Fine, P. E.; Godfrey-Faussett, P. & Beyers, N. (2007). The risk of disseminated Bacille Calmette-Guerin (BCG) disease in HIV-infected children. *Vaccine*, Vol. 25, No. 1, (January 2007), pp. 14-18.

Hill, P. C.; Brookes, R. H.; Adetifa, I. M.; Fox, A.; Jackson-Sillah, D.; Lugos, M. D.; Donkor, S. A.; Marshall, R. J.; Howie, S. R.; Corrah, T.; Jeffries, D. J.; Adegbola, R. A. & McAdam, K. P. (2006). Comparison of enzyme-linked immunospot assay and tuberculin skin test in healthy children exposed to *Mycobacterium tuberculosis*. *Pediatrics,* Vol. 117, No. 5, (May 2006), pp. 1542-1548.

Huygen, K. (2006). DNA vaccines against mycobacterial diseases. *Future Microbiology;* Vol. 1, No. 1, (June 2006), pp. 63-73.

Kurup, S. K.; Buggage, R. R.; Clarke, G. L.; Ursea, R.; Lim, W. K. & Nussenblatt, R. B. (2006). Gamma interferon assay as an alternative to PPD skin testing in selected patients with granulomatous intraocular inflammatory disease. *Canadian Journal of Ophthalmology,* Vol. 41, No. 6, (December 2006), pp. 737-740.

Leung, C. C.; Yew, W. W.; Tam, C. M.; Chan, C. K.; Chang, K. C.; Law, W. S.; Lee, S. N.; Wong, M. Y. & Au, K. F. (2005). Tuberculin response in BCG vaccinated school children and the estimation of annual risk of infection in Hong Kong. *Thorax,* Vol. 60, No. 2, (January 2005), pp. 124-129.

Mahadevan, B.; Mahadevan, S.; Serane, V. T. & Narasimhan, R. (2005). Tuberculin reactivity in tuberculous meningitis. *Indian Journal of Pediatrics,* Vol. 72, No. 3, (March 2005), pp. 213-215.

Mahairas, G. G.; Sabo, P. J.; Hickey, M. J.; Singh, D. C. & Stover, C. K. (1996). Molecular analysis of genetic differences between *Mycobacterium bovis* BCG and virulent *M. bovis. Journal of Bacteriology,* Vol. 178, No. 5, (March 1999), pp. 1274-1282.

Mitsuyama, M. & McMurray, D. N. (2007). Tuberculosis: vaccine and drug development. *Tuberculosis (Edinb),* Vol. 87, No. 1001, (August 2007), pp. S10-S13.

Moffitt, M. P. & Wisinger, D. B. (1996). Tuberculosis. Recommendations for screening, prevention, and treatment. *Postgraduate Medicine,* Vol. 100, No. 4, (October 1996), pp. 201-204.

Mustafa, A. S. (2000).   HLA-restricted immune response to mycobacterial antigens: relevance to vaccine design. *Human Immunology,* Vol. 61, No. 2, (February 2000), pp. 166-171.

Mustafa, A. S. (2001). Biotechnology in the development of new vaccines and diagnostic reagents against tuberculosis. *Current Pharmaceutical Biotechnology,* Vol. 2, No. 2, (June 2001), pp. 157-173.

Mustafa, A. S. (2002). Development of new vaccines and diagnostic reagents against tuberculosis. *Molecular Immunology,* Vol. 39, No. 2, (September 2002), pp. 113-119.

Mustafa, A. S. (2005a). Progress towards the development of new anti-tuberculosis vaccines. In: *Focus on Tuberculosis Research,* Inc. Lucy T. Smithe, (Ed.), PP. 47-76, New York. Nova Science Publishers.

Mustafa, A. S. (2005b). Mycobacterial gene cloning and expression, comparative genomics, bioinformatics and proteomics in relation to the development of new vaccines and diagnostic reagents. *Medical Principles and Practice,* Vol. 14, No. 1, (November 2005), pp. 27-34.

Mustafa, A. S. (2005c). Recombinant and synthetic peptides to identify *Mycobacterium tuberculosis* antigens and epitopes of diagnostic and vaccine relevance. *Tuberculosis (Edinb),* Vol. 85, No.6, (September 2005), pp. 367-376.

Mustafa, A. S. (2009). Vaccine potential of *Mycobacterium tuberculosis*-specific genomic regions: in vitro studies in humans. *Expert Review of Vaccines,* Vol. 8, No. 10, (October 2009), pp. 1309-1312.

Mustafa, A. S. (2010). Cell mediated immunity assays identify proteins of diagnostic and vaccine potential from genomic regions of difference of *Mycobacterium tuberculosis*. *Kuwait Medical Journal*, Vol. 42, No. 2, (June 2010), pp. 98-105.

Mustafa, A. S. & Al-Attiyah, R. (2003). Tuberculosis: looking beyond BCG vaccines. *Journal of Postgraduate Medicine*, Vol. 49, No. 2, (June 2003), pp. 134-140.

Mustafa, A. S. & Al-Attiyah, R. (2004). *Mycobacterium tuberculosis* antigens and peptides as new vaccine candidates and immunodiagnostic reagents against tuberculosis. *Kuwait Medical Journal*, Vol. 36, No. 3, (September 2004), pp. 71-176.

Mustafa, A. S. & Al-Attiyah, R. (2009). Identification of *Mycobacterium tuberculosis*-specific genomic regions encoding antigens those induce qualitatively opposing cellular immune responses. *Indian Journal of Experimental Biology*, Vol. 47, No. 6, (June 2009), pp. 498-504.

Mustafa, A. S.; Al-Attiyah. R.; Hanif, S. N. M. & Shaban, F. A. (2008). Efficient testing of large pools of *Mycobacterium tuberculosis* RD1 peptides and identification of major antigens and immunodominant peptides recognized by human Th1 cells. *Clinical and Vaccine Immunology*, Vol. 15, No. 6, (June 2008), pp. 916-924.

Mustafa, A. S.; Al-Saidi, F.; El-Shamy, A. S. M. & Al-Attiyah, R. (2011). Cytokines in response to proteins predicted in genomic regions of difference of *Mycobacterium tuberculosis*. *Microbiology and Immunology*, Vol; 55, No. 4, (April 2011), pp. 267-78.

Mustafa, A. S.; Cockle,P. J.; Shaban, F.; Hewinson, R.G. & Vordermeier, H. M (2002). Immunogenicity of RD1 region gene products in M. bovis infected cattle. *Clinical and Experimental Immunology*, Vol. 130, No. 1, (October 2002), pp. 37-42.

Mustafa, A. S.; Oftung F.; Amoudy, H. A.; Madi, N. M.; Abal,A. T.; Shaban, F.& Andersen, P. (2000). Multiple epitopes from the *M. tuberculosis* ESAT-6 antigen are recognized by antigen specific human T cell lines. *Clinical Infectious Diseases*, Vol. 30, No. 3, (June 2000), pp. S201-S205.

Mustafa, A. S. & Shaban, F. A. (2006). Propred analysis and experimental evaluation of promiscuous Th1 cell epitopes of three major secreted antigens of *Mycobacterium tuberculosis*. *Tuberculosis (Edinb)*, Vol. 86, No. 2, (March 2006), pp. 115-124.

Mustafa, A. S.; Shaban, F. A.; Al-Attiyah, R.; Abal, A.T.; El-Shamy A. M.; Andersen, P & Oftung, F. (2003). Human Th1 cell lines recognize the *Mycobacterium tuberculosis* ESAT-6 antigen and its peptides in association with frequently expressed HLA class II molecules. *Scandinavian Journal of Immunology*, Vol. 57, No. 2, (February 2003), pp. 125-134.

Narayanan, P. R. (2006). Influence of sex, age & nontuberculous infection at intake on the efficacy of BCG: re-analysis of 15-year data from a double-blind randomized control trial in South India. *Indian Journal of Medical Research*, Vol. 123, No. 2, (February 2006), pp. 119-124.

Okada, M. & Kita, Y. (2010). Tuberculosis vaccine development: The development of novel (preclinical) DNA vaccine. *Human vaccines*, Vol. 6, No.4, (April 2010), pp. 297-308.

Rodrigues, L. C.; Pereira, S. M.; Cunha, S. S.; Genser, B.; Ichihara, M. Y.; de Brito, S. C.; Hijjar, M. A.; Dourado, I.; Cruz, A. A.; Sant'Anna, C.; Bierrenbach, A. L. & Barreto, M. L. (2005). Effect of BCG revaccination on incidence of tuberculosis in school-aged children in Brazil: the BCG-REVAC cluster-randomised trial. *Lancet*, Vol. 366, No. 9493, (October 2005), pp. 1290-1295.

Romano, M. & Huygen, K. (2009). DNA vaccines against mycobacterial diseases. *Expert Review of Vaccines, Vol.* 8, No. 9, (September 2009), pp. 1237-1250.

Roth, A. E.; Benn, C. S.; Ravn, H.; Rodrigues, A.; Lisse, I. M.; Yazdanbakhsh, M.; Whittle, H. & Aaby, P. (2010). Effect of revaccination with BCG in early childhood on mortality: randomised trial in Guinea-Bissau. *British Medical Journal,* Vol. 340, No. 5, (March 2010), pp. c671.

Sander, C. & McShane, H. (2007). Translational mini-review series on vaccines: development and evaluation of improved vaccines against tuberculosis. *Clinical and Experimental Immunology,* Vol. 147, No. 3, (March 2007), pp. 401-411.

Semba, R. D.; Darnton-Hill, I.; de Pee, S. (2010). Addressing tuberculosis in the context of malnutrition and HIV coinfection. *Food and Nutrition Bulletin,* Vol. 31, No. 4, (December 2010), pp. S345-S364.

Trunz, B. B.; Fine, P. & Dye, C. (2006). Effect of BCG vaccination on childhood tuberculous meningitis and miliary tuberculosis worldwide: a meta-analysis and assessment of cost-effectiveness. *Lancet,* Vol. 367, No. 9517, (April 2006), pp. 1173-1180.

Ulmer, J. B.; Wahren, B. & Liu, M. A. (2006). Gene-based vaccines: recent technical and clinical advances. *Trends in Molecular Medicine, Vol.* 12, No. 5, (May 2006), pp. 216-221.

World Health Organization. (2007). Revised BCG vaccination guidelines for infants at risk for HIV infection. *Weekly Epidemiological Record,* 82: 193

World Health Organization. (2008). The top 10 causes of death. *WHO Fact Sheet No.* 310.

World Health Organization. (2009). Global Tuberculosis Control. Epidemiology, strategy, financing. *WHO report WHO/HTM/2009.411.*

World Health Organization. (2010). Global Tuberculosis Control. *WHO Report 2010, WHO/HTM/TB/2010.7.*

Part 4

Toxicology

10

Molecular Cloning, Expression, Function, Structure and Immunoreactivities of a Sphingomyelinase D from *Loxosceles adelaida*, a Brazilian Brown Spider from Karstic Areas

Denise V. Tambourgi et al.[*]
Immunochemistry Laboratory, Butantan Institute, São Paulo, Brazil

1. Introduction

Loxosceles is the most poisonous spider in Brazil and, at least, three different species of medical importance are known in Brazil (*L. intermedia, L. gaucho, L. laeta*), with more than 5000 cases of envenomation reported each year. In South Africa, *L. parrami* and *L. spinulosa* are responsible for cutaneous loxoscelism (Newlands et al., 1982). In Australia, a cosmopolitan species, *L. rufescens*, is capable of causing ulceration in humans. In the USA, at least five *Loxosceles* species, including *L. reclusa* (brown recluse), *L. apachea, L. arizonica, L. unicolor* and *L. deserta* are known to cause numerous incidents (Ginsburg &Weinberg, 1988; Gendron, 1990; Bey et al., 1997; Desai et al., 2000).

Several studies have indicated that sphingomyelinase D (SMase D) present in the venoms of *Loxosceles* spiders is the main component responsible for the local and systemic effects observed in loxoscelism (Forrester et al., 1978; Kurpiewski et al., 1981; Tambourgi et al., 1998, 2000, 2002, 2004, 2005, 2007; van den Berg et al., 2002, 2007; Fernandes Pedrosa et al., 2002; Paixão Cavalcanti et al., 2006, Tambourgi et al., 2010). SMases D hydrolyze sphingomyelin resulting in the formation of ceramide-1-phosphate and choline (Forrester et al., 1978; Kurpiewski et al., 1981; Tambourgi et al., 1998) and, in the presence of Mg^{2+}, are able to catalyze the release of choline from lysophosphatidylcholine (van Meeteren et al., 2004).

All spider venom SMases D sequenced to date display a significant level of sequence similarity and thus likely possess the same $(\alpha/\beta)8$ or TIM barrel fold (Murakami et al., 2005, 2006). Based on sequence identity, biochemical activity and molecular modelling, a scheme for classification of spider venom SMases D was proposed (Murakami et al., 2006). The class 1 enzymes include SMase I and H13, SMases D from *L. laeta*, which possess a single disulphide bridge and contain an extended hydrophobic loop or variable loop (Murakami et

---

[*]Giselle Pidde-Queiroz[1], Rute M. Gonçalves-de-Andrade[1], Cinthya K. Okamoto[1], Tiago J. Sobreira[2], Paulo S. L. de Oliveira[2], Mário T. Murakami[2] and Carmen W. van den Berg[3]
[1]*Immunochemistry Laboratory, Butantan Institute, São Paulo, Brazil*
[2]*National Laboratory for Biosciences, National Centre for Research in Energy and Materials, Campinas, Brazil*
[3]*Department of Pharmacology, Oncology and Radiology, School of Medicine, Cardiff University, Cardiff, UK*

al., 2006). All other SMases D, such as SMases P1 and P2 from *L. intermedia* (Tambourgi et al., 1998, 2004), Lr1 and Lr2 from *L. reclusa* and Lb1, Lb2 and Lb3 from *L. boneti* (Ramos-Cerrillo et al., 2004) belong to class 2, which contains an additional intra-chain disulphide bridge that links the flexible loop with the catalytic loop (Murakami et al., 2006). The class 2 enzymes can be further subdivided into class 2a and class 2b depending on whether they are capable of hydrolysing sphingomyelin or not, respectively (Murakami et al., 2006). One representative of class 2b is the isoform 3 from *L. boneti* an inactive SMase D isoform (Ramos-Cerrillo et al., 2004) (Table 1).

Studies on the effect of venoms from synanthropic species of *Loxosceles* spiders have been reported, however, analysis from those living in natural environment have been poorly performed. *Loxosceles* species are present in several different habitats, including the karstic environment, and in Brazil it is the most common troglophile arachnid. The spiders are commonly found on the walls at the entrance of caves, especially in the shady rocky areas.

In order to characterize the venom of *Loxosceles* species living in natural environment, and compare their venom to those of synanthropic species, we have explored the caves of 'Parque Estadual do Alto do Ribeira' (PETAR - Ribeira Valley, SP, Brazil), which is an area of importance to both tourism and scientific research, due to the combination of tropical forest and extensive cave systems. Spiders captured in the caves of PETAR were identified by morphological analysis as *Loxosceles adelaida* Gertsch (1967), which belongs to '*gaucho* group' (Gertsch, 1967). This group includes four species: *L. adelaida*, *L. gaucho*, *L. similis* and *L. variegata*. *L. gaucho* belongs to the synanthropic fauna of Brazilian arachnids and is considered an important cause of the loxoscelic accidents in the south–eastern region of the country. Thus, the aims of the present study were to clone and express SMases D from the spider gland of *L. adelaida*, captured in the caves of PETAR (Brazil), to compare the functional activities of the recombinant proteins with toxins from synanthropic species and to investigate the inter- and intra-species cross-reactivities of antibodies raised against the purified recombinant proteins.

SMase Class	*Loxosceles* sp	SMase D	GenBank	Reference
I	*L. laeta*	SMase I H13 (SMase II)	AY093599.1 AY093600.1	Fernades-Pedrosa et al., 2002; de Santi-Ferrata et al., 2009 Fernandes-Pedrosa et al., 2002
II	*L. intermedia*	P1 P2	AY304471.2 AY304472.2	Tambourgi et al., 2004 Tambourgi et al., 2004
	L. reclusa	Lr1 Lr2	AY559846.1 AY559847.1	Ramos-Cerrillo et al., 2004 Ramos-Cerrillo et al., 2004
	L. boneti	Lb1 Lb2 Lb3	AY559844.1 - AY559845.1	Ramos-Cerrillo et al., 2004 Ramos-Cerrillo et al., 2004 Ramos-Cerrillo et al., 2004

2. Material and methods

2.1 Chemicals, reagents and buffers

Tween 20, bovine serum albumin (BSA), sphingomyelin (SM), choline oxidase, horseradish peroxidase (HRPO) and 3-(4-hydroxy-phenyl) propionic acid were purchased from Sigma Co. (St. Luis, MO). 5-bromo-4-chloro-3-indolyl-phosphate (BCIP) and Nitroblue Tetrazolium (NBT) were from Promega Corp. (Madison, WI, USA). Goat anti-horse IgG-alkaline phosphatase (GAH/IgG-AP) was from Sigma Co. (St. Luis, MO). Horse serum against SMases D P1 and P2 from *L. intermedia* and SMase D I from *L. laeta* was obtained as previously described (De Almeida et al., 2008). Buffers were: Veronal-Buffered Saline (VBS^{2+}), pH 7.4: 10 mM NaBarbitone, 0.15 mM CaCl$_2$ and 0.5 mM MgCl$_2$; Phosphate-Buffered Saline (PBS), pH 7.2: 10 mM NaPhosphate, 150 mM NaCl; HEPES-Buffered saline (HBS), pH 7.4: 10 mM Hepes, 140 mM NaCl, 5 mM KCl, 1 mM CaCl$_2$, 1 mM MgCl$_2$.

2.2 Spiders and venoms

Loxosceles intermedia, L. laeta, L. gaucho and *L. adelaida* spiders were provided by Immunochemistry Laboratory, Butantan Institute, Brazil (permission N° 1/35/95/1561-2 was provided by the Brazilian Institute of Environment and Renewable Natural Resources - IBAMA - a Brazilian Ministry of the Environment's enforcement agency). The venoms were obtained (permission N° 01/2009/IBAMA) by electrostimulation by the method of Bucherl (1969) with slight modifications. Briefly, 15–20 V electrical stimuli were repeatedly applied to the spider sternum and the venom drops were collected with a micropipette in PBS, aliquoted and stored at -20°C. The protein content of the samples was evaluated using the BCA Protein Assay Kit (Pierce Biotechnology, MA, USA).

2.3 Animals

The adult New Zealand white rabbits weighing approximately 3 kg were supplied by Butantan Institute animal facilities, SP, Brazil. All the procedures involving animals were in accordance with the ethical principles in animal research adopted by the Brazilian College of Animal Experimentation.

2.4 cDNA generation

The venom glands were removed from 50 specimens of *Loxosceles adelaida* 5 days after venom extraction, when the maximum level of RNA synthesis was achieved. Then, the glands were quickly frozen in liquid nitrogen and stored at −80°C until used. The total RNA was extracted from the glands using the Trizol reagent, according to the manufacturer's instructions (Invitrogen, USA). The quality and yield of total RNA were verified by the integrity of 28S and 18S rRNA, through denaturing agarose gel electrophoresis, and spectrophotometrically using the ratio 260/280 nm. The cDNAs were synthesised from 5 μg of total RNA using the cDNA Cycle® Kit for RT-PCR (Invitrogen, USA).

2.5 PCR Amplification of Sphingomyelinase D homologue from *L. adelaida*

The SMase D cDNA was amplified by PCR using total reverse transcriptase-PCR products as template. Degenerate primers AdeF1 (5'-G(G/A)ACG(C/A)GC(G/T)GATAA(A/C) CGTCG(A/T)CC-3') and AdeR2 (5'–CTA(A/T/G)TT(C/T)TT(A/G)AA(A/T/G) GTCTCCCA(A/T)GG–3'), were designed according to the highly conserved 5'- and 3'-

SMases D P1 and P2 from *L. intermedia* (accession numbers AY304471 and AY304472, respectively) and Smase D I from *L. laeta* (accession number AY093599) (Tambourgi et al, 2004; Fernandes Pedrosa et al, 2002). The PCR protocol included denaturation at 96°C for 3 min, followed by 35 cycles of denaturation (30 s at 95°C), annealing (30 s at 58°C), and extension (2 min at 72°C), and a final extension for 7 min at 72°C. The amplified fragments were purified by low melting using the UltraPure™ Low Melting Point Agarose (Invitrogen, USA) and cloned in a pGEM-T easy vector (Promega, USA). *E. coli* XL1-Blue competent cells (Strategene, USA) were transformed following the manufacturer's instructions. Positive clones were selected by growing the transformed cells in Luria broth (LB) medium containing 100 µg/ml ampicilin and blue-white color screening. The nucleotide sequence was determined by the dideoxy chain-termination method using the BigDye™ Terminator Cycle Sequence Kit and the ABI 3100 automatic system (Applied Biosystems, USA).

2.6 Subcloning in expression vector
The cDNA corresponding to the mature *L. adelaida* SMase D was amplified from plasmids containing full length sequence using primer AdeF3 (5'-CTCGAGGCGGATAAACGTCGACCCATATG-3'), which contains a XhoI restriction site (underlined) and AdeR4 (5'-CCATGGTCAATTCTTGAAAGTCTCC-3'), which includes a restriction site for NcoI (underlined) and a stop codon (in italics and bold).
PCR fragments (approximately 900 bp) were digested with XhoI and NcoI and cloned into the corresponding sites in pRSET-B bacterial expression vector (Invitrogen, USA). The use of the pRSET-B bacterial expression system results in the expression of a recombinant fusion protein, including a 6xHis-tag at the N-terminus and a 26 amino acid linker followed by the mature protein (N-terminal amino acid sequence before the coding sequence of the mature protein: 'MRGSHHHHHHGMASMTGGQQMGRDLYDDDDKDPSSR').

2.7 Recombinant protein expression and purification
Recombinant proteins were produced as described previously (Tambourgi et al., 2004). In brief: pRSETB-*L. adelaida* SMase D cDNA transformed *Escherichia coli* BL21 (DE3) (Invitrogen, USA) cells were inoculated in 500 mL of 2YT/amp and grown overnight at 37°C and induced with IPTG. Recombinant proteins were harvested from the pellet by french-pressure and purified on a Ni (II) Chelating Sepharose Fast Flow column (Pharmacia, Sweden, 1.0×6.4 cm). The fractions containing the recombinant proteins were pooled and concentrated using Centricon-30 (30,000-mw cutoff; Amicon, Inc., USA). Smase D P1 from *L. intermedia* and SMase D I from *L. laeta* were obtained as previously described (Tambourgi et al., 2004; Fernandes-Pedrosa et al., 2002).;The protein content of the samples was evaluated by the BCA protein kit assay, following the manufacturer's protocol (Pierce, USA).

2.8 Enzymatic activity
The SMase D enzymatic activity was estimated by determining the choline liberated from lipid substrates, using a fluorimetric assay (Tokumura et al., 2002). Briefly, sphingomyelin (SM – 50 µg) was diluted in 1 mL HEPES-buffered saline (HBS). Samples of the recombinant proteins or venom (2.5 to 20 µg) was added and the reaction was developed for 20 min at 37°C. After incubation, a mixture composed by 1 unit/mL choline oxidase, 0.06 unit/mL of horseradish peroxidase and 50 µM of 3-(4-hydroxy-phenyl) propionic acid in HBS was added and incubated for 10 min. The choline liberated was oxidized to betaine and H_2O_2

and this product determined by fluorimetry at λem=405 nm and λex=320 nm, using 96-well microtiter plates, in a spectrofluorimeter (Victor 3™, Perkin-Elmer, USA).

2.9 Electrophoresis and western blotting

Samples of the recombinant SMases D (5 µg) or *Loxosceles* venoms (10 µg) were solubilised in non-reducing sample buffer, run on 12% SDS-PAGE (Laemmli, 1970) and silver stained. Alternatively, gels were blotted onto nitrocellulose (Towbin et al., 1979). After transfer, the membranes were blocked with PBS containing 5% BSA and incubated with horse serum anti-SMases D from *L. intermedia* (Smases D P1 and P2) and *L. laeta* (SMase D I) (diluted 1:1000) for 1 h at room temperature. Membranes were washed 3 times with PBS/0.05% Tween 20 for 5 min each wash, and incubated with GAH/IgG-AP (1:7500) in PBS/1% BSA for 1 h at room temperature. After washing 3 times with PBS/0.05% Tween 20 for 5 min each wash, blots were developed using NBT/BCIP according to the manufacturer's instructions (Promega).

2.10 Normal human serum and erythrocytes

Human blood was obtained from healthy donors. Blood samples drawn to obtain sera were collected without anticoagulant, allowed to clot for 4 h at 4°C and the normal human serum (NHS) was stored at -80°C. Blood samples drawn to obtain erythrocytes for subsequent use as target cells were collected in anticoagulant (Alsever's old solution: 114 mM citrate, 27 mM glucose, 72 mM NaCl, pH 6.1).

2.11 Treatment of erythrocytes with *Loxosceles* recombinant proteins/venom

Human erythrocytes were washed and resuspended at 1.5% in VBS^{2+} and incubated with different concentrations of the recombinant proteins or venom for 1 h at 37°C. Control samples were incubated with VBS^{2+}. The cells were washed, resuspended to the original volume in VBS^{2+} and analysed in a haemolysis assay as described (Tambourgi et al., 2000).

2.12 Dermonecrotic activity

Samples of *L. adelaida* venom (5 µg), SMase P1 from *L. intermedia* (5 µg) or SMase D from *L. adelaida* (5 µg and 10 µg), in PBS, were injected intradermally in the shaved back of adult rabbits. Control sites were injected with equal volume of PBS. The size of the lesions was measured over a period of 16, 24, 48 and 72 h. After 72 h, the animals were euthanized and skin specimens were obtained for histological examination.

2.13 Histological analysis

Skin samples were fixed in 4% buffered formalin solution, and then embedded in paraffin. Tissue sections were stained with haematoxylin and eosin and examined for the presence of epithelial necrosis, epithelial slough, dermal infiltrates, haemorrhage and level of collagen dissociation in the dermis and skin muscle fiber degeneration.

2.14 Homology molecular modelling and quality analysis

The atomic coordinates of SMase I, a SMase D from *L. laeta* (PDB accession code: 1XX1) was used as 3D-model for restraint-based modelling as implemented in the program MODELLER (Fiser and Sali, 2003). The overall model was improved enforcing the proper stereochemistry using spatial restraints and CHARMM energy terms, followed by conjugate gradient simulation based on the variable target function method (Fiser and Sali, 2003). Ten models were built for the *L. adelaida* SMase D sequence based on the (m)GenThreader alignment (Lobley et al., 2009) and the model with best global score was selected for explicit solvent MD simulations using the package Yasara (http://www.yasara.org) to check its stability and consistency. The overall and local quality analyses of the final model were assessed by VERIFY3D (Eisenberg et al., 1997), PROSA (Wiederstein and Sippl, 2007) and VADAR (Willard et al., 2003). Three-dimensional structures were displayed, analyzed and compared using the program COOT (Emsley and Cowtan, 2004).

2.15 Molecular dynamics simulation of the enzyme-substrate complex

The SMase D I structure (PDBid 1XX1) obtained from the Protein Data Base (www.rcsb.org) and a predicted structure of the sphingomyelin (SM) was manually docked using the sulfate ion from the experimental structure, as a reference of the position of the phosphate moiety of the SM.

The structure was prepared for energy minimization (EM) molecular dynamics (MD) simulation using YASARA program for building missing atoms and hydrogens in the model. The parameters for the force field were obtained from YAMBER3 (Krieger et al., 2004). The pKa values for Asp, Glu, His and Lys residues were predicted. Based on the pH 7.0, the protonation states were assigned according to convention: Asp and Glu were protonated if the predicted pKa was higher than the pH; His was protonated if the predicted pKa was higher than the pH and it did not accept a hydrogen bond, otherwise it was deprotonated; Cys was protonated; Lys was deprotonated if the predicted pKa was lower than the pH and; Tyr and Arg were not modified (www.yasara.org). A simulation box was defined at 15 Å around all atoms of each macromolecular complex. Then, the simulation box was filled with water molecules and Na/Cl counter ions, that were placed at the locations of the lowest/highest electrostatic potential, until the cell neutralization, and the requested NaCl concentration reached 0.9%. A short MD simulation was performed for the solvent adjust, and water molecules were subsequently deleted until the water density reached 0.997 g/mL. A short steepest descent EM was carried until the maximum atom speed dropped below 2200 m/s. Then 500 steps of simulated annealing EM were performed with a target temperature at 0 K. Finally, a 20 ns MD production simulation was performed at 298 K and a non-bonded cutoff of 7.86 A. A snapshot was saved every 7.5 ps (picoseconds). The graphical analysis was carried out using Visual Molecular Dynamics (VMD) software (Humphrey et al., 1996) and the plots were generated using R (cran.r-project.org).

Ligand binding analysis was carried out taking into account the potential energy obtained with the current force field for the complex and components:

$$E_{binding} = (E_{protein} + E_{ligant}) - E_{complex}$$

In our equation, the complex energy is placed after the energy of the components; therefore, higher energies equated to higher affinity between the protein and the substrate.

3. Results

3.1 Identification and characterization of L. adelaida recombinant protein

Analysis of L. adelaida SMase D clone revealed that the mature cDNA covers an open reading frame of 843 nucleotides encoding 280 amino acid residues (GenBank JN202927). The complete nucleotide sequence of this cDNA clone and the deduced amino acid sequence are shown in Figure 1A. Sequence analysis revealed that SMase D from L. adelaida exhibits similarity with previously characterized class 2 SMase D and shares 75% and 59% homology with the sequences of L. intermedia SMase P1 and L. laeta SMase I, respectively (Figure 1B).

```
1    GCG GAT AAA CGT CGA CCC ATA TGG ATC ATG GGA CAT ATG GTT AAC   45
1     A   D   K   R   R   P   I   W   I   M   G   H   M   V   N    15

46   GCC ATC TCT CAA ATA GAC GAA TTT GTC AAC CTT GGG GCG AAT TCC   90
16    A   I   S   Q   I   D   E   F   V   N   L   G   A   N   S    30

91   ATC GAA ACG GAC GTG GCT TTC GAT AAG CAA GCC AAT CCT CAA TAC  135
31    I   E   T   D   V   A   F   D   K   Q   A   N   P   Q   Y    45

136  ACG CAT CAC GGG GTC CCA TGC GAT TGT GGA AGG AAT TGT TGG AAA  180
46    T   H   H   G   V   P   C   D   C   G   R   N   C   W   K    60

181  AAG GAG AAC TTC CCA GAC TTC GTA AAA GCT CTA CGA AGT GCT ACA  225
61    K   E   N   F   P   D   F   V   K   A   L   R   S   A   T    75

226  ACC CCC AGT GAT TCC AAA TAT CAT GAA AAA CTG ATC TTA GTT GTG  270
76    T   P   S   D   S   K   Y   H   E   K   L   I   L   V   V    90

271  TTC GAC CTA AAA ACG GGT AGT CTT TCC AAC AAT CAA GCC TAC GAC  315
91    F   D   L   K   T   G   S   L   S   N   N   Q   A   Y   D   105

316  GCA GGA AAG AAT TTG GCA AAA AAT CTC CTT CAA AAT TAC TGG AAC  360
106   A   G   K   N   L   A   K   N   L   L   Q   N   Y   W   N   120

361  AAT GGC AAT AAT GGT GGA AGA GCA TAC ATA GTG TTA TCA GTA CCA  405
121   N   G   N   N   G   G   R   A   Y   I   V   L   S   V   P   135

406  TAC CTT GCA CAT TAT AAA TTG ATA ACA GGA TTT CAA GAA ACG CTT  450
136   Y   L   A   H   Y   K   L   I   T   G   F   Q   E   T   L   150

451  AAA AAC GAG GGA CAT CAA GAA TTG TTG GAG AAA GTT GGA TAC GAC  495
151   K   N   E   G   H   Q   E   L   L   E   K   V   G   Y   D   165

496  TTC TCC AGG AAT GAC TAC ATC AGC GAT GTT AAG GCA GCT TAT AGT  540
166   F   S   R   N   D   Y   I   S   D   V   K   A   A   Y   S   180

541  AGA GCT GGA GTA TCA AGT CCC GTG TGG CAA AGC GAC GGT GTC ACC  585
181   R   A   G   V   S   S   P   V   W   Q   S   D   G   V   T   195

586  AAC TGT TGG CTA CGT GGT TTT GGT CGT GTG AAG CAA GCT GTG GCA  630
196   N   C   W   L   R   G   F   G   R   V   K   Q   A   V   A   210

631  AAT AGA GAC TCC GCG GAC GGA TTC ATT AAC AAA GTA TAC TAT TGG  675
211   N   R   D   S   A   D   G   F   I   N   K   V   Y   Y   W   225

676  ACA GTG GAT AAA CGC GCA ACA ACT AGA AAA TCA CTT AAT GCT GGA  720
226   T   V   D   K   R   A   T   T   R   K   S   L   N   A   G   240

721  GTT GAC GGT ATA ATG ACC AAT TAC CCC GAT GTA ATT GCT AGT GTA  765
241   V   D   G   I   M   T   N   Y   P   D   V   I   A   S   V   255

766  CTC AAG GAA CCT GCT TAC AGT TCA AAA TTC AGG GTC GCC ACA TAC  810
256   L   K   E   P   A   Y   S   S   K   F   R   V   A   T   Y   270

811  ACG GAT AAT CCT TGG GAG ACT TTC AAG AAT TGA
271   T   D   N   P   W   E   T   F   K   N   *
```

A

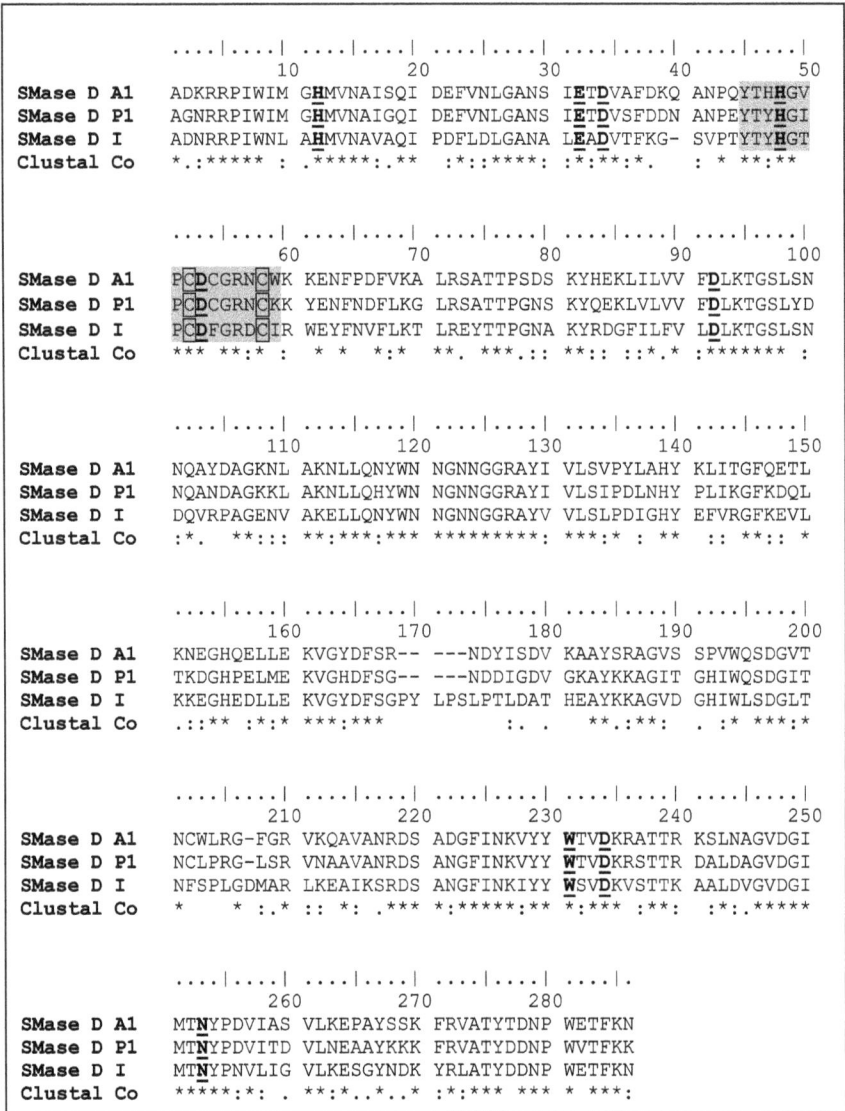

```
                ....|....| ....|....| ....|....| ....|....| ....|....|
                    10         20         30         40         50
SMase D A1      ADKRRPIWIM GHMVNAISQI DEFVNLGANS IETDVAFDKQ ANPQYTHHGV
SMase D P1      AGNRRPIWIM GHMVNAIGQI DEFVNLGANS IETDVSFDDN ANPEYTYHGI
SMase D I       ADNRRPIWNL AHMVNAVAQI PDFLDLGANA LEADVTFKG- SVPTYTYHGT
Clustal Co      *.:****** : .*****:.** :*::****: :*:**:*.    : * **:**


                ....|....| ....|....| ....|....| ....|....| ....|....|
                    60         70         80         90        100
SMase D A1      PCDCGRNCWK KENFPDFVKA LRSATTPSDS KYHEKLILVV FDLKTGSLSN
SMase D P1      PCDCGRNCKK YENFNDFLKG LRSATTPGNS KYQEKLVLVV FDLKTGSLYD
SMase D I       PCDFGRDCIR WEYFNVFLKT LREYTTPGNA KYRDGFILFV LDLKTGSLSN
Clustal Co      *** **:* :   *  * *:* **. ***.:: **:: ::*.* :*******  :


                ....|....| ....|....| ....|....| ....|....| ....|....|
                   110        120        130        140        150
SMase D A1      NQAYDAGKNL AKNLLQNYWN NGNNGGRAYI VLSVPYLAHY KLITGFQETL
SMase D P1      NQANDAGKKL AKNLLQHYWN NGNNGGRAYI VLSIPDLNHY PLIKGFKDQL
SMase D I       DQVRPAGENV AKELLQNYWN NGNNGGRAYV VLSLPDIGHY EFVRGFKEVL
Clustal Co      :*. **::: **:***:*** *********: ***:* : **  :: **:: *


                ....|....| ....|....| ....|....| ....|....| ....|....|
                   160        170        180        190        200
SMase D A1      KNEGHQELLE KVGYDFSR-- ---NDYISDV KAAYSRAGVS SPVWQSDGVT
SMase D P1      TKDGHPELME KVGHDFSG-- ---NDDIGDV GKAYKKAGIT GHIWQSDGIT
SMase D I       KKEGHEDLLE KVGYDFSGPY LPSLPTLDAT HEAYKKAGVD GHIWLSDGLT
Clustal Co      .::** :*:* ***:***       :.  **.:**:  . :* ***:*


                ....|....| ....|....| ....|....| ....|....| ....|....|
                   210        220        230        240        250
SMase D A1      NCWLRG-FGR VKQAVANRDS ADGFINKVYY WTVDKRATTR KSLNAGVDGI
SMase D P1      NCLPRG-LSR VNAAVANRDS ANGFINKVYY WTVDKRSTTR DALDAGVDGI
SMase D I       NFSPLGDMAR LKEAIKSRDS ANGFINKIYY WSVDKVSTTK AALDVGVDGI
Clustal Co      *    * :.* :: *: .*** *:*****:** *:*** :**:  :*:.*****


                ....|....| ....|....| ....|....| ....|.
                   260        270        280
SMase D A1      MTNYPDVIAS VLKEPAYSSK FRVATYTDNP WETFKN
SMase D P1      MTNYPDVITD VLNEAAYKKK FRVATYDDNP WVTFKK
SMase D I       MTNYPNVLIG VLKESGYNDK YRLATYDDNP WETFKN
Clustal Co      *****:*: . **:*..*..* :*:*** *** * ***:
```

B

Fig. 1. Sequence of *L. adelaida* SMase D A1. **[A]** Complete cDNA sequence of *L. adelaida* SMase D A1 and its deduced mature protein sequence. The primers used in the subcloning are underlined. The nucleotide residues are numbered in the 5′ to 3′ direction. **[B]** Alignment of the complete deduced amino acid sequences of *L. adelaida* SMase D A1 and SMase D P1 of *L. intermedia* and SMase D I of *L. laeta*. The conserved residues forming the catalytic pocket are underlined and in bold letters. Open boxes indicate the two conserved cysteine residues in SMases D. The catalytic loops (residues 45-59) are shaded gray. Asterisks indicate identical amino acid residues.

The theoretical molecular mass for the putative recombinant mature protein of *L. adelaida* is 31.545 kDa, with a pI = 8.85. Expression of *L. adelaida* SMase D resulted in an approximately 36 kDa recombinant His6-tagged protein, clearly visible by SDS-PAGE in the bacterial cell extracts (Figure 2A). The recombinant protein was purified from the soluble fraction of cell lysates by Ni^{+2}-chelating chromatography and eluted from the resin in extraction buffer containing 0.8 M imidazole at greater than 95% purity (Figure 2B).

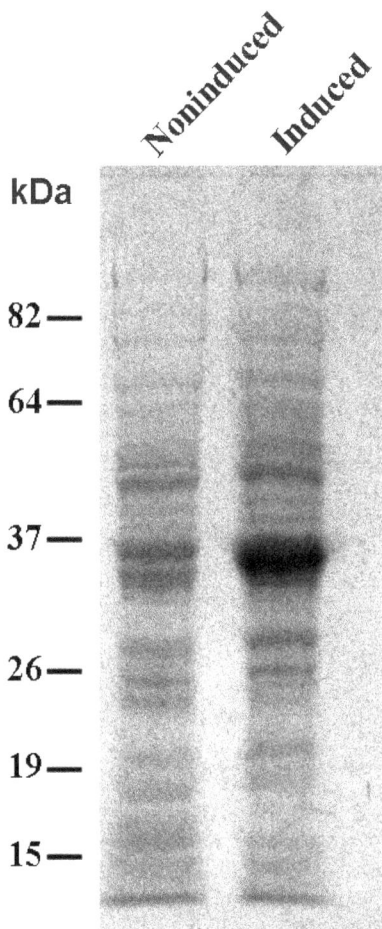

A

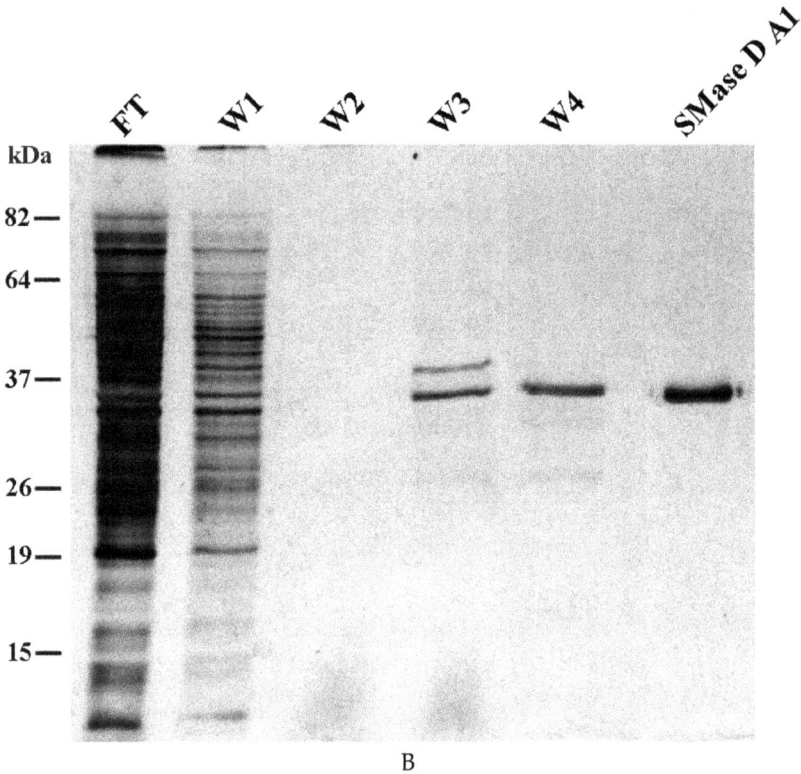

B

Fig. 2. Expression and purification of recombinant SMase D A1 of *L. adelaida*. **[A]** Extracts of *E. coli* BL21(DE3) transformed with the plasmid pRSET B clone A1 were induced or not with IPTG for 2 h. The cells were collected by centrifugation, resuspended in buffer extraction and disrupted by French-pressure. **[B]** The supernatant was loaded onto Ni(II)-Chelating Sepharose column, the flowthrough was collected (FT), the column was washed (W1 – W4), and the recombinant protein was eluted with the elution buffer. Samples were separated by SDS-PAGE (12% gel) under reducing conditions and stained by silver.

3.2 Enzymatic activity

The ability of the *L. adelaida* recombinant SMase D protein to degrade sphingomyelin was investigated and compared with the activity of the previously characterized active recombinant isoform from *L. intermedia* gland, SMase D P1, and with *L. adelaida* crude venom. Figure 3 shows that the crude *L. adelaida* venom, as well as the recombinant *L. adelaida* SMase D A1 and Smase D P1 proteins present significant sphingomyelinase activity as shown by the breakdown of the substrate, being the activity of *L. adelaida* SMase D protein lower than that of the *L. intermedia* SMase P1. However, *L. adelaida* recombinant SMase D activity was approximately ten times higher than that of the crude *L. adelaida* venom. Since the recombinant *L. adelaida* protein was endowed with sphingomyelinase activity, we named it as SMase D A1.

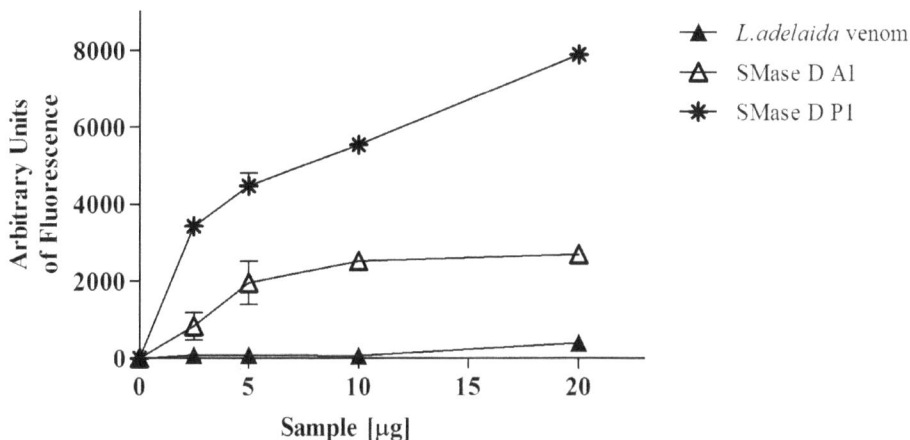

Fig. 3. Sphingomyelinase activity of *L. adelaida* SMase D recombinant protein.
Sphingomyelin (50 µg) was incubated with increasing amounts of buffer, *L. adelaida* or *L. intermedia* recombinant SMases D or crude *L. adelaida* venom. After 30 min at 37°C, the formed choline was oxidized to betaine and determined fluorimetricaly. Results are representative for two separate experiments expressed as mean of duplicates +/-SD.

3.3 Cross-reactivities of SMase D A1 of *L. adelaida*

The protein profile from the *Loxosceles* spp venoms and the recombinant SMases D were analyzed by SDS-PAGE followed by silver staining. Figure 4A shows that the venoms from *L. intermedia, L. laeta, L. gaucho* and *L. adelaida* differ in composition, number and intensity of bands. It can also be observed that the recombinant proteins were purified to homogeneity and that P1 and A1 exhibit similar molecular weight. SMase I presents Mr of approximately 33 kDa, and SMases P1 and A1, around 37 kDa and 36 kDa, respectively.

In order to analyze the inter- and intra-species cross-reactivities, horse polyclonal antiserum raised against a combination of the SMases D P1 and P2 from *L. intermedia* and Smase D I from *L. laeta,* were used in western blotting reactions. Figure 4B shows that the antiserum was able to recognize the purified recombinant proteins, SMases D A1, I and P1, as well as bands of approximately 35 kDa in the *Loxosceles* spp spider venoms. The slightly higher Mr of the recombinant proteins is attributed to the extra N-terminal tag, which increased the size of *L. adelaida* SMase A1, *L. intermedia* SMase P1 and *L. Laeta* SMase I proteins by, approximately, 4 kDa, 3 kDa and 1 kDa, respectively.

3.4 *L. adelaida* recombinant SMase D induces haemolysis and dermonecrosis

Sphingomyelinases D isolated from *Loxosceles* spider venoms have been shown to be able to transform human erythrocytes in activators of the complement system (Tambourgi et al., 1995, 1998, 2000). In order to assess whether the recombinant SMase A1 could also induce Complement-dependent haemolysis, erythrocytes were incubated with increasing amounts of the SMases A1, P1 or crude venom. Although the activity of SMase A1 was relatively lower compared with the SMase P1 and *L. adelaida* venom, as shown in Figure 5, *L. adelaida* recombinant SMase A1, as well as SMase P1 and *L. adelaida* venom were able to render erythrocytes susceptible to lysis by autologous Complement.

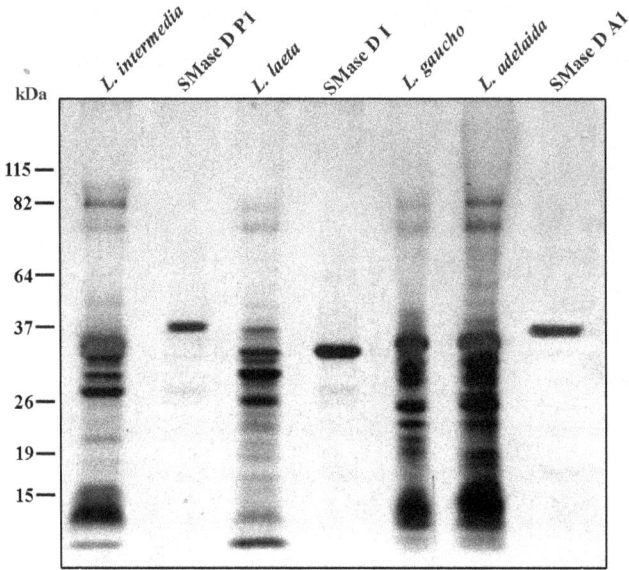

A

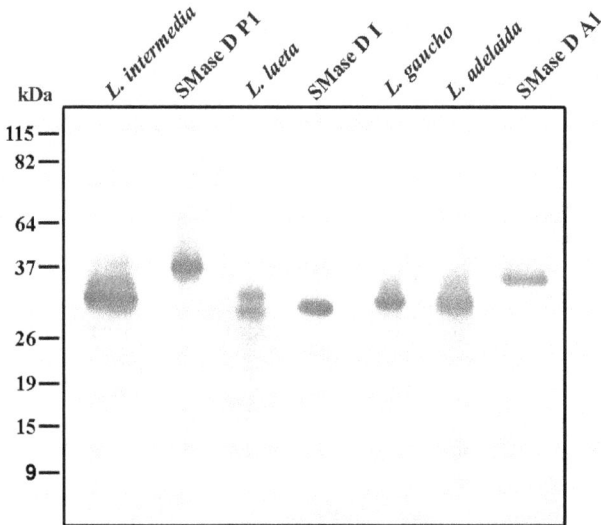

B

Fig. 4. SDS–PAGE and Western blotting analysis of the recombinant mature *L. adelaida* SMase D protein. Purified recombinant proteins, *L. intermedia* SMase D P1, *L. laeta* SMase D I and *L. adelaida* SMase D A1, and *L. intermedia, L. laeta, L. gaucho* and *L. adelaida* venoms were compared. **[A]** Samples were separated by SDS-PAGE (12% gel) under non-reducing conditions and silver stained. **[B]** Purified recombinant proteins and venoms were run on 12% SDS-PAGE gel under non-reducing conditions and western blotted using horse antiserum raised against recombinant SMases D P1 and P2 and anti-*L. laeta* SMase I.

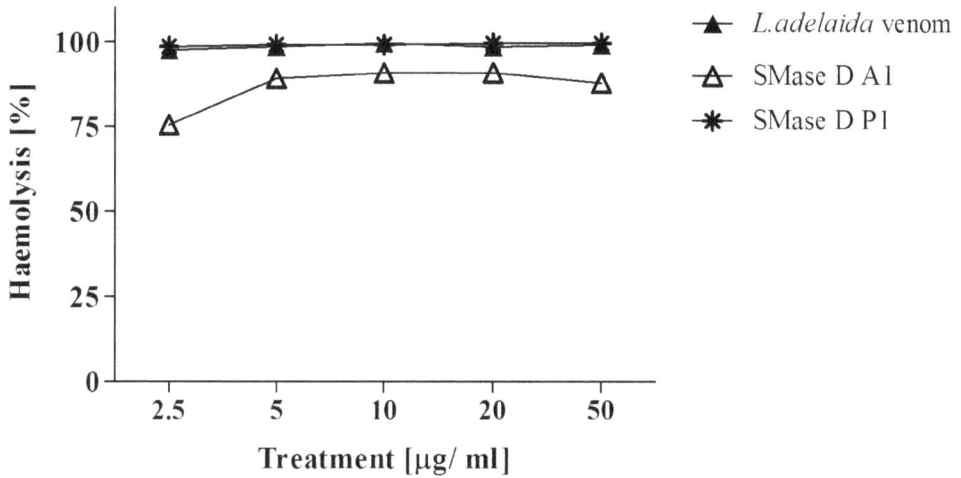

Fig. 5. Induction of haemolysis by *L. adelaida* SMase D A1. Erythrocyte pre-treated with different amounts of *L. adelaida* or *L. intermedia* recombinant SMases D or crude *L. adelaida* venom or VBS^{2+} were incubated with NHS. After incubation for 1h at 37°C, the absorbance of the supernatant was measured at 414 nm and expressed as percentage of lysis. Results are representative for three separate experiments and expressed as mean of duplicates +/-SD.

The ability of *L. adelaida* recombinant SMase A1 to induce dermonecrotic lesions was tested by injecting rabbits with 5 µg of the toxin. The animals received buffer, as negative control, and 5 µg of SMase D P1 or venom as positive controls. A typical loxoscelic lesion, as revealed by the presence of oedema, oerythema and mild tenderness, developed in the skin area injected with the recombinant proteins or venom within a few hours of injection and approximately 24 h post injection, necrosis with gravitational spread and scar were observed at the inoculation site (Figure 6A). Figure 6B shows that the dermonecrotic action of *L. adelaida* SMase D A1 is dose dependent, but that this activity was less intense than that exhibited by the crude venom and SMase D P1.

Histological analysis of the skin samples obtained from PBS inoculated animals showed a thin epidermis and a normal pattern for the collagenous area and muscle fibers (Figure 7, panels A/A1/A2). Despite the *L. adelaida* recombinant SMase D A1 has exhibited lower dermonecrotic activity than the Smase D P1, skin samples obtained from the recombinant proteins or venom inoculation sites showed a thin epidermis, dissociation of the collagenous fibers due to the oedema, degeneration of muscle fibers, moderate haemorrhage in the superficial dermis and intense neutrophil infiltration in the deep dermis and musculature (Figure 7).

[A]

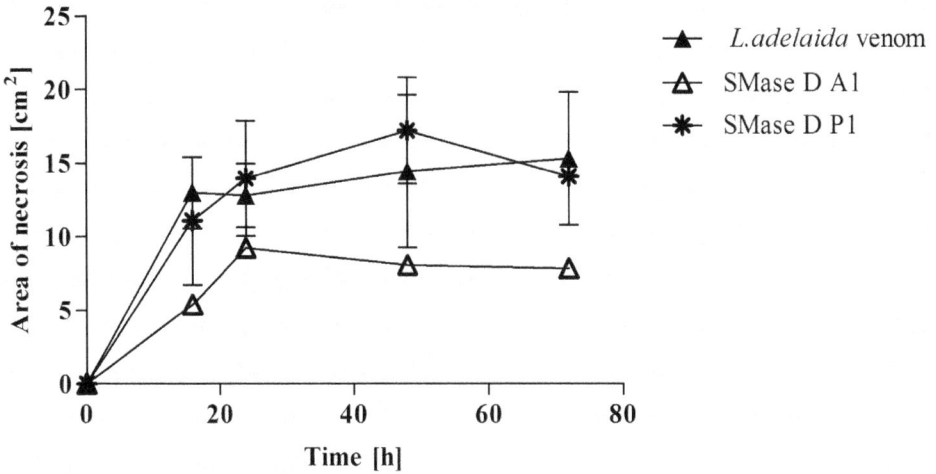

[B]

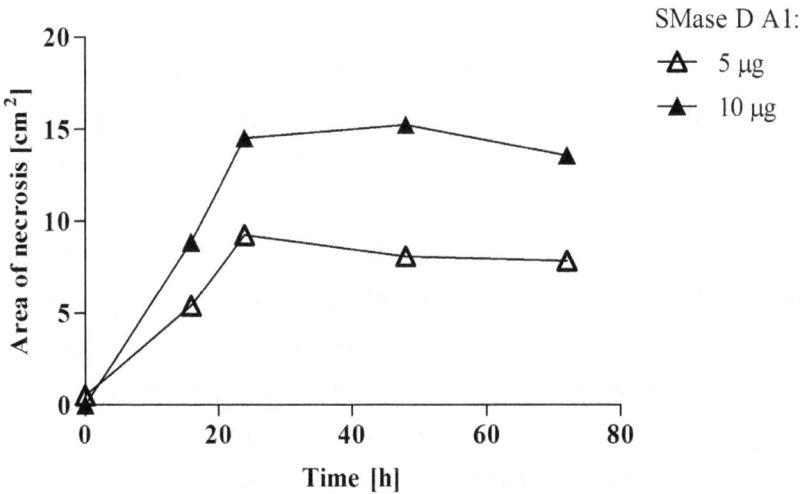

Fig. 6. Induction of dermonecrosis by *L. adelaida* SMase D recombinant protein. **[A]** Adult rabbits were injected intradermally with 5 µg of *L. adelaida* or *L. intermedia* recombinant SMases D or crude *L. adelaida* venom. The animals received buffer for negative control reactions. The areas of the dermonecrotic lesions were determined 16, 24, 48 and 72 h after injection. Results are representative for three separate experiments and expressed as mean of duplicates +/-SD. **[B]** Samples of 5 µg or 10 µg of SMase D A1 were injected intradermally in rabbits and the areas of the dermonecrotic lesions were determined 16, 24, 48 and 72 h after injection.

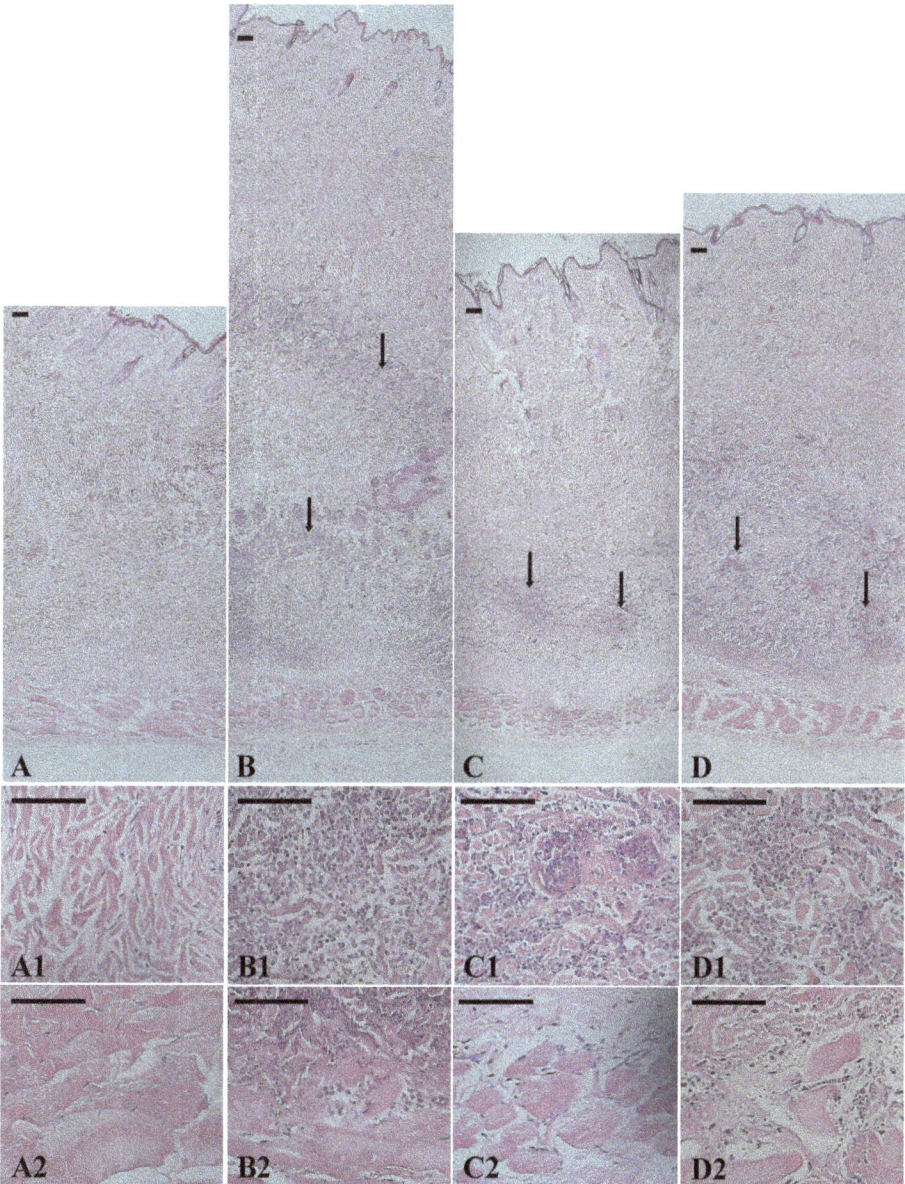

Fig. 7. Histological analysis. Analysis of skin of rabbits injected with 5 µg *L. adelaida* or *L. intermedia* recombinant SMases D or crude *L. adelaida* venom. Control sites were injected with an equal volume of phosphate-buffered saline (PBS). Panels correspond to the panoramic view of skin sections from rabbits injected with PBS **[A]**, *L. adelaida* venom **[B]**, purified *L. adelaida* SMase D A1 **[C]** and *L. intermedia* SMase D P1 **[D]**. Arrows indicate areas of leukocyte infiltration. Panels A1/2–D1/2 show details of the collagenous area of the dermis of the same sections. Bars at the top of each panel indicate 100 µm.

3.5 Overall structure description

MD analysis of *L. adelaida* SMase D A1 *in silico* model converged to a RMSD of 1.17 Å and showed a stable behavior over the simulation. Global and local stereochemical assessment indicated a very good model for comparative structural analyses. SMase D A1 displays a typical TIM $(\alpha/\beta)_8$-barrel fold and its active-site cleft, formed by the metal-binding site (Asp, Glu and Asp) and the two catalytic histidines, is furthermore surrounded by the catalytic loop (residues 46-60), variable loop (residues 167-175), flexible loop (residues 196-203) and other short hydrophobic loops (Figure 8). Based on the current classification of SMases D (Murakami et al., 2006), SMase A1 belongs to class II containing an additional disulphide bridge (Cys53–Cys201), which connects the catalytic loop to flexible loop (Figure 8). This feature, not present in class I SMases D, diminishes significantly the active-site volume and also alters the inherent flexibility exhibited by the flexible loop. Beyond that, all the structural features observed in class I SMases D are fully conserved and details concerning the action mechanism are well described in Murakami et al., 2005 and Murakami et al., 2006.

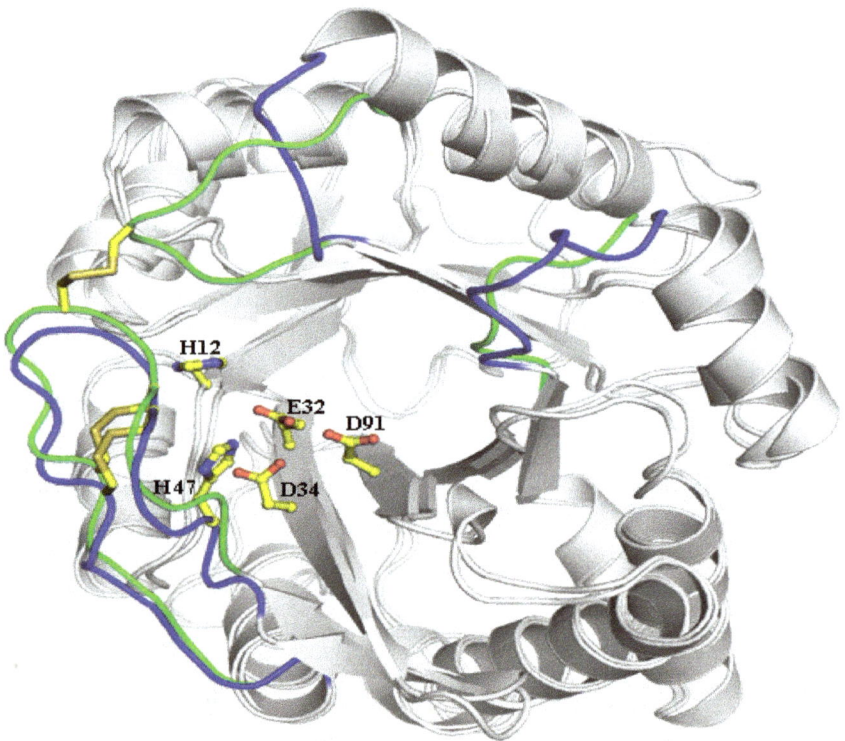

Fig. 8. Structure superposition of Smase D I (class I) and *L. adelaida* SMase D A1 (class II). The residues involved in metal-ion binding and catalysis are presented in atom colors (PDB code: 1XX1). Differences in the catalytic, flexible, and variable loops in *L. laeta* SMase D I (blue) and *L. adelaida* SMase D A1 (green).

3.6 Sphingomyelin-binding mode to SMases D

Despite the structure of a SMase D member has been solved, there is neither structural nor biophysical data relating the binding mode of the sphingomyelinase (SM) into the active-site cleft of SMases D. Thus, in order to shed light into the structural determinants for recognition and binding of SM by SMases D, a MD simulation was performed using the crystal structure of the SMase D I and the SM docked into enzyme taking into account the crystallographic position of the sulphate ion, which provides a good notion how the phosphate group from SM is oriented in the active-site cavity. As observed in Figure 9, the RMSD of the protein is low (~ 1.5 Å), whereas the SM shows a higher variation in the first 3000 ps (~ 5.15 Å) and after that become more stable (~ 3 Å). Although, the aliphatic tails of the substrate exhibit high flexibility as observed by RMSD analysis over the simulation, the polar head is stable. The binding analysis showed that the substrate-enzyme interaction increases during the simulation and it stabilizes around 291 Kcal/mol (Figure 9). In 10 ns of MD simulation, the model achieved a stable conformation with some fluctuations in the aliphatic tails as expected (Figure 9). The choline head is buried in the active-site cleft making van der Waals contacts with Trp230, which is conserved among SMases D (Figures 10A-B). The phosphate moiety is coordinated by the magnesium ion as expected, forming a tetrahedral cage of the ion along with the three acidic residues (Figure 10B). The residue Lys93, which is also highly conserved, is found in a distance range that permits to interact with the carbonyl oxygen of the sphingosine backbone. Val89, Ser132, Asp164, Ser166, Pro134, Pro168, Tyr169, Leu170, Leu198, Tyr228 and Met250 are also participating in the coordination of the substrate (Figure 10B). These data corroborates with previous crystallographic studies, whose suggested the importance of Lys93, Trp230 and other aromatic residues in the recognition and interaction of the substrate.

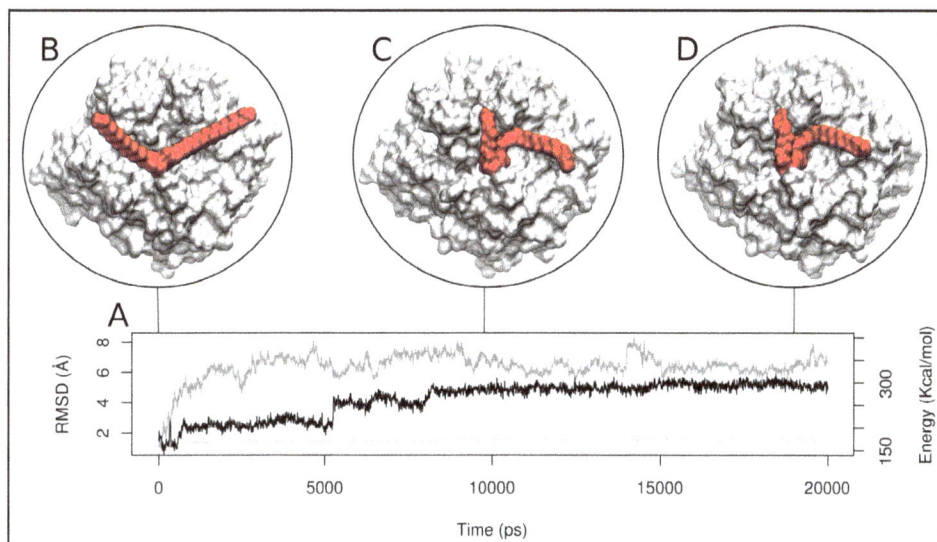

Fig. 9. MD simulation of SM/SMase D I complex. **[A]** RMSD of the protein (light gray) and the substrate (dark gray). MD structural frames in 0 ns (initial state) **[B]**, 10 ns **[C]** and 20 ns (final state) **[D].**

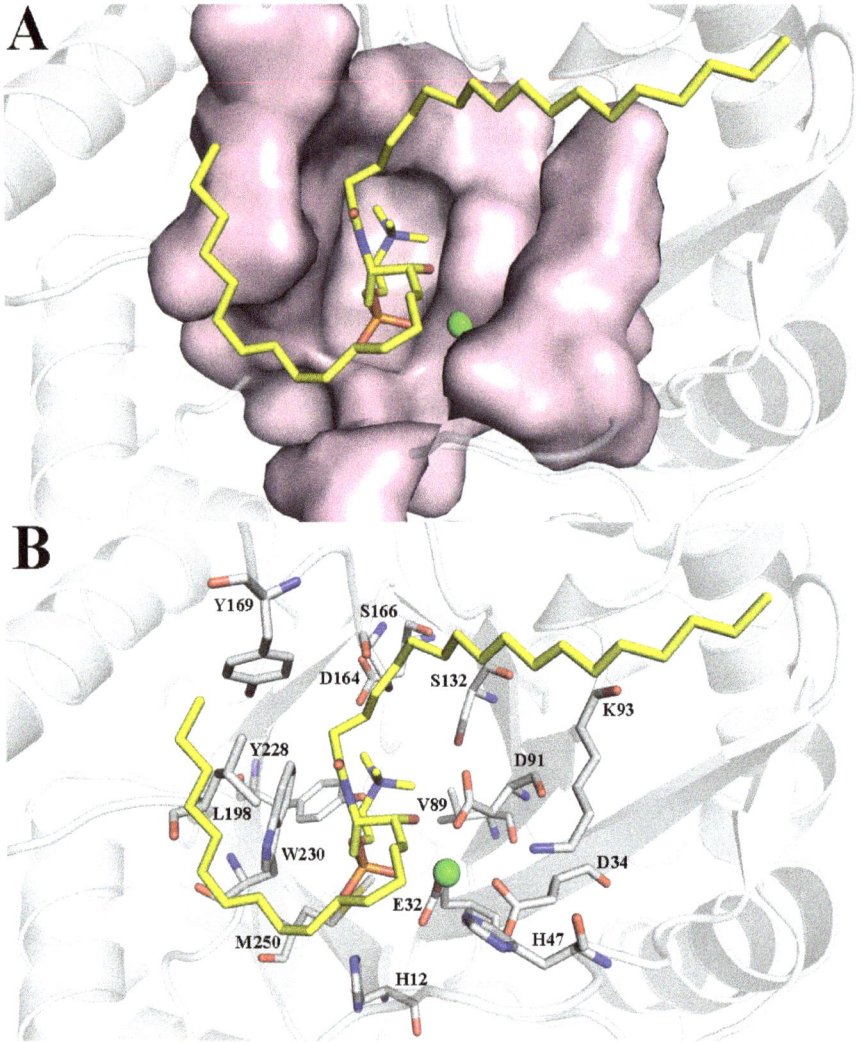

Fig. 10. Binding mode of sphingomyelin in SMase D I. **[A]** Surface representation of the active-site cavity with the substrate. **[B]** Schematic representation of the residues involved in the sphingomyelin (SM) interaction.

4. Discussion

Previously, we have characterized the biochemical and biological properties of *L. adelaida* venom and evaluate the toxic potential of envenomation by this non-synanthropic *Loxosceles* species (Pretel et al., 2005). The biological activities of the *L. adelaida* venom was compared to that of *Loxosceles gaucho*, a synanthropic species of medical importance in Brazil. *L. adelaida* venom showed a similar potential to induce haemolysis, dermonecrosis and lethality as *L. gaucho* venom. Thus, showing that the troglophile *Loxosceles* species, *L. adelaida*, commonly found in the complex of caves from PETAR, is potentially able to cause envenomation with the same gravity of those produced by synanthropic species.

Since various studies have indicated that sphingomyelinase D present in the venoms of *Loxosceles* spiders is the main component responsible for the local and systemic effects observed in loxoscelism, we have cloned and expressed a SMase D from the spider gland of *L. adelaida* to compare the functional activities of the recombinant protein with toxins from synanthropic species and to investigate the inter- and intra-species cross-reactivities of antibodies raised against the purified recombinant proteins.

The *L. adelaida* SMase D A1 cDNA exhibits similarity with previously characterized class 2 SMase D, *i.e.*, the SMase P1 from *L. intermedia*. Both sequences show the residues of the active site pocket, *i.e*, His^{12}, Glu^{32}, Asp^{34}, Asp^{91}, His^{47}, Asp^{52}, Trp^{230}, Asp^{233}, and Asn^{252}, which are essential for the metal-ion binding of SMases D (Mg^{2+} is coordinated by Glu^{32}, Asp^{34}, Asp^{91}), and for acid-base catalytic mechanisms (His^{12} and His^{47} play key roles and are supported by a network of hydrogen bonds between Asp^{34}, Asp^{52}, Trp^{230}, Asp^{233}, and Asn^{252}) (Murakami et al., 2005, and Figures 1, 8 and 10). The importance of histidine residues was also demonstrated by Lee et al. (2005) through site-directed mutagenesis of a *Loxosceles reclusa* recombinant SMase D isoform.

Antiserum produced against the recombinant SMases D P1, P2, from *L. intermedia,* and I, from *L. laeta* was highly cross-reactive against *L. adelaida* SMase A1, and also exhibit a high level of recognition to SMases present in *Loxosceles adelaida* and *L. gaucho* venoms. These data suggest that SMases D from *Loxosceles* species analized share the main immunogenic epitopes. This also means that this antivenom is of potential benefit to patients being bitten, not only by the spiders of the *Loxosceles* species, which the antiserum was raised against, but also by *L. adelaida*.

We show here that the SMases D A1 and P1 in spite of their cross-reactivity, being able to induce a typical dermonecrotic reaction, exhibit differences in their toxic potential, being the lesions produced by *L. adelaida* SMase A1 smaller than that induced by SMase P1. These enzymes were also able to transform erythrocytes into activators of the autologous complement system, as demonstrated by increase of lysis susceptibility in the presence of complement. But, again, SMase D P1 was more active than A1.

Based on sequence and structural similarities, the SMases D can be grouped into two classes depending on the presence of an additional disulphide bridge between the catalytic loop and flexible (Murakami et al., 2006). *L. adelaida* SMase D A1 is a class II member and conserves all structural features for catalytic activity upon sphingomyelin. MD simulations indicated the binding mode of SM in the SMase I, a class I member that already has its crystallographic structure solved (Murakami et al., 2005) and they demonstrated the role exerted by Trp230 in the orientation of choline head, the magnesium ion in the coordination of the phosphate group and other aliphatic residues in the stabilization of substrate at the active site.

In conclusion, we have cloned, expressed and biochemically and structurally characterized a new sphingomyelinase D from *Loxosceles adelaida* spider and shown that it displays all the functional characteristics of whole venom. The recombinants toxins, representing different classes of SMase D molecules, will allow us to further characterize the functionally important domains of these proteins. The identification of the active site(s) would aid in the design and testing of suitable anti-sphingomyelinase compounds in the development of novel therapies to treat loxoscelism.

5. Acknowledgements

This work was supported by FAPESP, CNPq, INCTTOX.

6. References

Bey, T.A., Walter, F.G., Lober, W., Schmidt, J., Spark, R. & Schlievert, P.M. 1997. *Loxosceles arizonica* bite associated with shock. Ann Emerg Med. 30(5), 701-703.

Bucherl, W. 1969. Biology and venoms of the most important South American spiders of the genera *Phoneutria, Loxosceles, Lycosa,* and *Latrodectus.* Am. Zool. 9, 157-159.

de Almeida, D.A., Fernandes-Pedrosa, M.F., de Andrade, R.M.G., Marcelino, J.R., Gondo-Higashi, H., Junqueira-de-Azevedo, I.L.M., Ho, P.L., van den Berg, C. & Tambourgi, D.V. 2008. A new Anti-loxoscelic serum produced against recombinant Sphingomyelinase D: results of preclinical trials. Am. J. Trop. Med. Hyg., 79(3), 463-470.

de Santi Ferrara GI, Fernandes-Pedrosa M de F, Junqueira-de-Azevedo I de L, Gonçalves-de-Andrade RM, Portaro FC, Manzoni-de-Almeida D, Murakami MT, Arni RK, van den Berg CW, Ho PL, Tambourgi DV. 2009. SMase II, a new sphingomyelinase D from Loxosceles laeta venom gland: molecular cloning, expression, function and structural analysis. Toxicon 53, 743-753.

Desai, A., Lankford, H.A. & Warren, J.S 2000. *Loxosceles deserta* spider venom induces the expression of vascular endothelial growth factor (VEGF) in keratinocytes. Inflammation. 24(1), 1-9.

Eisenberg, D., Luthy, R. & Bowie, J.U. 1997. VERIFY3D: assessment of protein models with three-dimensional profiles. Methods Enzymol 277, 396–404.

Emsley, P. & Cowtan, K. 2004. Coot: model-building tools for molecular graphics. Acta Crystallogr D Biol Crystallogr 60, 2126–2132.

Fernandes-Pedrosa, M.F., Junqueira de Azevedo, I.L.M., Gonçalves de Andrade, R.M., van den Berg, C.W., Ramos, C.R.R., Ho, P.L. & Tambourgi, D.V. 2002. Molecular cloning and expression of a functional dermonecrotic and haemolytic factor from *Loxosceles laeta* venom. Biochem. Biophys. Research Commun. 298, 638-645.

Fiser, A. & Sali, A. 2003. Modeller: generation and refinement of homology based protein structure models. Methods Enzymol 374, 461–491.

Forrester, L.J., Barrett, J.T. & Campbell, B.J. 1978. Red blood cell lysis induced by the venom of the brown recluse spider. The role of sphingomielinase D. Arch. Biochem. Biophys. 187, 355-365.

Gendron, B.P. 1990. *Loxosceles reclusa* envenomation. Am J Emerg Med. 8(1):51-4.

Ginsburg, C.M. & Weinberg, A.G. 1988. Hemolytic anemia and multiorgan failure associated with localized cutaneous lesion. J Pediatr. 112(3), 496-499.

Humphrey, W., Dalke, A. and Schulten, K., "VMD - Visual Molecular Dynamics". J. Molec. Graphics 1996, 14,33-38.

Krieger, E., Darden, T., Nabuurs, S.B., Finkelstein, A. & Vriend, G. 2004. Making optimal use of empirical energy functions: force-field parameterization in crystal space. Proteins 57, 678–683.

Kurpiewski, G., Forrester, L.J., Barrett, J.T. & Campbell, B.J. 1981. Platelet aggregation and sphingomyelinase D activity of a purified toxin from the venom of *Loxosceles reclusa.* Biochem. Biophys. Acta 678, 467-476.

Laemmli, U.K. 1970. Cleavage of structural proteins during the assembly of the head of bacteriophage T4. Nature 227, 680-685.

Lee, S. & Lynch, K.R. 2005. Brown recluse spider (*Loxosceles reclusa*) venom phospholipase D (PLD) generates lysophosphatidic acid (LPA). Biochem J. 391,317-323.

Lobley, A., Sadowski, M.I., Jones, D.T., 2009. pGenTHREADER and pDomTHREADER: new methods for improved protein fold recognition and superfamily discrimination. Bioinformatics 25,1761-1767.

Murakami, M.T., Fernandes-Pedrosa, M.F., Andrade, S.A., Gabdoulkhakov, A., Betzel, C., Tambourgi, D.V.& Arni, R.K. 2006. Structural insights into the catalytic mechanism of sphingomyelinases and evolutionary relationship to glycerophosphodiester phosphodiesterases. Biochem. Biophys. Research Commun. 342, 323–329.

Murakami, M.T., Fernandes-Pedrosa, M.F., Tambourgi, D.V. & Arni, R.K. 2005. Structural basis for metal-ion coordination and the catalytic mechanism of sphingomyelinases D. J. Biol. Chem. 280, 13658-13664.

Newlands, G., Isaacson, C. & Martindale, C. 1982. Loxoscelism in the Transvaal, South Africa. Trans. R. Soc. Trop. Med. Hyg. 76(5), 610-615.

Paixão-Cavalcante, D., van den Berg, C.W., Fernandes-Pedrosa, M.F., Gonçalves-de-Andrade, R.M. & Tambourgi, D.V. 2006. Role of matrix metalloproteinases in HaCaT keratinocytes apoptosis induced by *Loxosceles* venom sphingomyelinase D. J. Invest. Dermatol. 126, 61-68.

Pretel, F., Gonçalves-de-Andrade, R.M., Magnoli, F.C., da Silva, M.E., Ferreira, J.M Jr, van den Berg, C.W. & Tambourgi, D.V. 2005. Analysis of the toxic potential of venom from *Loxosceles adelaida*, a Brazilian brown spider from karstic areas. Toxicon. 45(4), 449-58.

Ramos-Cerrillo, B., Olvera, A., Odell, G.V., Zamudio, F., Paniagua-Solis, J., Alagon, A. & Stock, R.P. 2004. Genetic and enzymatic characterization of sphingomyelinase D isoforms from the North American fiddleback spiders *Loxosceles boneti* and *Loxosceles reclusa*. Toxicon 44, 507-514.

Tambourgi, D.V., Fernandes-Pedrosa, M.F., Gonçalves de Andrade, R.M., Billington, S.J., Griffiths, M. & van den Berg, C.W. 2007. Sphingomyelinases D induce direct association of C1q to the erythrocyte membrane causing complement mediated autologous haemolysis. Mol. Immunol. 44, 576-582.

Tambourgi, D.V., Fernandes-Pedrosa, M.F., van den Berg, C.W., Gonçalves-de-Andrade, R.M., Ferracini, M., Paixão-Cavalcante, D., Morgan, B.P.& Rushmere, N.K. 2004. Molecular cloning, expression, function and immunoreactivities of members of a gene family of sphingomyelinases from *Loxosceles* venom glands. Mol. Immunol. 41, 831–840.

Tambourgi, D.V., Gonçalves-de-Andrade, R.M. & van den Berg, C.W. 2010. Loxoscelism: From basic research to the proposal of new therapies. Toxicon 56, 1113-1119.

Tambourgi, D.V., Magnoli, F.C., van den Berg, C.W., Morgan, B.P., de Araujo, P.S., Alves, E.W. & Dias da Silva,W. 1998. Sphingomyelinases in the venom of the spider *Loxosceles intermedia* are responsible for both dermonecrosis and complement-dependent hemolysis. Biochem. Biophys. Res. Commun. 251, 366-373.

Tambourgi, D.V., Morgan, B.P., Gonçalves-de-Andrade, R.M., Magnoli, F.C. & van den Berg, C.W. 2000. *Loxosceles intermedia* spider envenomation induces activation of an endogenous metalloproteinase, resulting in cleavage of glycophorins from the erythrocyte surface and facilitating complement-mediated lysis. Blood 95, 683-691.

Tambourgi, D.V., Paixão-Cavalcante, D., Gonçalves de Andrade, R.M., Fernandes Pedrosa, M.F., Magnoli, F.C., Morgan, B.P.& van den Berg, C.W. 2005. *Loxosceles* Sphingomyelinase induces Complement dependent dermonecrosis, neutrophil infiltration and endogenous gelatinase expression. J. Invest. Dermatol. 124, 725-731.

Tambourgi, D.V., Sousa da Silva, M., Billington, S.J., Gonçalves-de-Andrade, R.M., Magnoli, F.C., Songer, J.G. & van den Berg, C.W. 2002. Mechanism of induction of complement susceptibility of erythrocytes by spider and bacterial sphingomyelinases. Immunology 107, 93-101.

Tokumura, E., Majima, Y., Kariya, K., Tominaga, K., Kogure, K., Yasuda, K.& Fukuzawa, K. 2002. Identification of human plasma lysophospholipase D, a lysophosphatidic acid-producing enzyme, as autotaxin, a multifunctional phosphodiesterase. J. Biol. Chem. 277, 39436-39442.

Towbin, H., Staeehelin, T. & Gordon, J. 1979. Electrophoretic transfer of proteins from acrylamide gels to nitrocellulose sheets: Procedure and some applications. Proc. Natl. Acad. USA 76, 4350-4354.

van den Berg, C.W., Gonçalves de Andrade, R.M., Magnoli, F.C., Marchbank, K.J. & Tambourgi, D.V. 2002. *Loxosceles* spider venom induces metalloproteinase mediated cleavage of MCP/CD46 and MHCI and induces protection against C-mediated lysis. Immunology 107, 102-110.

van den Berg, C.W., Gonçalves-de-Andrade, R.M., Magnoli, F.C.& Tambourgi, D.V. 2007. *Loxosceles* spider venom induces the release of thrombomodulin and endothelial protein C receptor: implications for the pathogenesis of intravascular coagulation as observed in loxoscelismo. J. Thromb. Haem. 5, 989-995.

van Meeteren, L.A., Frederiks, F., Giepmans, B.N., Fernandes-Pedrosa, M.F., Billington, S.J., Jost, B.H., Tambourgi, D.V. & Moolenaar, W.H. 2004. Spider and bacterial sphingomyelinases D target cellular lysophosphatidic acid receptors by hydrolyzing lysophosphatidylcholine. J. Biol Chem. 279, 10833-10836.

Wiederstein, M.& Sippl, M.J. 2007. ProSA-web: interactive web service for the recognition of errors in three-dimensional structures of proteins. Nucleic Acids Res 5, 407–410.

Willard, L., Ranjan, A., Zhang, H., Monzavi, H., Robert, F.B., Syakes, B.D., Wishart, D.S. 2003. VADAR: a web server for quantitative evaluation of protein structure quality. Nucleic Acids Res 31, 3316–3319.

11

Molecular Toxinology – Cloning Toxin Genes for Addressing Functional Analysis and Disclosure Drug Leads

Gandhi Rádis-Baptista

Institute of Marine Sciences, Federal University of Ceará

Brazil

1. Introduction

The revolution in Biology started earlier with the genetic works of Gregor Mendel (1866), who, through his work with pea breeding, observed the phenomena of dominance and segregation of traits and discovered several laws of heredity. The pioneered endeavor of deciphering the linkage between transmission of heredity and a biomolecule was succeeded by the works of Griffith (1928), Oswald Avery, Colin McLeod, and Marylin McCarty (1944), who demonstrated that the instruction for virulence traits in bacteria was contained in the deoxyribonucleic acid (DNA) molecule, as well as Alfred Hershey and Martha Chase (1952), who elegantly proved that the progeny of bacteriophage is propagated after the injection of the bacteriophage's genetic material into the host cell. The hallmark of molecular biology arose in 1953 with the description of the double-helix backbone of DNA by James Watson and Francis Crick, who also described that such a structure may suggest a mechanism of DNA replication (Watson and Crick, 1953). Notably, the deduction of the DNA structure was based on the data of other works by Erwin Chargaff, who determined the base correspondence and ratios in nucleic acid, and by Rosalind Franklin, who obtained DNA fiber images from X-ray diffraction. The elegant genetic experiments from François Jacob and Jacques Monod (1961) established the concept of *cis*-acting elements and the mechanism by which the operator and repressor regulate β-galactosidase expression in *Escherichia coli* and sugar metabolism (Jacob and Monod, 1961). At same time, Marshal Nirenberg was leading the race to decipher the genetic code (Nirenberg, 2004). By the 1970s, Fred Sanger and Walter Gilbert developed two distinct methodologies for DNA sequencing, which culminated with automated high-throughput DNA sequence analysis, thus opening the door for the genomic revolution and the publication of hundred of genomes, including the human genome. The central dogma of molecular biology, which postulated that DNA directs its own replication and its transcription to yield RNA, which in turn directs its translation to form protein, was wisely proposed by F. Crick in 1958 (Crick, 1970). This also included the "probable transference of information", which we know now as RNA replication and reverse transcription, after the seminal works of D. Baltimore (1970) and J.M. Bishop on the molecular virology of retroviruses and oncogenes (1973). The 'biological revolution' continued forward with important discoveries such as the mechanism by which chromosomes are protected at their ends (the telomeres) against degradation and the involvement of the enzyme (telomerase) that forms these ends by Elizabeth H. Blackburn, Carol W. Greider, and Jack W. Szostaktelomere (Blackburn, 2005) as well as the novel mechanism of gene regulation mediated by double-stranded RNA (interference RNA) that triggers suppression of gene activity by Andrew Z. Fire and Craig C. Mello (Mello and Conte, 2004; Fire, 2007). Most of these scientific conquests have been nicely compiled in the writing of Lander and Weinberg (2002), where the initial events of the Biological Revolution are outlined in more detail. Other actors have played a significant role in the field, but a review of this short time frame of scientific conquests may give us a taste of how the field of molecular biology experienced periods of excitement and 'epic' concretizations, culminating with what we know today. New avenues are being explored with the advent of two new disciplines that incorporate paradigmatic concepts and approaches to interrogate the complexity of life, namely, systems biology and synthetic biology. In the first case, the focus is on the analysis of complex biological systems from a holistic point of view, i.e., to study how individual genes and proteins interact to build entire organisms (single cell or multicellular) and allow them to operate properly. In the field of synthetic biology, principles of engineering are used to build from scratch living systems able to perform alternative functions not found in nature (Bader, 2011).

Currently, recombinant DNA technology and DNA cloning represent an indispensable tool box for the research of distinct life science fields as diverse as environmental science, evolutionary biology, cell biology, microbiology, molecular medicine and pharmacology, and structural and systems biology.

A particular field of interest that involves recombinant DNA technology and covers the study of ecological, biochemical, pharmacological and structural aspects of animal toxins is referred to in the scientific community as Molecular Toxinology. From an ecological point of view, animal venom might be considered an arsenal of organic and protein substances capable of immobilizing the competitor or prey by interfering with specific molecular targets in their cells and tissues. Therefore, animals equipped with venom glands and an inoculating apparatus have a significant fitness advantage. Biochemically and pharmacologically, chemical and protein diversities correlate with biodiversity, i.e., diverse indigenous animals inhabiting a given biome may offer novel compounds and bioactive molecules. Before the advent of molecular cloning and recombinant DNA technology, only the major components of venom were purified in high yield using protein chemistry techniques suited for functional characterization. Consequently, milligrams of purified toxin were required for studies by classical tissue-based assays, which limited the analysis of toxin biological activity to the molecular level. However, with the refinement of instrumentation in the context of 'omics' (e.g., genomics, transcriptomics, proteomics, and interactomics), toxins expressed in the venom even in minute amounts could be thoroughly examined.

In this chapter, I wish to introduce some selected molecular biology techniques that can be applied to investigate the diversity of polypeptide molecules present in animal venoms.

2. Animal venom peptides and proteins as therapeutics

The global market for peptide and protein therapeutics was valued at over US$ 57 Billion in 2006 and estimates suggest that it will grow continuously in the next years at a 9.7% compound annual growth rate (CAGR) and reach a value of $103 billion by 2014 (Nair, 2011). According to these analyses, there is a high demand in the therapeutic proteins market for engineered monoclonal antibodies (MAb), insulins, cytokines, interferons and related immune modulators, enzymes, hematopoietic growth factors (erythropoietins) and coagulation factors. These polypeptide drugs are designed for the treatment of autoimmune and cardiovascular diseases, diabetes and cancer. The increasing demanding for therapeutic polypeptides is due in part to the excellent affinity and selectivity for the disease target displayed by such molecules, as well as biological compatibility. Currently, therapeutic peptides and proteins are manufactured by means of synthetic peptide chemistry and recombinant technology, respectively. Consequently, polypeptide drugs offer unparalleled opportunities for innovation in molecular design, improved pharmacokinetics, and disease target-delivery of therapeutics.

Animal venom is a collection of molecules selected during millions of years of metazoan evolution with which specific tissue and target preferences are invariably observed among numerous families of toxins. In fact, snakes (Birrel et al., 2007), scorpions (Bringans et al., 2008), spiders (Estrada et al., 2007), and sea snails (Becker et al., 2008), to name just a few, produce and secrete a valuable diversity of toxins capable of interacting with distinct molecular targets within the cells of their prey or victims. Importantly, from the point of view of molecular evolution and medical biotechnology, using phylogenetic analysis, Fry (2005) inferred that polypeptide toxins secreted in a given venom evolved from endogenous bioactive protein genes, which were expressed early in tissues other than venom glands. For example, snake venom three finger toxins (3FTs) appear to have evolved from a common protein ancestor, such as the nicotinic acetylcholine receptor-binding LYNX, which is expressed in large projection neurons in the hippocampus, cortex and cerebellum. Other toxins that were related to ancestral proteins, for which the genes were preserved during the evolution of snake venom gland, include acetylcholinesterases, disintegrins/metalloproteinases, C-type lectins, complement C3, crotasin/defensin-like peptides, cystatins, endothelins, factor V, factor X, kallikrein, kunitz-type protease inhibitors, LYNX/SLUR, L-amino oxidase, natriuretic peptides/bradykinin potentiating peptides, nerve growth factor, phospholipase A2, and vascular endothelial growth factor.

Considering the evolutionary conservation of polypeptides toxins and the excellent market opportunity for use a therapeutic peptide and proteins, animal toxins represent one unique source of ready-to-use engineered polypeptides capable of modulating vital human physiological and pathological processes. Numerous examples of the use of animal toxin in medicine have been reported. In their articles "Bugs as Drugs" (part I and II), E. P. Cherniack (2010; 2011) reviews the use of a number of different 'bugs' (worms, leeches, snail, ticks, centipedes, spider) and their metabolic products (endogenous or secreted) in medicine and describes the clinical benefits of such biological/pharmacological resources. As a particular biological effect is the result of a combination of a specific activity intrinsically contained in a single molecule, a myriad of new pharmacologically active compounds can be isolated from animal venom. For example, a dozen snake venom toxins belonging to several protein families such as C-type lectin, metalloprotease, phospholipase A2, and three-finger toxin display anticoagulant activities and a high potential for therapeutic use in preventing pathological clot formation (Kini, 2006). The snake venom components that act on the vertebrate blood coagulation cascade can be categorized, depending on their hemostatic action, as follows: enzymes that clot fibrinogen; enzymes that degrade fibrin (ogen); plasminogen activators; prothrombin activators; factor V activators; factor X activators; anticoagulants (inhibitors of prothrombinase complex formation, inhibitors of thrombin, phospholipases, and protein C activators); enzymes with hemorrhagic activity; enzymes that degrade plasma serine proteinase inhibitors; and platelet aggregation inducers (direct acting enzymes, direct acting non-enzymatic components, and inhibitors of platelet aggregation) (Markland, 1998). Some of these toxins have been clinically used as therapeutics or diagnostic reagents, while others are under pre-clinical trials (Fox and Serrano, 2007). For example, Exendin-4, a 39-amino acid peptide from the saliva of the lizard *Heloderma suspectum*, is able to improve blood sugar control in adults with type 2 diabetes mellitus, and it has been commercially registered as Byetta® (Exenatide). Furthermore, a number of toxins isolated from aquatic (including marine) and terrestrial animals represent the ultimate resource of novel molecules to treat highly prevalent cardiovascular diseases, such as high blood pressure and arrhytmias (Hodgson & Isbister, 2009), human neurological disturbances (Mortari et al., 2007), and cancer (Molinski et al., 2009). In fact, some of the best selling drugs used to treat high blood pressure, namely captopril and analogues, which act by inhibiting the angiotensin converting enzyme (ACE), were developed using a rational chemical synthesis approach using a pentapeptide toxin expressed in the venom of the Brazilian pit viper *Bothrops jararaca* as a model (Ferreira, 1965; 1985; Ondetti et al., 1977; Cushman and Ondetti; 1991). The therapeutic potential of venom peptides have also been investigated with regards to their pharmacology effects on ion channels and neural receptors (Lewis and Garcia, 2003). For example, an N-type voltage-sensitive calcium channel blocker peptide, isolated from the marine mollusk Conus (*Conus magus*), was recently approved by the US Food and Drug Administration as a drug (ziconotide/ Prialt®) for the treatment of severe chronic pain (Schmidtko et al., 2010).

To study the mechanistic specificities of a given toxin toward a molecular target, milligrams of pure polypeptides are usually required, which is not always easy to obtain from certain venomous animals. This is particularly true for purified molecules from the venom and saliva of small creatures such as scorpions, spiders, wasps, poison worms, and hematophagous animals. Consequently, a powerful approach used to investigate the therapeutic potential of animal venoms is based on proteomics and molecular cloning techniques. Information about proteomic-based characterization of animal venom (venomics) can be found elsewhere (Calvete et al., 2007; 2009; Escoubas et al., 2008; 2009). This chapter reviews select topics on molecular toxinology, which include not only the cloning and recombinant expression of a single toxin but also the receptor-guided high-throughput screening of polypeptide venom libraries.

2.1 Molecular cloning of animal toxin genes

DNA cloning, or molecular cloning, is the process of constructing recombinant DNA molecules, transferring them in a given host cell and making copies of the inserted DNA, usually genes or the product of their transcription, i.e., messenger RNAs (Watson et al., 2008). For cloning, all that is necessary for propagation of cloned DNA is the piece of DNA of interest (i.e., insert DNA) from a particular source, a vector (small molecule of DNA capable of self-replication and containing a selectable marker), and restriction and modifying enzymes used to cut and join the insert and vector DNA together. Once recombinant DNA molecules are prepared *in vitro*, a host cell, usually Gram-negative bacteria *Escherichia coli* is transformed with the engineered vector and propagated millions of times to produce large quantities of cloned DNA. When a collection of thousands of DNA sequences is cloned instead of a single piece of DNA, it is referred to as a library. Essentially, two main types of libraries have been prepared and utilized in toxin research:

complementary or copied DNA (cDNA) and genomic libraries. The cDNA library is designed to represent the transcriptome, i.e., the pool of messenger RNAs (mRNA) molecules that is produced by a cell type, tissue or organ in given time or metabolic condition. Therefore, it represents a snapshot of cell status, even if some transcripts (mRNAs) may be constitutively expressed. Thus, libraries of cDNAs or expressed sequence tags (ESTs) can be highly useful when the focus of a particular research project encompasses interrogative studies of gene expression profiles or differential gene expression. Using cDNA and EST sequencing, the gene expression profiles of the venom glands from several species of poison animals have been analyzed, including snakes (Zhang et al., 2006; Wagstaff and Harrison, 2006; Pahari et al., 2007; Cassewell et al., 2009; Neiva et al., 2009; Georgieva et al., 2010; Durban et al., 2011; Jiang et al., 2011, Rokyta et al., 2011), scorpions (Schwartz et al., 2007; Ma et al., 2009; Ruiming et al., 2010; Morgenstern et al., 2011), spiders (Chen et al., 2008 Fernandes-Pedrosa et al., 2008; Gremski et al., 2010; Jiang et al., 2010), platypus (Whittington et al., 2099; 2010), conus (Pi et al., 2006; Hu et al., 2011), and jellyfish (Yang et al., 2003).

Differential gene expression of venom gland libraries has been used to investigate the molecular diversity of venom polypeptides. For example, Morgenstern and collaborators (2011) have reported that a significant difference exists in the transcriptome of resting venom glands from the buthid scorpion *Hottentotta judaicus* in comparison with the gland that is actively engaged in regenerating its venom. The transcriptome profile of a replete (resting) venom gland is rich in low-abundance toxin transcripts and tends to predominantly consist of open reading frame (ORF) sequences encoding toxins acting on voltage- and calcium-activated potassium ion channels. To perform such a study, a cDNA library was prepared from the venom gland of scorpions that had not been milked or induced to produce venom and was then compared with a cDNA library prepared from the venom gland of milked specimens in which the venom glands were committed to replenishing the venom pool of transcripts and toxins. This technique, termed cDNA subtraction, which will be described in the next section, and the resulting subtracted cDNA library, are generally used for the screening of cDNAs corresponding to mRNAs differentially expressed or regulated. The differential pattern of transcripts in resting and regenerating venom glands revealed an important aspect to be considered when studying transcriptomes of venomous animals and suggests which strategy should be taken into account to prepare cDNA libraries. In practical terms, most cDNA libraries are prepared after milking the venom from poisonous animals to empty the venom gland and induce the synthesis of total RNA prior to mRNA purification (Rottenberg et al., 1971; Rádis-Baptista, 1999). Thus, by examining full or subtracted cDNA libraries, as exemplified for scorpion venom glands, qualitative and quantitative differences are detected, particularly when one wishes to know the constitutive or induced venom transcriptomes for comparison with the corresponding proteomes (Ma et al., 2010). In fact, a combination of transcriptome analysis, i.e., cDNA sequencing, with mass spectrometry represents a useful alternative to characterize the animal venom, as was utilized for Cone marine snail venom analysis, and compare the inter- and intra-species variation that exists among venom peptide libraries (Gowd et al., 2008).

In molecular toxinology, the data obtained from genomic DNA library analysis seems, at first glance, less informative than those obtained from cDNA libraries due to the relative static nature of genomes. However, genomic libraries provide information about gene number, diversity and organization. For example, using a PCR homology screening method, a crotamine paralogous toxin gene, crotasin, which is 2.5 kilobase (kb) long and organized in three exons intervened by two introns, was identified in a rattlesnake genomic DNA bacteriophage library; this suggested that gene duplication and accelerated independent evolution operated in the diversification of crotamine/crotasin genes (Rádis-Baptista et al., 2004). Such phenomena of gene evolution is a recurrent theme in toxinology (Kordis and Gubensek, 2000; Fry et al., 2009) and genomic libraries configure a good technical resources to retrieve such informations. Genomic DNA libraries constructed in bacterial artificial chromosomes (BACs) for *Bungarus multicinctus* and *Naja naja*, two old world toxic elapid snakes, were screened with probes for four major families of toxins, three-finger toxin (3FTx), phospholipase A2 (PLA2), Kunitz-type protease inhibitor (Kunitz) and natriuretic peptide (NP), and results showed 3FTx as the major toxin gene in elapidae venom (Jiang et al., 2011). These 3FTx genes are also composed of three exons and two introns in a region of approximately 2.5 kb (Tamiya and Fujimi, 2006), and they were shown to represent five putative tandem duplicates in *B. multicinctus* and seven in *N. atra*, thereby suggesting that tandem duplications has also contributed to the expansions of toxin multigene families in these two elapids (Jiang et al., 2011). Interesting, genome analysis of platypus (*Ornithorhynchus anatinus*) coupled with the transcriptome of its venomous apparatus revealed 83 novel putative platypus venom genes, distributed among 13 toxin families, which are homologous to known toxins from a wide range of vertebrates (fish, reptiles, insectivores) and invertebrates (spiders, sea anemones, starfish). A number of these platypus venom toxin families are expressed in tissues other than the venom gland, as observed earlier with snake venom toxins (Whittington et al., 2010).

A glimpse into the realm of molecular toxinology reveals that recombinant DNA technology is essential for the analysis of the complex pharmacological effects of venom toxins and their potential biomedical and clinical applications. In the next section, the current and potential molecular techniques used for investigating toxins in a high-throughput manner are described in further detail.

2.1.1 Construction of a venom gland cDNA library

The basic steps for constructing a cDNA library include the following: (1) excision of venom glands from the poisonous animal of interest, (2) preparation of total and messenger RNA for cDNA cloning, (3) synthesis of cDNA from mRNA by reverse transcription, (4) selection of a vector, plasmid or bacteriophage (phage) for cloning and propagation of cDNA libraries, (5) sequencing of all cDNA libraries or screening for desired clones before sequence analysis, and (6) validation of cDNA clones for functional analysis. Synthesis kits for convenient construction of cDNA libraries are commercially available, and detailed information of a particular system can be obtained from sales representatives at companies such as Clontech Laboratories, Inc (Mountain View, CA-U.S.A.) and Stratagene (presently, Agilent Technologies, Inc., Santa Clara, CA-U.S.A.).

Total RNA is typically purified from the tissue of choice, from which the sequences of interest are more abundant, using the single-step acid guanidinium thiocyanate-phenol-chloroform method of Chomczynski and Sacchi (1987), by which cells are quickly disrupted, their components are solubilized, and the endogenous RNase is simultaneously denatured using guanidinium salt. Although such methods of RNA extraction and purification are very effective and reliable, caution should be taken when the sources of RNA extraction are tissues rich in lipids, polysaccharides and proteoglycans. In such cases, modification of the method, such as by introducing an organic extraction step and changing in the condition of RNA precipitation, can counteract the interference of these contaminants that would otherwise inhibit reactions of reverse-transcription. Total RNA purified by this single-step method is not only used for mRNA - or poly(A)$^+$ RNA - purification and for cDNA synthesis but also for applications such as northern hybridization, RNase protection assay and dot/slot blotting (Sambrook and Russel, 2001). In general, tissues such as snake venom glands and sea anemone tentacles yield between 3 and 10 μg/mg of RNA. Once the total RNA has been purified, the quality and integrity must be analyzed. The quality is assessed spectrophotometrically by analyzing the A_{260}/A_{280} ratio, which should be between 1.8 and 2.0. RNA quality is confirmed by running an aliquot of total RNA preparation using denaturing formaldehyde agarose gel electrophoreses (stained with ethidium bromide) to observe the ratio of 28S and 18S ribosomal RNA. The theoretical 28S:18S ratio for eukaryotic RNA is approximately 2:1, but this ratio might be different with RNA extracted from tissues of other organisms. An alternative to denaturing formaldehyde agarose gel electrophoresis is the analysis of total RNA in a microfluidic device, by which RNA integrity and concentration are automatically assessed. However, as it is more convenient and less laborious and inexpensive, gel-based analysis of total RNA quality can be still useful, when formamide is used with RNA samples and TAE agarose gel instead of formaldehyde as denaturing agent (Masek et al., 2004).

Traditionally, cDNA libraries are prepared using mRNA, and not total RNA, to promote the reaction of reverse transcription. Thus, methods of preparing high-quality mRNA should be applied. These methods are based on the presence a 3' end polyadenosine tail found in most eukaryotic mRNAs. Short oligonucleotides of deoxythymidine (18 to 30 nucleotides in length) - oligo(dT)$_{18-30}$, immobilized on cellulose or linked to biotin, form a stable hybrid with the poly(A)+ tail of mRNA in the presence of a high concentration of salt. The polyadenylated RNAs are denatured (at 70-72°C for 5-10 min), allowed to hybridize with the oligo(dT) and separated by affinity (column chromatography) or captured (with streptavidin-coated paramagnetic beads) from the other RNAs (ribosomal RNA, transfer RNA), which are washed away, and the RNA is eluted with a low-salt buffer. The poly(A)$^+$ RNA pool is quantified by ultraviolet (UV) spectrophotometry and then used for cDNA synthesis. It is important to note that only 1 to 5% of total RNA constitute poly(A)$^+$ RNA, and RNAs that represent less than 0.5% of the total mRNA population of the cell are referred as 'rare' mRNAs. Therefore, when a cDNA library is being prepared, it should be comprehensive enough to include clones that represent 'low abundance' or 'rare' transcribed sequences (mRNAs). This is estimated the following formulas:

$$N= \ln (1 - P)/\ln (1-[1/n]),$$

where N is the number of clones required, P is the probability (usually 0.99), and $1/n$ is the fraction of the total mRNA that is represented by a single type of rare mRNA (Sambrook and Russel, 2001).

$$Or, \quad q = 1 - P = [1 - (n/T)]^B$$

where n is the number of molecules of the rarest mRNA in a cell and T is the total number of mRNA molecules in a cell. The desired base (B) is the number of clones that should be screened to achieve a 99% probability that a cDNA clone will exist in the library (Ausubel et al., 1998).

Often, when 500,000 to 1,000,000 independent cDNA clones are present in an unamplified library, at least one copy of every mRNA should be present in the library. In more practical terms, to obtain a representative cDNA library with a high probability of finding a clone harboring a rare transcribed sequence, 1 to 5 μg (or less) of poly(A)+ mRNA is usually sufficient.

To synthesize cDNAs, commercially available RNA-dependent DNA polymerases (reverse transcriptase, RT), derived from avian or murine retroviruses, catalyze the addition of deoxyribonucleotides to the 3'-hydroxyl terminus of a primed RNA-DNA hybrid, for which a deoxyoligonucleotide, such as an oligo(dT), a pool of random hexamers or a specific sequence, is used as a primer and the mRNA is used as template. This action of extension is referred to as reverse transcriptase 5'→3' DNA polymerase activity, and the product is a hybrid molecule composed of a single-stranded RNA (the mRNA) and a single-stranded cDNA (ss-cDNA). The non-engineered RTs have two additional catalytic activities and are consequently multifunctional enzymes. These RTs display a low level of DNA polymerase activity but a considerable 3'→5' or 5'→3' exonuclease processivity of RNA degradation in an RNA:DNA hybrid, or RNase H activity (Ausubel et al., 1998). Although the RNase H activity of RTs is useful in molecular biology (e.g., selective destruction of parts of an RNA molecule), it has been eliminated from most recombinant engineered enzymes commercialized for research to avoid the degradation of an mRNA template and improve the yield of synthetic cDNA. It is important to understand the mechanism by which RT functions, given the final product of this step. ss-cDNA is a useful starting material for other specific techniques of cloning, such as 3'- and 5'-RACE (rapid amplification of cDNA ends), reverse transcription coupled with PCR (RT-PCR), and quantitative real-time PCR (qPCR). One critical step in the preparation of ds-cDNA for cloning is the choice of primers for synthesis of first-strand cDNA. Primers used for cDNA synthesis include (1) oligo (dT), a 12-18 nucleotide oligo that binds to the poly(A) tail at the 3' end of mRNA; (2) primer-adaptors that contain a homopolymeric tract at the 3' end and a restriction site; (3) primers linked to a plasmid; and (4) random primers (Sambrook and Russel, 2001). However, the most popular methods for cDNA library construction are based on priming cDNAs with an oligo(dT) adaptor for directional cloning and homopolymeric priming of second-strand cDNA; this latter technique is commercially known as "switching mechanism at 5' end of RNA template – SMART".

Once the first-strand cDNA synthesis is accomplished (typically for 60 min, at 37-42°C), other modifying enzymes are employed, such as RNase H for introducing nicks into the RNA molecule of the mRNA:cDNA hybrids; E. coli DNA polymerase I for extending the 3' end of RNA primers (generated by RNase H activity on mRNA moiety) and replacing the fragments of mRNA in the mRNA:cDNA hybrids; bacteriophage T4 DNA polymerase for polishing the double-stranded cDNA (ds-cDNA); and T4 polynucleotide kinase for phosphorylating 5'-hydroxyl groups on the ends of the ds-cDNA for ligation of linkers or adaptors, which is accomplished by T4 DNA ligase. Methylases (e.g., Eco RI methylase) that catalyze the methylation of internal cleavage sites in ds-cDNA and thus protect synthesized cDNA from restriction enzymes are generally used for linker-adaptor digestion in the last steps of preparation of cDNAs for cloning. It is interesting to note that steps of phenol-chloroform extraction and ethanol precipitation are necessary for cleaning up the ds-cDNA by removing enzymes, buffers and reaction components (Ausubel et al., 1998; Sambrook and Russel, 2001).

Another important step when preparing a cDNA library is the size fractionation of cDNA molecules. Size exclusion chromatography (e.g., Sepharose CL-4B) is used for fractionation of cDNAs and allows for the elimination of linkers or adaptors, enzymes, and buffer components, which would be detrimental in the subsequent steps of cloning. Moreover, with this procedure, truncated cDNAs that arise from incomplete first- or second-strand cDNA synthesis are discarded and sequences >400-500 nucleotides are conveniently selected for construction of 'high molecular' weight libraries. Researchers dealing with venom gland cDNA libraries should note that a high number of toxins (e.g., cardiotoxin, crotamine-like and sarafotoxin) are encoded by short nucleotide sequences, and thus, it is advisable to prepare a low molecular weight sub-library from venom gland cDNAs. As a rule of thumb, the step of size fractionation is accomplished after the cDNA synthesis has been completed and just before the ds-cDNA is cloned into a vector. Figure 1 displays a typical agarose gel electrophoresis analysis of size fractionation of ds-cDNA from an animal venom gland.

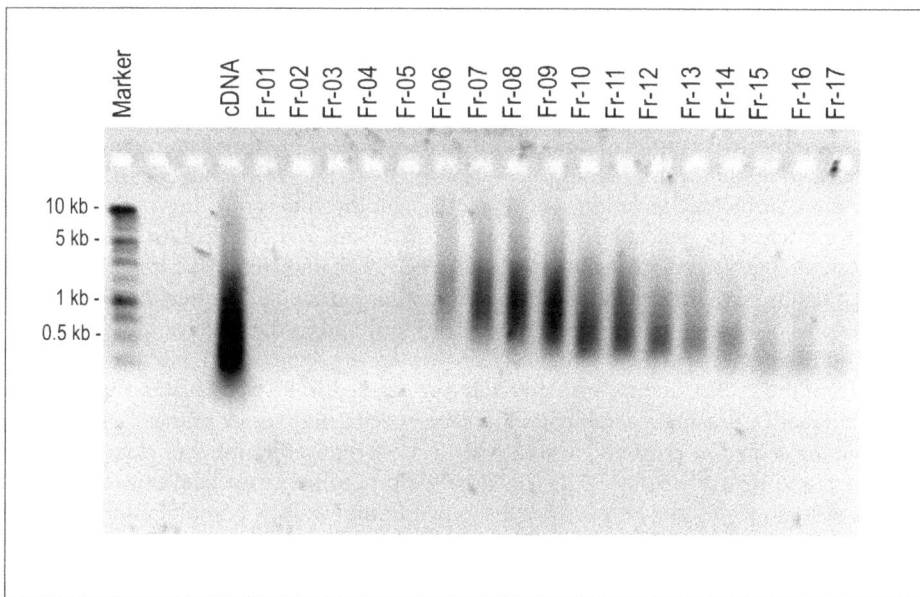

Fig. 1. Size fractionation of cDNA prepared with mRNAs from sea anemone (*Anthopleura cascaia*) tentacles. A doubled-stranded cDNA pool, prepared using SMARTER technology according the recommended protocol (Clontech Labs), was size fractionated with a CHROMA Spin DEPC-400 column (bed volume = 1 ml). Aliquots of each fraction (total = 17, Fr-01 to Fr-17) were analyzed on a 1.0% agarose gel and visualized after ethidium bromide staining. Marker = 1 kb Plus ladder (Promega Corp., WI-USA). 'cDNA' = unfractionated cDNA pool.

Both popular methods of cDNA synthesis for library construction, priming mRNA with an oligo(dT) adaptor and homopolymeric priming of second-strand cDNA, produce cDNAs for directional cloning. Thus, the choice of vectors and strategy for cDNA cloning is selected based on the approach for ds-cDNA synthesis. With these two methods, adaptors and linkers at the 5' and 3' ends of cDNA molecules, containing restriction sites, are catalytically cleaved with two different enzymes, thereby producing cDNAs with distinct cohesive stick terminals. Vectors that are used for cloning cDNA include bacteriophage (phage) and plasmids. Today, either vector is appropriate for the preparation of libraries in the range of 10^6 to 10^7 independent recombinant clones when high-quality phage packaging extract and high-efficiency electrocompetent *E. coli* are used for the construction of phage and plasmid cDNA libraries. Examples of vectors currently used are the phage-engineered plasmids (phagemids) λZAP, λZAPII and λZAP Express (Stratagene/Agilent Technologies, Inc., Santa Clara, CA-U.S.A.) and the plasmids pcDNA3.1, pDNR and pSMARTer (Clontech Laboratories, Inc., Mountain View, CA-U.S.A.). The phagemid vectors are composed from the genomes of high-efficiency infectious bacteriophage particles, from which a high number of recombinant clone can be obtained, and allow for the *in vivo* recovery of plasmids harboring the inserted cDNAs. These vectors are commercially available either linearized or as phage arms, and the linear plasmid can be prepared in advance with the appropriate restriction enzymes, although the process of vector preparation is relatively laborious. Rádis-Baptista and collaborators (1999, Kassab et al., 2004, Neiva et al., 2009) have successfully prepared venom gland cDNA libraries in both λphage and plasmids with a high number of independent recombinant clones and satisfactory insert size averages. Libraries prepared in these ways are then amplified, and the clones are pooled or stored in microwell plates and maintained at -80°C for screening.

Several molecular biology techniques are typically employed to screen for a specific cDNA sequence of interest. In molecular toxinology, a convenient and practical protocol involves the 'PCR homology screening' technique (Israel 1993; 1995; Radis-Baptista, 1999). Given that families of toxins have divergent members with conserved motifs, the nucleotide or amino acid sequences of known toxin representatives are multi-aligned, the consensus segment is selected, and sequence-specific oligonucleotide primers are synthesized. Hundreds of small aliquots from the cDNA library are separately propagated in 96-wells plates and analyzed by conventional PCR using a combination of gene-specific and vector primers. The pools of clones that are scored as positive are subdivided into a reduced number of clones per well, propagated, and re-screened by PCR. This iterative procedure is repeated until 100% of all clones in a single well score positive for the gene of interest. PCR homology screening not only facilitates the identification of clones of interest but also generates amplicons that can be easily and directly cloned for sequencing analysis. Moreover, PCR homology screening seems to be less laborious and tedious than screening cDNA libraries with labeled oligonucleotide probes, which requires replica plates, support membranes, and radioactive or fluorescent probes.

As most vectors incorporate elements for functional analysis of cloned inserts, cDNAs from a library can be screened by detecting a biological activity as a result of recombinant protein expression. However, this approach is not commonly used for screening toxin cDNA clones because of the low efficiency of toxin refolding *in vitro*, which is due to the high number of disulfide bonds that toxin molecules contain. Phenotype coupled to genotype-based screening of cDNA libraries is best achieved by alternative molecular techniques of protein-protein interaction, as described in the next sections. In figure 2, the main steps involved in the construction of cDNA and genomic DNA libraries are summarized.

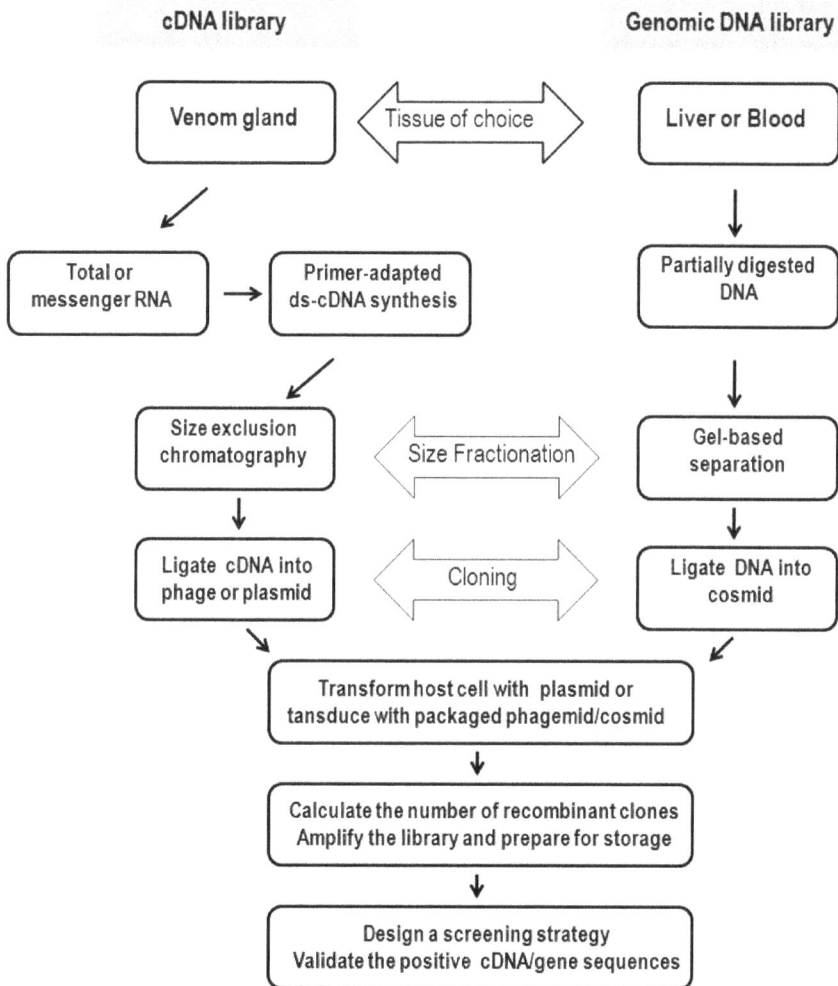

Fig. 2. Overview of main steps for preparing venom gland cDNA and genomic DNA libraries

2.1.2 Subtractive and genomic libraries in animal toxin research

The subtracted cDNA library is very convenient for comparative studies of transcript expression. For example, when comparing the pattern of expression in venom glands of two distinct poisonous animals from specific geographical localities or the expression profiles of resting (replete) and regenerating venom glands, the production of a subtracted cDNA library is applicable (and recommended). The principle of subtractive cloning relies on nucleic acid hybridization, by which nucleotide sequences differentially expressed in one cell or tissue type (the tracer) are hybridized to a complementary nucleic acid pool isolated from the cell or tissue that is not expected to express the sequence of interest (the driver) (Sagerströmet al., 1997; Ausubel et al., 1998). An excess of driver, at least 10-fold higher than the tracer, is prepared from the cell or tissue lacking the sequences of interest. The driver (also designed [-]) and tracer (bait or [+]) nucleic acid pools are hybridized, and only nucleotide sequences common to driver and tracer form hybrids. Driver-tracer hybrids are removed, and unhybridized sequences are utilized for successive rounds of subtraction for maximal removal of sequences that are common to both tracer and driver. The enriched unhybridized sequences are prepared for cloning, and the tracer-specific clones subsequently constitute the subtracted library.

There are at least four strategies applied for subtractive cloning for which both DNA and RNA can be used as the driver and tracer. However, each subtractive cloning strategy has associated advantages and drawbacks, which should be considered when designing the experiment. For example, it is important to consider the source and amount of material that will serve as the driver and tracer sequences, whether an amplification step is necessary, the kind of molecules that will serve as tracer (e.g., first-strand cDNA, ss-DNA or ds-DNA) and as driver (poly(A)+ RNA, RNA, ss-DNA or ds-DNA). Taking these points into consideration, the strategy of subtraction can be chosen from the usual schemes, categorized as the basic PCR-based technique, library-library hybridization and positive selection. With the basic subtractive cloning technique, cDNA serves as the tracer and mRNA as the driver. A large amount of tissue is required when performing this procedure, for which the subtractive hybridization can be performed only twice. With the PCR-based strategy, a small amount of starting material (ss-cDNA or ds-cDNA) can be used, and multiple rounds of subtraction are feasible. With library-library subtraction, collections of ss-cDNAs are used, and full-length sequences are obtained; however, the procedure is not easy to repeat. In positive subtractive selection, also known as the cohesive restriction sites method, an excess of [-] cDNA sequences is digested with restriction enzymes to produce blunt-ended fragments and then mixed with sonicated [+] cDNAs. After hybridization, only clonable cDNAs represent the tracer-specific sequences (Sagerströmet al., 1997). The literature describing the application of subtractive cloning for the investigation of differentially expressed genes is plentiful (for example, Lockyer et al., 2008; Zhang et al., 2008; Lefèvre and Murphy, 2009; Li et al., 2010; Matsumoto et al., 2011; Chengxiang et al., 2011; Liu et al., 2011; Padmanabhan et al., 2011), while only a few examples have been reported in the field of molecular toxinology (Baek et al., 2009; Baek and Lee, 2010; Morgenstern et al., 2011).

As mentioned before, another type of library typically used for studies of toxin genes is the genomic DNA library (Figure 2). As with the first step in library construction, the tissue is surgically excised from anesthetized animals (e.g., liver from venomous animals) or cells (e.g., blood), which are collected from the organism of interest without killing it, and high molecular weight genomic DNA is then purified. Genomic DNA inserts are then prepared by controlled digestion with rare cutting restriction enzymes, producing blunted-end DNA fragments, which are then linked to a selected vector. Vectors for cloning genomic DNA should have a high capacity to accommodate large fragments (in the range of 20 to 40 kb or more). These high-capacity vectors include, listed in order of capacity, cosmid (30-45 kb), bacteriophage P1 (70-100 kb), bacterial artificial chromosome, BACs (120-300 kb) and yeast artificial chromosomes (YACs; 0.25 to 1.2 Mb) (Ausubel et al., 1998, Sambrook and Russel, 2001).

Recombinant YACs are produced by digesting the vector with selected restriction enzymes and ligating the restriction-digested (and size-selected) genomic DNA into the left and right arms of the YAC vector. The YAC libraries are then transformed into yeast (*Saccharomyces cerevisiae*) and then utilized for the genetic screening of clones of interest. The YAC vectors contain restriction sites, a centromere (*CEN4*), an autonomously replicating sequence (*ARS1*), two selective markers, and telomeric sequences (*TEL*) at the vector terminus. Such features allow recombinant YACs to be autonomously replicated, segregate between daughter yeast cells and, due to their stability, integrate into host chromosome.

Bacterial artificial chromosomes are circular, double-stranded DNA, low copy number vectors derived from *E. coli* fertility (F) factor, a large plasmid responsible for carrying genes encoding proteins involved in replication, partition and conjugation. Genetic engineered BACs are commercially available, and manipulation of this vector for preparing genomic libraries is straightforward in comparison to YAC libraries. In such cases, restriction-digested genomic DNA fragments, selected by size, are ligated into linearized BAC vectors and then transformed into *E. coli* by electroporation. BAC libraries are then stored frozen until arrayed for clone screening by hybridization with oligonucleotide probes or PCR primers.

Cosmids are conventional plasmids that have been engineered to allow for cloning of large pieces of DNA. Cosmid vectors contain one or two selectable marker (e.g., neo^r and amp^r), a plasmid origin of replication (*ori*), restriction cloning sites, and one or two cohesive end sites (*cos*) from phage λ, which are essential for packaging of recombinant viral genome into infective phage particles. Restriction-digested genomic DNA and the vector are ligated, and the resultant linear concatenated recombinant molecules are *in vitro* packaged into phage heads. The infectious recombinant λphage injects the cosmid DNA into susceptible *E. coli* host cells, where the host's ligase covalently join the complementary sticky ends of the cosmid vector, thereby producing circular molecules that replicate as plasmids. Methods of screening genomic DNA cosmid libraries include hybridization of colonies plated on membranes (replica filters) and PCR-based techniques. As cosmids are λ phage-derived vectors that possess high cloning efficiency and high capacity for incorporating relatively large DNA fragments, they represent excellent vectors for the construction of complex genomic DNA libraries from venomous organisms (e.g., Nobuhisa et al., 1997; Beye et al., 1998; Fujimi et al., 2003; Rádis-Baptista et al., 2004). These libraries have been used for the identification of toxin genes and have led to reports describing their organization, comprehension of molecular evolution and diversification of toxin genes, as well as studies of comparisons of genomes and basic genetics. Genomic DNA libraries of venomous animals also constitute real archives of a particular biological resource, by which genomic information can be retrieved at any time regardless of how difficult it is to find and capture the organism in nature.

The first step for constructing a genomic DNA library involves the preparation of high molecular weight (HMW) genomic DNA. In the basic protocol, the tissues of interest are quickly frozen in liquid nitrogen and pulverized. The tissue powder is then transferred to a solution containing proteinase K and sodium dodecyl sulfate (SDS) and incubated until the cellular protein is completely degraded and the nucleic acid is released. The digest solution is extracted by phenol/chloroform/isoamyl alcohol, and the HMW genomic DNA is precipitated with ethanol, fished (to avoid sharing forces and DNA rupture), dried, and finally resuspended in aqueous buffer. Genomic DNA prepared in this manner is sufficient (size in the range of 100-150 kb) for cloning into phage and cosmids when precautions to minimize DNA degradation are taken (Ausubel et al., 1998; Sambrook and Russel, 2001). After purifying HMW genomic DNA, cosmid inserts are partially digested with rare cutting restriction enzymes (e.g., *Sal*I and *Spe*I), dephosphorylated with calf intestinal phosphatase (CIP), and fractionated by size using agarose gel electrophoresis. Standard agarose gels are convenient when separating genomic DNA fragments in the range of 0.5 to 25 kb, whereas pulsed-field agarose gels are required to resolve molecules ranging from 10 kb to 2 Mb. Partially digested, dephosphorylated genomic DNA is ligated to cosmid arm DNA using bacteriophage T4 DNA ligase, and the ligated genomic DNA-cosmid vector is then incubated with packaging extract, which contains all proteins required to produce infectious λ phage. *E. coli* host cells are transduced with the packaged cosmids and then spread onto selective agar plates. The number of λ phage plaques is counted, the efficiency of cloning is calculated, and the average size of the inserts is analyzed by restriction endonucleases or agarose gel electrophoresis. In general, the efficiency of cloning is in the order or 10^5 to 10^7 plaque forming units (pfu)/μg of genomic DNA. The genomic DNA cosmid library is titered by plaque dilution assay and amplified for storage. Cosmid libraries are stored in aliquots and preserved at -80°C in either 7% dimethyl sulfoxide or 15-20% sterile glycerol. Additional information regarding the cloning of large genomic DNA fragments for preparation of libraries can be obtained from dedicated companies such as Agilent Technologies, Inc. (Santa Clara, CA-U.S.A.). In addition to the construction of genomic DNA libraries, there are other simpler techniques for cloning toxin genes and examining their structural organization, such as direct clone by PCR-mediated amplification based on specific primers probes for cDNA sequences (Jiqun et al., 2004; Zhijian et al., 2006, Mao et al., 2007) and 'genome (DNA) walking' (Gendeh et al., 1997; Legros et al., 1997; Afifiyan et al., 1999; Jeyaseelan et al., 2003).

2.2 Molecular techniques of protein interaction for target discovery of animal toxins

Polypeptide toxins initially act on target cells and tissues by interacting with a particular biomolecule (e.g., membrane lipids, proteoglycans, ion-channels, glycoprotein and integrin receptors). Protein-protein interaction techniques designed to assess the association of proteins in mammalian cells can be used to analyze the mechanism by which toxins act to intoxicate organisms at the molecular (protein) level and to identify ligands with high specificity and selectivity for a given partner (receptor). These techniques used for the identification of protein-protein interactions include the fusion of a protein of interest with glutathione-S-transferase (GST fusions) and analysis by far-western or pull-down (Einarson & Orlinick, 2002), co-immunoprecipitation and mass spectrometry-based protein characterization (Adams et al., 2002), yeast and bacterial two hybrid selection systems (Serebriiski & Joung, 2002), or phage display (Goodyear & Silverman, 2002). These represent robust and promising techniques for the identification of therapeutic polypeptides due to the direct linkage between the molecular cloning of genes (genotype) and analysis of the biological activity of expressed peptides and proteins (phenotype). As described by E. Golemis (2002), using such techniques, three types of information may be obtained: (1) the identification of every possible set of interacting proteins for a target (protein of interest); (2) the physiological significance of such interactions once partner proteins have been identified; and (3) the validation of the physiological role of interacting proteins with the systematic use of modulators. By making use of this protein-protein interaction technique, target-driven identification of novel animal polypeptide toxins in the context of clinical application is enhanced. With this technique, instead of cloning a single DNA sequence into a vector, which produces the bait chimera, a library of thousands of cloned DNA is best suited to be scrutinized though rounds of target-driven (bait) specific binding. Consequently, methods used to examine protein-protein interactions are extremely promising for the screening of animal toxins in a customized approach, i.e., based on target-driven selection.

2.2.1 Phage display and its potential to interrogate animal toxin libraries

In phage display, or peptide display, the polypeptide library of interest is fused to capsid protein and expressed on the surface of a bacteriophage (or phage), which becomes 'decorated' with the recombinant coat protein and is therefore available for analysis of receptor and ligand binding. The "phage display" technology is a robust strategy used to identify and investigate highly specific protein-protein interactions and to find or model novel ligand receptors, as there is a direct physical linkage between the polypeptide expressed in the capsid (phenotype) and the genetic information (genotype) (Smith, 1985). The physical link between genetic information and recombinant protein expression in each single phage particle allows for iterative rounds of selection of clones with particular capabilities, i.e., panning, coupled with steps of amplification of a clone of interest or sub-libraries. Rare ligand-binding clones are enriched based their specificity and rescued from complex libraries (over 10^{10} recombinant phage particles) (Goodyear & Silverman, 2001). Polypeptide molecules of diverse types and properties have been successfully displayed on the surface of filamentous phage and selected by different means. As reported by Goodyear and Silverman (2002), these polypeptides include enzymes (McCafferty et al., 1991), antibodies (McCafferty et al., 1990; Burton & Barbas, 1993; Winter et al., 1994, Zheng et al., 2005), short peptides and protein fragments (Smith, 1985; Cwirla et al., 1990, Scott and Smith, 1990, Petersen et al., 1995), cytokines (Gram et al., 1993), antigens (Crameri et al., 1994), and extracellular receptor domains (Chiswell and McCafferty, 1992; Wu et al., 1995). The surface display technology is also a robust molecular tool for the purpose of investigating the proteomic complexity of protein-protein interactions mediated by natural or artificial ligands. Natural ligands displayed on the phage surface include allergen libraries (Crameri & Walter, 1999, Crameri et al., 1994), carbohydrate and polysaccharides (Deng et al., 1994; Yamamoto et al., 1999), RNA binding proteins (Danner & Belasco, 2001), fatty acids and membrane lipids (Qiu and Marcus, 1999; Gargir et al., 2002; Nakai et al., 2005) and ligands for Gal80p (Hertveldt et al., 2003). Moreover, phage display technology can be used for the identification and characterization of novel ligand receptor-binding interactions in studies of structure activity relationships (SARs) (Qiu et al., 1999; Li et al., 2009; Bannister et al., 2011; Garbelli et al., 2011).

In phage display (or peptide display) technology, segments of genes (selected or generated at random) or full-length cDNAs of interest are inserted in frame with a gene encoding one of the capsid proteins from lytic (T4, T7 or lambda phage) or non-lytic filamentous bacteriophages (M13, f1 or fd) (Goodyear & Silverman, 2002; Li et al., 2010). In the case of filamentous phage, the adopted strategy for phage display relies on the cloning of the nucleotide segment encoding the peptide sequence of interest in fusion with the genes for one of phage capsid proteins, typically protein III or VIII (pIII and pVIII, respectively). The choice of the display protein, whether it be pIII, pVIII or another bacteriophage coat protein, will influence the panning outcome in terms of either ligand avidity or binding affinity (Qiu et al., 1999; Fagerlund et al., 2008). The recombinant virus genome is then packed, and the library proteins are expressed as fusion (chimera) capsid proteins on the phage surface. The phage display library is typically screened by bait selection on immobilized supports. These supports include functionalized beads (chromatographic resins), multi-well plates or membranes, in which the bait protein is attached. As the partner protein (bait) can be localized in complex biological systems (cells, tissues and organs), the panning procedure can be performed *in vivo* with experimental animals as well as *in vitro* with isolated proteins, cells and tissues in culture (Michon et al., 2002; Kehoe and Kay, 2005; Valadon et al., 2006; Li et al., 2006; Zahid et al., 2010; Bahudhanapati et al., 2011; Kanki et al., 2011).

Although there are numerous advantages associated with this technique, one potential limitation of phage display is the difficulty of correctly expressing cDNA libraries fused to capsid phage proteins. Technically, nucleotide-coding sequences are cloned into the 5' end of capsid genes to produce virus particles decorated with heterologous proteins fused to the N-terminus of coat protein. However, the insertion of stop codons and unpredictable reading frame shifts in the fused gene constructs cause interference with coat protein expression and decrease the efficiency of polypeptide surface display. Strategies have been developed by several researchers to overcome such technical issues and improve phage display technology. For example, one direct alternative is to express library polypeptides at the carboxyl terminus of filamentous phage coat protein and avoid interruption of the translated chimera (Jespers et al., 1995; Hufton et al., 1999; Brunet et al., 2002). Using a similar strategy, fusion of cDNAs to the 5'-end of 10B gene of T7 phage allows for the expression of the cDNA phage display library at the C-terminus of the capsid protein (Danner and Belasco, 2001). Using a coupled version to produce chimeric cDNA libraries and coat proteins for display, Crameri and Suter (1993) prepared a phagemid vector in which cDNA libraries were cloned into the 3' end of a c-Fos leucine zipper domain gene segment to produce polypeptide libraries fused to the carboxyl terminus of c-Fos, and the C-jun leucine zipper domain is expressed in fusion with lambda phage pIII. As a result, during phage assembly, a c-Jun and c-Fos heterodimer is formed, and polypeptide libraries are displayed at the C-terminus of pIII. Another strategy used to display cDNA libraries, which was developed by Caberoy and collaborators (2009), employs a modified version of T7 phage display. In this version, referred to as the T7Bio3C vector, a cleaved motif of human rhinovirus 3C protease was fused to the C-terminus of the capsid 10B protein, two GS flexible linkers and a biotin tag. Such vectors accommodate cDNA libraries in all three possible reading frames, improving the recovery of recombinant full-length cDNAs.

Regardless of the polypeptide phage display library preparation method, panning (affinity selection) and clone enrichment represent critical aspects of the method. Although some technical problems may arise when cloning cDNA sequences for polypeptide display on bacteriophage capsids, they can be potentially solved by changing the strategies of DNA construction, and phage display is still an efficient, sensitive and indispensable method to investigate the diversity of peptide molecules in natural libraries expressed in the glands of venomous animals. Moreover, phage display is a powerful component from the arsenal of functional proteomics dedicated to elucidate protein-protein interactions, novel receptor-binding peptide ligands, and discovery of peptide drug leads, among other applications of protein chemistry. With such technical characteristics, the surface display technology can be easily adapted and automated to a array format of high-throughput screening, conditions that meet the productivity seen in genomic and proteomics approaches (Walter et al., 2001; Georgieva and Konthur, 2011) and applications in drug discovery and medicine (Sergeeva et al., 2006). Phage display platforms for cDNA cloning based on T7 phage biology are commercially available from Novagen/EMD4Biosciences (USA), by which technical literature can be consulted.

2.2.2 Yeast two hybrid system for investigating animal toxin interactions

Another powerful molecular biology technique suitable for characterization of specific protein-protein interactions and potentially useful for binding studies of animal toxins and their cellular or tissue targets is known as the yeast two-hybrid system (Y2H). The Y2H is an interaction trap technology designed for the analysis of the interaction between two known cloned proteins of interest or to screen a library for a gene encoding an unknown protein that interacts with a specific known target (the bait). The system works with chimeric proteins, which are prepared by fusing nucleotide sequences encoding a polypeptide of interest (e.g., a given animal toxin) with a DNA binding domain (DBD), resulting in the chimera DBD-toxin on one side, and the putative partner protein (or library) and transcriptional activation domain (AD), generating the AD-partner on the other side. The fused DNA sequences coding for both constructs are co-transformed into yeast (*Saccharomyces cerevisiae*), and the chimeric proteins are then expressed and transported to the nucleus. When two proteins interact, the DNA binding domain and transcriptional activation domain are brought together and then activate the transcription of the two reporter genes (Serebriiski and Joung, 2002). The DBD used is the LexA protein from bacteria that interacts in the nucleus with *lexA* operators located upstream of β-galactosidase and *LEU2* (or *HIS3*) genes (the first and second reporters of the system). Positive yeast two-hybrid clones with trapped interacting protein, which are plated on selective medium lacking leucine (or histidine) and containing X-Gal, are identified by the blue color and are selected for validation. A more detailed assessment of Y2H technology can be found in several articles published during the last two decades (Luban and Goff, 1995; Miller and Stagljar 2004; Lentze and Auerbach, 2008; Ratushny and Golemis, 2008; Suter et al., 2009; Brückner et al., 2009; Fields et al., 2009). Examples in the literature of yeast or bacterial two-hybrid systems used for toxin research are still rare. Most studies in which experiments of interaction trap with two-hybrid systems involve toxins from organisms other than venomous animals, such as plant and microorganisms. For example, in wheat, a necrotizing toxin produced by *P. tritici-repentis* (Ptr ToxA) was shown to interact with a chloroplast protein involved in endocytosis in both ToxA-sensitive and ToxA-resistant plant cultivars (Manning et al., 2007).

Using Y2H, Rádis-Baptista and collaborators (personal communication) analyzed the interaction of crotamine with neural protein targets. Crotamine is a low molecular weight cationic polypeptide found in the venom of the South American rattlesnake (*Crotalus durissus terrificus*) that, despite its toxic effect on animal tissues, has arisen as a promising template for drug development and biomedical applications (Kerkis et al., 2010; Radis-Baptista, 2011). When injected i.p. into mice, crotamine causes rapid (< 10 min) and specific hind limb paralysis. Thus, it was first hypothesized that a neural receptor was a target of crotamine, thereby triggering a toxic response. To address this hypothesis, a mouse nervous system cDNA library was constructed into a pB42AD plasmid vector, producing fusions between the transcriptional activation domain (AD) and library sequences, while the bait DBD-crotamine fusion was cloned into a pLexA plasmid. Both plasmids were obtained from the MATCHMAKER LexA Two-Hybrid System (Clontech Laboratories, Inc, Mountain View, CA-USA). After co-transformation of a suitable yeast strain (EGY48 [p8op-lacZ]) with the fusion plasmids, tests were made to ascertain that some protein partners interact. The preliminary results displayed that DBD-crotamine self-activated the expression of the β-galactosidase gene reporter. Initially, this mechanism was unclear, but later, Kerkis and collaborators (2004, 2010) proved that crotamine was able to enter eukaryotic cells and bind specifically to the chromosomes, thereby validating a portion of the data obtained from the analysis with Y2H and crotamine. A scheme illustrating the cloning strategy for Y2H screening is presented in Figure 3.

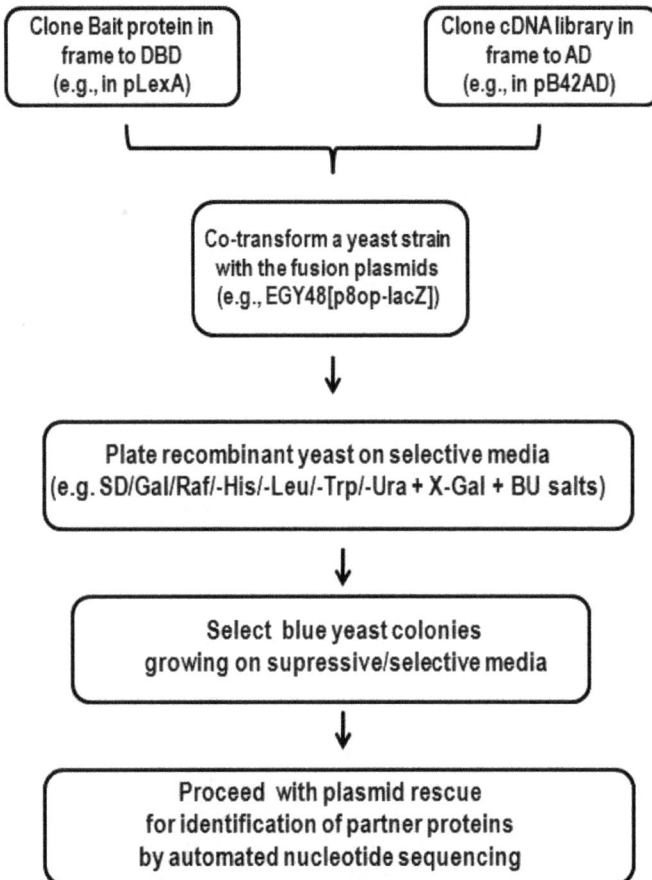

Fig. 3. Basic steps involved in the preparation of an interaction trap assay with Y2H.

The bacterial two-hybrid (B2H) system is analogous to the Y2H system , except that one polypeptide of interest (e.g., from a library) is linked to a subunit of the *E. coli* RNA polymerase (RNA pol), while the other protein of interest (for example, the bait) is fused to a DNA-binding domain, but the other. Similar to Y2H, when a protein-protein interaction does occur, the DNA-binding domain (in bait-DBD fusion) recruits a RNA pol moiety to a weak promoter in the host *E. coli*, and transcription of a reporter gene is activated, thereby indicating a positive interaction trap (Goodyear and Silverman, 2001; Dove and Hochschild, 2004).

2.2.3 *In vitro* display techniques for investigation of animal toxins

One concern regarding phage display and two-hybrid systems is the involvement of living organisms (bacteria, phage and yeast) in the process of library preparation and selection. Therefore, true *in vitro* selection technologies have been developed by which the number of molecules that can be handled are not limited by cellular transformation efficiencies and directed protein evolution can be achieved independently of successive rounds of randomization (Amstutz et al., 2001). Two of these techniques are the ribosome display and directed evolution coupled to cDNA display. Like phage display and Y2H (or B2H), a physical link between genotype (RNA and DNA) and phenotype (expressed/displayed protein) exists in ribosome and cDNA display. In ribosome display, non-covalent ternary complexes, consisting of mRNA, ribosome and nascent polypeptide, which can fold correctly while still attached to ribosomes, are formed, which demonstrates the coupling of genotype and phenotype (Hanes and Plückthum, 1997). DNA library coding for particular proteins of interest is transcribed *in vitro*. The mRNA is then purified and used for *in vitro* translation. Because the stop codon has been removed from the protein encoding sequences in the DNA library, the ribosome stalls at the 3' end of the mRNA during *in vitro* translation, giving rise to a ternary complex of mRNA, ribosome, and encoded protein. In general, the protein is able to fold correctly on the ribosome because a carboxyl-terminal spacer had been genetically fused to it, thus allowing the protein of interest to fold outside of the ribosomal tunnel. High concentrations of magnesium and low temperature further stabilize the ternary complex. These complexes, which are formed during *in vitro* translation, can directly be used to select for the properties of the displayed protein. After affinity selection and elution from a ligand (immobilized or in solution), the mRNA is purified, reverse-transcribed, and amplified by PCR. Following successive rounds of selection, which enriches the ligands at a rate of 100-1000-fold per cycle, the pooled DNA can be cloned in an expression vector for identification and large-scale preparation of selected ligand (Hanes and Plückthum, 1997; Amstutz et al., 2001; Schaffitzel et al., 2002). Ribosome display has been used for *in vitro* selection of biologically relevant macromolecules such as antigenic epitopes (Yau et al., 2003; Lee et al., 2004; Yang et al., 2007), cell-surface receptor modulators (Milovnik et al., 2009), and enzymes (Amstutz et al., 2006; Quinn et al., 2008).

cDNA display is a technology similar to ribosome display in which the ternary complexes are formed by the covalent coupling of mRNA, bearing puromycin at the 3' end via an oligonucleotide linker to the carboxyl terminus of nascent protein (Roberts and Szostak, 1997; Nemoto et al., 1997; Miyamoto-Sato et al., 2000). In cDNA display, as seen in ribosome display, the DNA library coding for particular polypeptides of interest is first transcribed *in vitro*. The mRNA is purified and ligated to the puromycin linker in the presence of T4 RNA ligase. The mRNA-puromycin linker is then translated in a cell-free system (e.g., reticulocyte lysate), and when the ribosome reaches the RNA-DNA junction, translation stops, and the puromycin moiety enters the peptidyl transferase site of the ribosome, thereby allowing for the formation of a covalent link between the puromycin linker and the nascent polypeptide. The covalent linked polypeptide and mRNA are rapidly purified from the ribosome by biotin-streptavidin capture, and cDNA is synthesized by reverse transcription. The purified complex, composed of the hybrid mRNA-cDNA and polypeptide, is then ready for affinity selection using the target molecule of interest (Yamaguchi et al., 2009).

In addition to affinity selection, both technologies of ribosome and cDNA display allow for *in vitro* directed protein evolution (Yanagida et al., 2010; Dreier and Plückthun, 2011). In each round of selection, conventional or error-prone PCR with non-proofreading *Taq* polymerase can introduce a number of mutations and consequently increase the diversity of nucleotide sequences (DNA library) and displayed proteins (Cadwell and Joyce, 1992; Schaffitzel and Plückthun, 2001). Recently, Naimuddin and collaborators (2011) applied cDNA display technology and directed protein evolution to engineer an elapidae snake three-finger toxin (3FTx) scaffold designed to identify modulators of interleukin-6 receptor (IL-6R). The three-fingers toxins are well conserved protein structures characterized by β-sheets and three protruding loops (loops I to III) and are slightly distinct among different snake toxins and responsible, to a certain degree, for diverse biological activity and toxicity (Endo and Tamyia, 1987; Kini and Doley, 2010). Based on the work by Naimuddin and colleagues (2011) and previous studies (Yamaguchi et al., 2009), they were able to generate a 3FTx library containing 1.2×10^{11} molecules by randomization of DNA sequences encoding all three loops of the *Micrurus corallinus* (coral snake) α-neurotoxin (MicTx3) as template and discover 3FTx-derived peptide ligand of interleukin-6 receptor. This reinforces the potential benefits of employing cDNA display for *in vitro* protein evolution and target-driven selection. Figure 4 depicts a schematic view of the process of cDNA display.

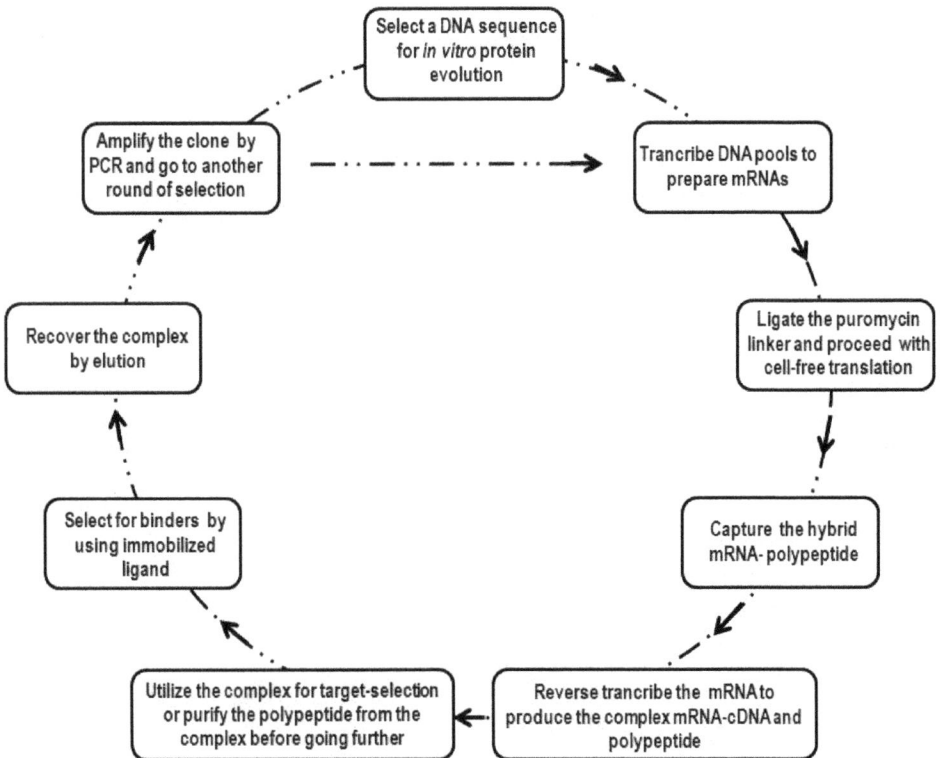

Fig. 4. Depiction of the reiterative steps involved in cDNA display and main procedure for generating libraries for selective screening.

2.3 Heterologous expression of toxin genes in eukaryotic and prokaryotic systems

Although numerous classes of polypeptide toxins identified are expressed in relatively minute amounts in the venom glands of several animals, milligrams of such bioactive molecules are required for accurate biochemical and pharmacological characterization. Recombinant DNA technology can be utilized to prepare vectors for cloning and expression of the toxin gene of interest. In most cases, the expression vectors are plasmids that contain promoters that direct the synthesis of large amounts of mRNA (cDNA), sequences that encode genetic traits that allow vector-containing cells to be selected and sequences that increase the efficiency with which the mRNA is translated (Ausubel et al., 1995). For recombinant expression of proteins and polypeptide toxins, researchers can make use of prokaryotic and eukaryotic systems. The most utilized prokaryotic host is the bacteria *E. coli,* and the expression is directed by T7 promoters inserted into the vectors, as originally developed by Studier and collaborators (Studier and Moffatt, 1986; Studier et al., 1990). In this T7 promoter-driven system, a relatively small amount of T7 RNA polymerase provided from a cloned copy of T7 gene 1 is sufficient to direct high-level transcription from a T7 promoter in a multicopy plasmid, thereby producing the recombinant protein in a short amount of time (< 3 h) and in a quantity higher than 50% of the total cell protein. To improve the solubility of recombinant protein produced in *E. coli* and to facilitate the downstream process of purification, vectors are available that allow for the expression of fusion proteins, such as maltose-binding protein, glutathione-S-transferase, hexa-histidine and thioredoxin fusions (Ausubel et al., 1995). Recombinant fusion proteins are easily purified by affinity chromatography, and their extra portion ('tags' and 'carrier' protein) can be chemically or enzymatically removed via cleavage signals present in the protein sequence (Sambrook and Russel, 2001). Overexpression of recombinant protein can be toxic and detrimental to the host bacterial cell; therefore insoluble intracellular aggregates, known as inclusion bodies, are often formed. These inclusion bodies are composed of almost pure unfolded protein, which can be properly refolded after disruption by different means (Marston 1986; Marston and Hartley, 1990). Despite the numerous parameters that have been tested to avoid IB formation, such as promoter strength, codon usage and gene dosage, and the temperature of induced expression (Martinez-Alonso et al., 2009), Garcia-Fruitos and collaborators (2005) have shown that overexpressed protein aggregation as inclusion bodies does not cause inactivation of enzymes and fluorescent proteins. Given this, when designing the experimental strategy, it is important to take into consideration not only the election of an appropriate lysis method but also the design of the necessary washing steps to isolate native protein and recover undisturbed active protein (Garcia-Fruitos, 2010). Prokaryotic expression systems are commercially available from several biotech companies, and their respective technical resources are easily assessed for additional information (e.g., EMD4Biosciences USA, Life Technologies/Invitrogen USA, GE Healthcare USA and New England Biolabs USA). A system that combines an insect virus (baculovirus) as a vector and cultured insect cells as a host has been utilized for the expression and production of heterologous protein. The baculovirus-insect cell system has proved to be an excellent choice for protein expression as it has several advantages, including the production the protein in high yield with the appropriate eukaryotic post-translational modifications (Luque and O'Reilly, 1999). The biological activity and similarity to native proteins offer a great advantage over conventional bacterial expression systems (Patterson et al., 1995). Vectors for expression of heterologous proteins in eukaryotic cells have also been developed, which allow for the production of recombinant polypeptides in mammalian-derived cells, such as African green monkey kidney fibroblast-like (COS) cells (Warren and Shields, 1984) and Chinese hamster ovary (CHO) cells (Cockett et al., 1990; Kaufman et al., 1991), budding yeast (*S. cerevisiae*), fission yeast (*Schizosaccharomyces pombe*) and methylotrophic yeast (*Pichia pastoris*) (Trueman, 1995; Gellissen and Hollenberg, 1997; Li et al., 2005; Takegawa et al., 2009). A study was performed analyzing the performance of the five principal eukaryotic expression systems, including the stable expression of transfected adherent CHO cells, transient expression in mammalian COS cells, and baculovirus expression in invertebrate insect cells (Geisse et al., 1996). Each expression system has advantages and disadvantages that should be considered when selecting a method to prepare recombinant proteins. The expected yield, time required for production of the desired protein, necessity of protein refold and post-translational modifications, possibility of scale-up, and costs are examples of what should be taken into account when expressing a biologically active protein. When dealing with animal toxin, one should remember that most toxin families contain members with three, four or five disulfide bonds, which confer to each toxic polypeptide high stability and structural rigidity– properties that are necessary for proteins secreted into the venom. Therefore, systems or conditions that allow disulfide bond formation are much more advantageous. The successful heterologous production of animal toxin using the systems mentioned above has been reported. For example, with *Pichia pastoris*, several functional snake venom proteins have been expressed, such as the cystatin-like cysteine-protease inhibitors from the elapidae *Austrelaps superbus* (Richards et al., 2010); a venom P-II metalloproteinase (Jerdonitin) composed of metalloproteinase and disintegrin domains from *T. jerdonii* venom (Zhu et al., 2010); a thrombin-like enzyme (gloshedobin) from the venom of *Gloydius shedaoensis* in fusion with HSP-70 (Yang et al., 2009); the disintegrin domain of a metalloprotease from the green pit viper *Trimeresurus albolabris* (Singhamatr and Rojnuckarin, 2007); a thrombin-like enzyme (Ancrod) from the venom of *Calloselasma rhodostoma* (Yu et al., 2007); and a fibrinogenolytic serine protease from *T. albolabris* venom (Muanpasitporn and Rojnuckarin, 2007). With bacterial expression systems, recent reports have described a disintegrin (r-mojastin 1) from the venom of the mohave rattlesnake, *Crotalus scutulatus scutulatus* (Sánchez et al., 2010); gloshedobin (separately fused at its N terminus with three fusion partners) (NusA, GST, and TrxA) (Jiang et al., 2010); C-type lectins (BML-1 and BML-2) from the venom of *Bungarus multicinctus* (Lin et al., 2007); a c-type lectin (BJcuL) from the *Bothrops jararacussu* venom (Kassab et al., 2004); eretrin (an active spider toxin with penile erectile function) from the venom of *Phoneutria nigriventer* (Torres et al., 2010); and Huwentoxin-I, a small neurotoxin (33 amino acid in length) from the venom of the Chinese bird spider *Ornithoctonus huwena* (Che et al., 2009). Scorpion toxins have been expressed in baculovirus for the purpose of insect control (Gershburg et al., 1998; Rajendra et al., 2006). Yonamine and collaborators (2009) have transiently expressed a thrombin-like enzyme (gyroxin) with esterase activity from the venom of the South American rattlesnake *C. durissus terrificus* using COS-7 cells. This article was the first to report the functional expression of a snake toxin in a mammalian expression system.

Such works provide guidance for researchers choosing a system and using a vector to express a toxin of interest for the characterization of a biological activity at the molecular level.

3. Conclusion

The scientific achievements that outline the biochemical basis of the transmission of genetic information culminated with a revolution in Biology with the advent of Molecular Biology, genetic engineering (or recombinant DNA technology) and revolutionized the field of Life Sciences. Influenced by genetic engineering, molecular techniques have been employed in fields as diverse as Environmental Science and Medicine, although all these fields work towards common goal of improving the quality of human life. The applications of recombinant DNA technology range from food analysis and process to clinical diagnostics and therapy. The holistic view that genome, transcriptome and the other associate 'omics' (encompassing metabolome, glycome, peptidome and pharmacogenomics) are responsible for the characteristics of a biological system and the status of an organism in conditions of good health or disease has radically influenced the way we are using chemicals, including medicines. In addition, new disciplines have emerged from the sequential 'revolutions' in the fields of Biology as well as Systems and Synthetic Biology. Methods that are based on the linkage between genotype (DNA, RNA and cDNA) and phenotype (expressed polypeptides), which are suitable for high-throughput evaluation, have become essential for yielding timely results for current challenges. Therefore, gene sequence libraries prepared and maintained in different formats (e.g., cDNA, genomic, and subtractive) are of paramount significance to understand cell function and to comprehend the entire organism. Moreover, focusing on healthy and wealthy humans, the society (and scientific community) may take advantage of the information contained in such libraries with the aim of producing better biotechnological goods and drugs. Moreover, genotype-linked phenotype-based libraries, such as phage/surface display, yeast and bacterial two-hybrid systems, and ribosome and cDNA display offer unprecedented technical capabilities to rapidly identify specific target-binding ligands with potential drug applications. Because polypeptides are involved in such interaction trap technologies, the gene (mRNA) of interest can be sub-cloned into a given expression vector, and the (therapeutic) peptide and protein of interest can be prepared in sufficient amounts required for clinical research and medical practice. In this context, animal toxins found in nature as complex mixtures in the venom of numerous species of organisms inhabiting distinct geographical location and belonging to various biomes, constitute the ultimate biological resources for drug discovery and development. Families of animal toxins contain proteins that evolved for millions of years as a result of positive Darwinian selection (accelerated evolution), thereby generating conserved protein scaffolds with distinct biological and pharmacological activities. In this way, a dozen animal toxins have been converted either into drugs and diagnostic tools or have served as templates for drug design. Consequently, the combined use of refined and robust molecular techniques, designed to assess the biotechnological potential of venom polypeptides and their precursors (genes and mRNA), will offer priceless rewards concerning scientific endeavors in toxin research. Thus far, as presented in this chapter, the application of molecular cloning techniques in toxin studies can link basic research of natural compounds to the applied research from pharmaceutical industries, which might be ultimately translated into the practical scientific answers for an inquiring biological world.

4. Acknowledgments

I am grateful to the National Brazilian Council for the Scientific and Technological Development (CNPq), the Ministry of Science and Technology, and the Federal Government of Brazil for administrative and financial support. G.R.-B. is a member of the Scientific Committee of CNPq.

5. References

Adams, P.D.; Seeholzer, S. & Ohh, M. (2002). Identification of associated proteins by coimmunoprecipitation. In: E. Golemis. *Protein-Protein Interactions. A Molecular Cloning Manual*, CSHL Press, New York.

Afifiyan, F.; Armugam, A.; Tan, C.H.; Gopalakrishnakone, P. & Jeyaseelan, K. (1999). Postsynaptic alpha-neurotoxin gene of the spitting cobra, *Naja naja sputatrix*: structure, organization, and phylogenetic analysis. *Genome Res.* 9(3):259-66.

Amstutz, P.; Forrer, P.; Zahnd, C. & Plückthun, A. (2001). *In vitro* display technologies: novel developments and applications. *Curr Opin Biotechnol.* 12(4):400-5.

Amstutz, P.; Koch, H.; Binz, H.K.; Deuber, S.A. & Plückthun, A. (2006). Rapid selection of specific MAP kinase-binders from designed ankyrin repeat protein libraries. *Protein Eng Des Sel.* 19(5):219-29.

Ausubel, F.M., Brent, R., Kingston, R.E., Moore, D.D., Seidman, J.G., Smith, J.A., Struhl, K. (Eds.) (1998). *Current Protocols in Molecular Biology*, Wiley, New York.

Bader, J.S.(2011). Grand network convergence. *Genome Biol.* 12(6):306.

Baek, J.H. & Lee, S.H. (2010). Differential gene expression profiles in the venom gland/sac of *Eumenes pomiformis* (Hymenoptera: Eumenidae). *Toxicon.* 55(6):1147-56.

Baek, J.H.; Woo, T.H.; Kim, C.B.; Park, J.H.; Kim, H. Lee, S. & Lee, S.H. (2009). Differential gene expression profiles in the venom gland/sac of *Orancistrocerus drewseni* (Hymenoptera: Eumenidae). *Arch Insect Biochem Physiol.* 71(4):205-22.

Bahudhanapati, H.; Zhang, Y.; Sidhu, S.S. & Brew, K. (2011). Phage display of tissue inhibitor of metalloproteinases-2 (TIMP-2): Identification of selective inhibitors of collagenase-1 (MMP-1). J Biol Chem. 2011 Jun 29. Epub ahead of print.

Baltimore D. RNA-dependent DNA polymerase in virions of RNA tumour viruses. *Nature.* 1970 Jun 27;226(5252):1209-11.

Bannister, D.; Popovic, B.; Sridharan, S.; Giannotta, F.; Filée, P.; Yilmaz, N. & Minter, R. (2011). Epitope mapping and key amino acid identification of anti-CD22 immunotoxin CAT-8015 using hybrid β-lactamase display. *Protein Eng Des Sel.* 24(4):351-60.

Becker, S. & Terlau H. (2008). Toxins from cone snails: properties, applications and biotechnological production. *Appl. Microbiol. Biotechnol.* 79:1-9.

Beye, M.; Poch, A.; Burgtorf, C.; Moritz, R.F. & Lehrach, H. (1998). A gridded genomic library of the honeybee (*Apis mellifera*): a reference library system for basic and comparative genetic studies of a hymenopteran genome. *Genomics.* 49(2):317-20.

Birrell, G.W.; Earl, S.T.; Wallis, T.P.; Masci, P.P.; de Jersey J.; Gorman, J.J. & Lavin, M.F. (2007). The diversity of bioactive proteins in Australian snake venoms. Mol. Cell. *Proteomics* 6: 973-86.

Bishop, J. M.; Deng, C.T.; Faras, A. J.; Goodman, H. M.; Levinson, W. E.; Taylor, J. M. & Varmus., H. E. (1973). Transcription of the *Rous sarcoma* virus genome by RNA-directed DNA polymerase. In C. F. Fox (ed.), *Virus research, proceedings of second ICN-UCLA symposium.*

Blackburn, E.H. (2005). Telomeres and telomerase: their mechanisms of action and the effects of altering their functions. *FEBS Lett.* 579(4):859-62

Bringans, S.; Eriksen, S.; Kendrick, T.; Gopalakrishnakone, P.; Livk, A.; Lock, R. & Lipscombe, R. (2008). Proteomic analysis of the venom of *Heterometrus longimanus* (Asian black scorpion). *Proteomics* 8:1081-96.

Brückner, A.; Polge, C.; Lentze, N.; Auerbach, D. & Schlattner, U. (2009) Yeast two-hybrid, a powerful tool for systems biology. *Int J Mol Sci.* 10(6):2763-88.

Brunet, E.; Chauvin, C.; Choumet, V. & Jestin, J.L. (2002). A novel strategy for the functional cloning of enzymes using filamentous phage display: the case of nucleotidyl transferases. *Nucleic Acids Res.* 30(9):e40.

Burton, D.R. & Barbas, C.F. (1993). Human antibodies to HIV-1 by recombinant DNA methods. *Chem. Immunol.* 56:112-126.

Caberoy, N.B.; Zhou, Y.; Alvarado, G.; Fan, X. & Li, W. (2009). Efficient identification of phosphatidylserine-binding proteins by ORF phage display. *Biochem Biophys Res Commun.* 386(1):197-201.

Cadwell, R.C. & Joyce, G.F. (1992). Randomization of genes by PCR mutagenesis. *PCR Methods Appl.* 2(1):28-33.

Calvete, J.J.; Juárez, P. & Sanz, L. (2007). Snake venomics. Strategy and applications. *J Mass Spectrom.* 42(11):1405-14

Calvete, J.J.; Sanz, L.; Angulo, Y.; Lomonte, B. & Gutiérrez, J.M. (2009). Venoms, venomics, antivenomics. *FEBS Lett.* 583(11):1736-43.

Casewell, N.R.; Harrison, R.A.; Wüster, W. & Wagstaff, S.C. (2009). Comparative venom gland transcriptome surveys of the saw-scaled vipers (Viperidae: Echis) reveal substantial intra-family gene diversity and novel venom transcripts. *BMC Genomics.* 10:564.

Che, N.; Wang, L.; Gao, Y. & An, C. (2009). Soluble expression and one-step purification of a neurotoxin Huwentoxin-I in *Escherichia coli. Protein Expr Purif.* 65(2):154-9.

Chen, J.; Zhao, L.; Jiang, L.; Meng, E.; Zhang, Y.; Xiong, X. & Liang, S. (2008). Transcriptome analysis revealed novel possible venom components and cellular processes of the tarantula *Chilobrachys jingzhao* venom gland. *Toxicon.* 52(7):794-806.

Chengxiang, H.; Guangxing, Q.; Ting, L.; Xinglin, M.; Rui, Z.; Pan, Z.; Zhongyuan, S. & Xijie G. (2011). Differential gene expression in silkworm in response to *Beauveria bassiana* infection. *Gene.* Jun 1. Epub ahead of print.

Cherniack, E.P. (2010). Bugs as drugs, Part 1: Insects: the "new" alternative medicine for the 21st century? *Altern Med Rev.* 15(2):124-35.

Cherniack, E.P. (2011). Bugs as drugs, part two: worms, leeches, scorpions, snails, ticks, centipedes, and spiders. *Altern Med Rev.* 16(1):50-8.

Chiswell, D.J. & McCafferty, J. (1992). Phage antibodies: will new 'coliclonal' antibodies replace monoclonal antibodies? *Trends Biotechnol.* 10(3):80-4

Chomczynski, P. & Sacchi, N. (1987). Single-step method of RNA isolation by acid guanidinium thiocyanate-phenol-chlorofom extraction. *Anal. Biochem.* 162:156-159.

Cockett, M.I.; Bebbington, C.R. & Yarranton, G.T. (1990). High level expression of tissue inhibitor of metalloproteinases in Chinese hamster ovary cells using glutamine synthetase gene amplification. *Biotechnology (NY).* 8(7):662-7.

Crameri, R. & Suter, M. (1993). Display of biologically active proteins on the surface of filamentous phages: a cDNA cloning system for selection of functional gene products linked to the genetic information responsible for their production. *Gene*. 137(1):69-75.

Crameri, R. & Walter, G. (1999). Selective enrichment and high-throughput screening of phage surface-displayed cDNA libraries from complex allergenic systems. *Comb Chem High Throughput Screen*. 2(2):63-72

Crameri, R.; Jaussi, R.; Menz, G. & Blaser, K. (1994). Display of expression products of cDNA libraries on phage surfaces. A versatile screening system for selective isolation of genes by specific gene-product/ligand interaction. *Eur J Biochem*. 226(1):53-8.

Crick, F. (1970). Central dogma of molecular biology. *Nature*. 227(5258):561-3.

Cushman, D.W. & Ondetti, M.A. (1991). History of the design of captopril and related inhibitors of angiotensin converting enzyme. *Hypertension*. 17(4):589-92.

Cwirla, S. E.; Peters, E. A.; Barrett, R. W. & Dower, W. J. (1990). Peptides on phage: a vast library os peptides for identifying ligands. *Proc. Natl Acad. Sci. USA* 87:6378-6382.

Danner, S. & Belasco, J.G. (2001). T7 phage display: a novel genetic selection system for cloning RNA-binding proteins from cDNA libraries. *Proc Natl Acad Sci U S A*. 98(23):12954-9.

Deng, S.J.; MacKenzie, C.R.; Sadowska, J.; Michniewicz, J.; Young, N.M.; Bundle, D.R. & Narang, S.A. (1994). Selection of antibody single-chain variable fragments with improved carbohydrate binding by phage display. *J Biol Chem*. 269(13):9533-8.

Dove, S.L. & Hochschild, A. (2004). A bacterial two-hybrid system based on transcription activation. *Methods Mol Biol*. 261:231-46.

Dreier, B. & Plückthun, A. (2011). Ribosome display: a technology for selecting and evolving proteins from large libraries. *Methods Mol Biol*. 687:283-306.

Durban, J.; Juárez, P.; Angulo, Y.; Lomonte, B.; Flores-Diaz, M.; Alape-Girón, A.; Sasa, M.; Sanz, L.; Gutiérrez, J.M.; Dopazo, J.; Conesa, A. & Calvete, J.J. (2011). Profiling the venom gland transcriptomes of Costa Rican snakes by 454 pyrosequencing. *BMC Genomics*. 12:259.

Einarson, M.B. & Orlinick, J.R. (2002). Identification of protein-protein interaction with glutathione-S-transferase fusion proteins. In: E. Golemis. *Protein-Protein Interactions. A Molecular Cloning Manual*, CSHL Press, New York.

Endo, T. & Tamiya, N. (1987). Current view on the structure-function relationship of postsynaptic neurotoxins from snake venoms. *Pharmacol Ther*. 34(3):403-51.

Escoubas, P. & King, G.F. (2009). Venomics as a drug discovery platform. *Expert Rev Proteomics*. 6(3):221-4.

Escoubas, P.; Quinton, L. & Nicholson, G.M. (2008). Venomics: unravelling the complexity of animal venoms with mass spectrometry. *J. Mass. Spectrom*. 43(3):279-95.

Estrada, G.; Villegas, E. & Corzo, G. (2007). Spider venoms: a rich source of acylpolyamines and peptides as new leads for CNS drugs. *Nat. Prod. Rep*. 24:145-61.

Fagerlund, A.; Myrset, A.H. & Kulseth M.A. (2008). Construction and characterization of a 9-mer phage display pVIII-library with regulated peptide density. *Appl Microbiol Biotechnol*. 80(5):925-36.

Fernandes-Pedrosa, M. F.; Junqueira-de-Azevedo, I.; Gonçalves-de-Andrade, R.M.; Kobashi, L.S.; Almeida, D.D.; Ho, P.L. & Tambourgi, D.V. (2008). Transcriptome analysis of

Loxosceles laeta (Araneae, Sicariidae) spider venomous gland using expressed sequence tags. *BMC Genomics.* 9:279.

Ferreira, S.H. (1965). A Bradykinin-potentiating factor (BPF) present in the venom of *Bothrops jararaca. Br J Pharmacol Chemother.* 24:163-9.

Ferreira, S.H. (1985). History of the development of inhibitors of angiotensin I conversion. *Drugs.* 30 Suppl 1:1-5.

Fields, S. (2009). Interactive learning: lessons from two hybrids over two decades. *Proteomics.* 9(23):5209-13

Fire, A.Z. (2007). Gene silencing by double-stranded RNA. *Cell Death Differ.* 14(12):1998-2012.

Fox, J.W. & Serrano, S.M. (2007). Approaching the golden age of natural product pharmaceuticals from venom libraries: an overview of toxins and toxin-derivatives currently involved in therapeutic or diagnostic applications. *Curr Pharm Des.* 13(28):2927-34

Fry, B.G. (2005). From genome to "venome": molecular origin and evolution of the snake venom proteome inferred from phylogenetic analysis of toxin sequences and related body proteins. *Genome Res* 15(3):403-420.

Fry, B.G.; Vidal, N.; van der Weerd, L.; Kochva, E. & Renjifo, C. (2009). Evolution and diversification of the Toxicofera reptile venom system. *J Proteomics* 72(2):127-136.

Fujimi, T.J.; Nakajyo, T.; Nishimura, E.; Ogura, E.; Tsuchiya, T. & Tamiya, T. (2003). Molecular evolution and diversification of snake toxin genes, revealed by analysis of intron sequences. *Gene.* 313:111-8.

Garbelli, A.; Beermann, S.; Di Cicco, G.; Dietrich, U. & Maga, G. (2011). A Motif Unique to the Human Dead-Box Protein DDX3 Is Important for Nucleic Acid Binding, ATP Hydrolysis, RNA/DNA Unwinding and HIV-1 Replication. *PLoS One.* 6(5):e19810.

García-Fruitós, E. (2010). Inclusion bodies: a new concept. *Microb Cell Fact.* 9:80.

García-Fruitós, E.; Gonzalez-Montalban, N.; Morell. M.; Vera, A.; Ferraz, R.M.; Aris, A.; et al. (2005). Aggregation as bacterial inclusion bodies does not imply inactivation of enzymes and fluorescent proteins. *Microb Cell Fact* 4:27.

Gargir, A.; Ofek, I.; Meron-Sudai, S.; Tanamy, M.G.; Kabouridis, P.S. & Nissim, A. (2002). Single chain antibodies specific for fatty acids derived from a semi-synthetic phage display library. *Biochim Biophys Acta.* 1569(1-3):167-73.

Geisse, S.; Gram, H.; Kleuser, B.; Kocher, H.P. (1996). Eukaryotic expression systems: a comparison. *Protein Expr Purif.* 8(3):271-82.

Gellissen, G. & Hollenberg, C.P. (1997). Application of yeasts in gene expression studies: a comparison of *Saccharomyces cerevisiae, Hansenula polymorpha* and *Kluyveromyces lactis* - a review. *Gene.* 190(1):87-97

Gendeh, G.S.; Chung, M.C. & Jeyaseelan, K. (1997). Genomic structure of a potassium channel toxin from *Heteractis magnifica. FEBS Lett.* 418(1-2):183-8.

Georgieva, D.; Ohler, M.; Seifert, J.; von Bergen, M.; Arni, R.K.; Genov, N. & Betzel, C. (2010). Snake venomic of *Crotalus durissus terrificus*--correlation with pharmacological activities. *J Proteome Res.* 9(5):2302-16.

Georgieva, Y. & Konthur, Z. (2011). Design and screening of M13 phage display cDNA libraries. *Molecules.* 16(2):1667-81.

Gershburg, E.; Stockholm, D.; Froy, O.; Rashi, S.; Gurevitz, M. & Chejanovsky, N. (1998). Baculovirus-mediated expression of a scorpion depressant toxin improves the insecticidal efficacy achieved with excitatory toxins. *FEBS Lett.* 422(2):132-6.

Golemis, E. (2002). Toward and understanding of protein interactions. In: E. Golemis. *Protein-Protein Interactions. A Molecular Cloning Manual,* CSHL Press, New York.

Goodyear, C.S. & G.J. Silverman. (2002). Phage-display approached for the study of protein-protein interactions. In: E. Golemis. *Protein-Protein Interactions. A Molecular Cloning Manual,* CSHL Press, New York.

Gowd, K.H.; Dewan, K.K.; Iengar, P.; Krishnan, K.S. & Balaram, P. (2008).Probing peptide libraries from Conus achatinus using mass spectrometry and cDNA sequencing: identification of delta and omega-conotoxins. *J Mass Spectrom.* 43(6):791-805.

Gram, H.; Strittmatter, U.; Lorenz, M.; Glück, D. & Zenke, G. (1993). Phage display as a rapid gene expression system: production of bioactive cytokine-phage and generation of neutralizing monoclonal antibodies. *J Immunol Methods.* 161(2):169-76.

Gremski, L.H.; da Silveira, R.B.; Chaim, O.M.; Probst, C.M.; Ferrer, V.P.; Nowatzki, J.; Weinschutz, H.C.; Madeira, H.M.; Gremski, W.; Nader, H.B.; Senff-Ribeiro, A. & Veiga, S.S. (2010). A novel expression profile of the *Loxosceles intermedia* spider venomous gland revealed by transcriptome analysis. *Mol Biosyst.* 6(12):2403-16.

Hanes, J. & Plückthun, A. (1997). *In vitro* selection and evolution of functional proteins by using ribosome display. *Proc Natl Acad Sci U S A.* 94(10):4937-42.

Hertveldt, K.; Dechassa, M.L.; Robben, J. & Volckaert, G. (2003). Identification of Gal80p-interacting proteins by *Saccharomyces cerevisiae* whole genome phage display. *Gene.* 307:141-9.

Hodgson, W.C. & Isbister, G.K. (2009). The application of toxins and venoms to cardiovascular drug discovery. *Curr Opin Pharmacol.* 9(2):173-6.

Hu, H.; Bandyopadhyay, P.K.; Olivera, B.M. & Yandell, M. (2011). Characterization of the *Conus bullatus* genome and its venom-duct transcriptome. *BMC Genomics.* 12:60.

Hufton, S.E.; Moerkerk, P.T.; Meulemans, E.V.; de Bruïne, A.; Arends, J.W. & Hoogenboom H.R. (1999). Phage display of cDNA repertoires: the pVI display system and its applications for the selection of immunogenic ligands. *J Immunol Methods.* 231(1-2):39-51.

Israel, D.I. (1993). A PCR-based method for high stringency screening of DNA libraries. *Nucl. Ac. Res.* 21:2627-2631.

Israel, D.I. (1995). A PCR-based method for screening DNA libraries. In: Diffenbach, C.W., Dveksler, G.S. (Eds.), *PCR Primer. A Laboratory Manual.* Cold Spring Harbor Laboratory Press, New York.

Jacob, F. & Monod, J. (1961). Genetic regulatory mechanisms in the synthesis of proteins, *J. Mol. Biol.* 3:318–356.

Jespers, L.S.; Messens, J.H.; de Keyser, A.; Eeckhout, D.; Van dB., I.; Gansemans, Y.G.; Lauwereys, M.J.; Vlasuk, G.P. & Stanssens, P.E. (1995). Surface expression and ligand-based selection of cDNAs fused to filamentous phage gene VI. *Biotechnology (NY)* 13:378–382

Jeyaseelan, K.; Poh, S.L.; Nair, R. & Armugam, A. (2003). Structurally conserved alpha-neurotoxin genes encode functionally diverse proteins in the venom of *Naja sputatrix. FEBS Lett.* 553(3):333-41.

Jiang, L.; Zhang, D.; Zhang, Y.; Peng. L.; Chen, J. & Liang, S. (2010). Venomics of the spider *Ornithoctonus huwena* based on transcriptomic versus proteomic analysis. *Comp Biochem Physiol Part D Genomics Proteomics.* 5(2):81-8

Jiang, X.; Xu, J. & Yang, Q. (2010). Soluble expression, purification, and characterization of *Gloydius shedaoensis* venom gloshedobin in *Escherichia coli* by using fusion partners. *Appl Microbiol Biotechnol.* 85(3):635-42.

Jiang, Y.; Li, Y.; Lee, W.; Xu, X.; Zhang, Y.; Zhao, R.; Zhang, Y. & Wang, W. (2011). Venom gland transcriptomes of two elapid snakes (*Bungarus multicinctus* and *Naja atra*) and evolution of toxin genes. *BMC Genomics.* 12:1.

Jiqun, S.; Xiuling, X.; Zhijian, C.; Wanhong, L.; Yingliang, W.; Shunyi, Z.; Xianchun, Z.; Dahe, J.; Xin, M.; Hui, L.; Wenxin, L. & Teng, W. (2004). Molecular cloning, genomic organization and functional characterization of a new short-chain potassium channel toxin-like peptide BmTxKS4 from *Buthus martensii* Karsch(BmK). *J Biochem Mol Toxicol.* 18(4):187-95.

Kanki, S.; Jaalouk, D.E.; Lee, S.; Yu, A.Y.; Gannon, J. & Lee, R.T. (2011). Identification of targeting peptides for ischemic myocardium by in vivo phage display. *J Mol Cell Cardiol.* 50(5):841-8.

Kassab, B.H.; de Carvalho, D.D.; Oliveira, M.A.; Rádis-Baptista, G.; Pereira, G.A. & Novello, J.C. (2004). Cloning, expression, and structural analysis of recombinant BJcuL, a c-type lectin from the *Bothrops jararacussu* snake venom. *Protein Expr Purif.* 35(2):344-52.

Kaufman, R.J.; Davies, M.V.; Wasley, L.C. & Michnick, D. (1991). Improved vectors for stable expression of foreign genes in mammalian cells by use of the untranslated leader sequence from EMC virus. *Nucleic Acids Res.* 19(16):4485-90.

Kehoe, J.W. & Kay, B.K. (2005). Filamentous phage display in the new millennium. *Chem Rev.* 105(11):4056-72.

Kerkis, A.; Kerkis, I.; Rádis-Baptista, G.; Oliveira, E.B.; Vianna-Morgante, A.M.; Pereira, L.V. & Yamane, T. (2004). Crotamine is a novel cell-penetrating protein from the venom of rattlesnake *Crotalus durissus terrificus*. *FASEB J.* 18(12):1407-9.

Kerkis, I.; Silva, F.S.; Pereira, A.; Kerkis, A. & Rádis-Baptista, G. (2010). Biological versatility of crotamine - a cationic peptide from the venom of a South American rattlesnake. *Expert Opin Investig Drugs.* 19(12):1515-25.

Kini, R.M. & Doley, R. (2010). Structure, function and evolution of three-finger toxins: mini proteins with multiple targets. *Toxicon.* 56(6):855-67.

Kini, R.M. (2006). Anticoagulant proteins from snake venoms: structure, function and mechanism. *Biochem J.* 397(3):377-87

Kodzius, R.; Rhyner, C.; Konthur, Z.; Buczek, D.; Lehrach, H.; Walter, G. & Crameri, R. (2003). Rapid identification of allergen-encoding cDNA clones by phage display and high-density arrays. *Comb Chem High Throughput Screen.* 6(2):147-54.

Kordis, D. & Gubensek, F. (2000). Adaptive evolution of animal toxin multigene families. *Gene* 261(1):43-52.

Lander, E.S. & Weinberg, R.A. (2002). Genomics: Journal to the center of Biology. *In*: Amato, I. Science Pathways of Discovery, John Wiley & Sons, Inc., New York. p. 57-72.

Lee, M.S.; Kwon, M.H.; Kim, K.H.; Shin, H.J.; Park, S. & Kim, H.I. (2004). Selection of scFvs specific for HBV DNA polymerase using ribosome display. *J Immunol Methods.* 284(1-2):147-57.

Lefèvre, P.L. & Murphy, B.D. (2009). Differential gene expression in the uterus and blastocyst during the reactivation of embryo development in a model of delayed implantation. *Methods Mol Biol.* 550:11-61

Legros, C.; Bougis, P.E. & Martin-Eauclaire, M.F. (1997). Genomic organization of the KTX2 gene, encoding a 'short' scorpion toxin active on K+ channels. *FEBS Lett.* 402(1):45-9.

Lentze, N. & Auerbach, D. (2008). The yeast two-hybrid system and its role in drug discovery. *Expert Opin Ther Targets.* 12(4):505-15.

Lewis, R.J. & Garcia, M.L. (2003). Therapeutic potential of venom peptides. *Nat Rev Drug Discov.* 2(10):790-802.

Li, B.; Xi, H.; Diehl, L.; Lee, W.P.; Sturgeon, L.; Chinn, J.; Deforge, L.; Kelley, R.F.; Wiesmann, C.; van Lookeren Campagne, M. & Sidhu, S.S. (2009). Improving therapeutic efficacy of a complement receptor by structure-based affinity maturation. *J Biol Chem.* 284(51):35605-11.

Li, J.X.; Chen, R.; Fu, C.L.; Nie, J.H. & Tong, J. (2010). Screening of differential expressive genes in murine cells following radon exposure. *J Toxicol Environ Health A.* 73(7):499-506.

Li, P.; Anumanthan, A.; Gao, X.G.; Ilangovan, K.; Suzara, V.V.; Düzgüneş, N. & Renugopalakrishnan, V. (2007). Expression of recombinant proteins in *Pichia pastoris. Appl Biochem Biotechnol.* 142(2):105-24.

Li, W. & Caberoy, N.B. (2010). New perspective for phage display as an efficient and versatile technology of functional proteomics. *Appl Microbiol Biotechnol.* 85(4):909-19.

Lin, L.P.; Lin, Q. & Wang, Y.Q. (2007). Cloning, expression and characterization of two C-type lectins from the venom gland of *Bungarus multicinctus. Toxicon.* 50(3):411-9.

Liu, H.P.; Chen, R.Y.; Zhang, Q.X.; Peng, H. & Wang, K.J. (2011). Differential gene expression profile from haematopoietic tissue stem cells of red claw crayfish, *Cherax quadricarinatus,* in response to WSSV infection. *Dev Comp Immunol.* 35(7):716-24.

Lockyer, A.E.; Spinks, J.; Kane, R.A.; Hoffmann, K.F.; Fitzpatrick, J.M.; Rollinson, D.; Noble, L.R. & Jones, C.S. (2008). *Biomphalaria glabrata* transcriptome: cDNA microarray profiling identifies resistant- and susceptible-specific gene expression in haemocytes from snail strains exposed to Schistosoma mansoni. *BMC Genomics.* 9:634.

Luban, J. & Goff, S.P. (1995). The yeast two-hybrid system for studying protein-protein interactions. *Curr Opin Biotechnol.* 6(1):59-64.

Luque, T. & O'Reilly, D.R. (1999). Generation of baculovirus expression vectors. *Mol Biotechnol.* 13(2):153-63.

Ma, Y.; Zhao, R.; He, Y.; Li, S.; Liu, J.; Wu, Y.; Cao, Z. & Li, W. (2009). Transcriptome analysis of the venom gland of the scorpion *Scorpiops jendeki*: implication for the evolution of the scorpion venom arsenal. *BMC Genomics.* 10:290.

Ma, Y.; Zhao, Y.; Zhao, R.; Zhang, W.; He, Y.; Wu, Y.; Cao, Z.; Guo, L. & Li, W. (2010). Molecular diversity of toxic components from the scorpion *Heterometrus petersii* venom revealed by proteomic and transcriptome analysis. *Proteomics.* 10(13):2471-85.

Manning, V.A.; Hardison, L.K. & Ciuffetti, L.M. (2007). Ptr ToxA interacts with a chloroplast-localized protein. *Mol Plant Microbe Interact.* 20(2):168-77.

Mao, X.; Cao, Z.; Yin, S.; Ma, Y.; Wu, Y. & Li, W. (2007). Cloning and characterization of BmK86, a novel K⁺-channel blocker from scorpion venom. *Biochem Biophys Res Commun.* 360(4):728-34.

Markland, F.S. (1998). Snake venoms and the hemostatic system. Toxicon. 36(12):1749-800.

Marston, F.A. & Hartley, D.L. (1990). Solubilization of protein aggregates. *Methods Enzymol.* 182:264-76.

Marston, F.A. (1986). The purification of eukaryotic polypeptides synthesized in *Escherichia coli. Biochem J.* 240(1):1-12.

Martinez-Alonso, M.; Gonzalez-Montalban, N.; Garcia-Fruitos, E. & Villaverde A (2009). Learning about protein solubility from bacterial inclusion bodies. *Microb Cell Fact* 8:4.

Masek, T.; Vopalensky, V.; Suchomelova, P. & Pospisek, M. (2005). Denaturing RNA electrophoresis in TAE agarose gels. *Anal Biochem.* 336(1):46-50.

Matsumoto, T.; Ishizaki, S. & Nagashima Y. (2011). Differential gene expression profile in the liver of the marine puffer fish *Takifugu rubripes* induced by intramuscular administration of tetrodotoxin. *Toxicon.* 57(2):304-10.

McCafferty, J.; Griffiths, A.D.; Winter, G. & Chiswell, D.J. (1990). Phage antibodies: filamentous phage displaying antibody variable domains. *Nature* 348:552-554.

McCafferty, J.; Jackson, R.H. & Chiswell, D.J. (1991). Phage-enzymes: expression and affinity chromatography of functional alkaline phosphatase on the surface of bacteriophage. *Protein Eng.* 4:286.

Mello, C.C. & Conte Jr., D. (2004). Revealing the world of RNA interference. *Nature* 431:338-342.

Michon, I.N.; Hauer, A.D.; von der Thüsen, J.H.; Molenaar, T.J.; van Berkel, T.J.; Biessen, E.A. & Kuiper J. (2002). Targeting of peptides to restenotic vascular smooth muscle cells using phage display in vitro and in vivo. *Biochim Biophys Acta.* 1591(1-3):87-97.

Miller, J. & Stagljar, I. (2004). Using the yeast two-hybrid system to identify interacting proteins. *Methods Mol Biol.* 261:247-62.

Milovnik, P.; Ferrari, D.; Sarkar, C.A. & Plückthun, A. (2009). Selection and characterization of DARPins specific for the neurotensin receptor 1. *Protein Eng Des Sel.* 22(6):357-66.

Miyamoto-Sato, E.; Nemoto, N.; Kobayashi, K. & Yanagawa, H. (2000) Specific bonding of puromycin to full-length protein at the C-terminus. *Nucleic Acids Res.* 28(5):1176-82.

Molinski, T.F.; Dalisay, D.S.; Lievens, S.L. & Saludes, J.P. (2009). Drug development from marine natural products. *Nat Rev Drug Discov.* 8(1):69-85.

Morgenstern, D.; Rohde, B.H.; King, G.F.; Tal, T.; Sher, D. & Zlotkin, E. (2011). The tale of a resting gland: transcriptome of a replete venom gland from the scorpion *Hottentotta judaicus. Toxicon.* 57(5):695-703.

Mortari, M.R.; Cunha, A.O.; Ferreira, L.B. & dos Santos, W.F. (2007). Neurotoxins from invertebrates as anticonvulsants: from basic research to therapeutic application. *Pharmacol Ther.* 114(2):171-83.

Muanpasitporn, C. & Rojnuckarin, P. (2007). Expression and characterization of a recombinant fibrinogenolytic serine protease from green pit viper (*Trimeresurus albolabris*) venom. *Toxicon.* 49(8):1083-9.

Naimuddin, M.; Kobayashi, S.; Tsutsui, C.; Machida, M.; Nemoto, N.; Sakai, T & Kubo, T. (2011). Directed evolution of a three-finger neurotoxin by using cDNA display yields antagonists as well as agonists of interleukin-6 receptor signaling. *Mol Brain.* 4:2.

Nair, S. (2011). Global Protein Therapeutics Market Analysis, Bharat Book Bureu, India.

Nakai, Y.; Nomura, Y.; Sato, T.; Shiratsuchi, A. & Nakanishi, Y. (2005). Isolation of a *Drosophila* gene coding for a protein containing a novel phosphatidylserine-binding motif. *J Biochem.* 137(5):593-9.

Neiva, M.; Arraes, F.B.; de Souza, J.V.; Rádis-Baptista, G.; Prieto da Silva, A.R.; Walter, M.E.; Brigido, M.M.; Yamane, T.; López-Lozano, J.L. & Astolfi-Filho S. (2009). Transcriptome analysis of the Amazonian viper *Bothrops atrox* venom gland using expressed sequence tags (ESTs). *Toxicon.* 53(4):427-36.

Nemoto, N.; Miyamoto-Sato, E.; Husimi, Y. & Yanagawa, H. (1997). *In vitro* virus: bonding of mRNA bearing puromycin at the 3'-terminal end to the C-terminal end of its encoded protein on the ribosome in vitro. *FEBS Lett.* 414(2):405-8.

Nirenberg, M. (2004). Historical review: Deciphering the genetic code-a personal account. *Trends Biochem Sci.* 29(1):46-54

Nobuhisa, I.; Deshimaru, M.; Chijiwa, T.; Nakashima, K.; Ogawa, T.; Shimohigashi, Y.; Fukumaki, Y.; Hattori, S.; Kihara, H. & Ohno, M. (1997). Structures of genes encoding phospholipase A2 inhibitors from the serum of *Trimeresurus flavoviridis* snake. *Gene.* 191(1):31-7.

Ondetti, M.A.; Rubin, B. & Cushman, D.W. (1977). Design of specific inhibitors of angiotensin-converting enzyme: new class of orally active antihypertensive agents. *Science.* 196(4288):441-4.

Padmanabhan, P. & Sahi, S.V. (2011). Suppression subtractive hybridization reveals differential gene expression in sunflower grown in high phosphorus. *Plant Physiol Biochem.* 49(6):584-91.

Pahari, S.; Mackessy, S.P. & Kini, R.M. (2007). The venom gland transcriptome of the Desert *Massasauga rattlesnake (Sistrurus catenatus edwardsii)*: towards an understanding of venom composition among advanced snakes (Superfamily Colubroidea). *BMC Mol Biol.* 8:115.

Patterson, R.M.; Selkirk, J.K. & Merrick, B.A. (1995). Baculovirus and insect cell gene expression: review of baculovirus biotechnology. *Environ Health Perspect.* 103(7-8):756-759

Petersen, G.; Song, D.; Hügle-Dörr, B.; Oldenburg, I. & Bautz, E.K. (1995). Mapping of linear epitopes recognized by monoclonal antibodies with gene-fragment phage display libraries. *Mol Gen Genet.* 249(4):425-31.

Pi, C.; Liu, Y.; Peng, C.; Jiang, X.; Liu, J.; Xu, B.; Yu, X.; Yu, Y.; Jiang, X.; Wang, L.; Dong, M.; Chen, S. & Xu, A.L. (2006). Analysis of expressed sequence tags from the venom ducts of *Conus striatus*: focusing on the expression profile of conotoxins. *Biochimie.* 88(2):131-40.

Qiu, J.X.; Kai, M.; Padlan, E.A. & Marcus, D.M. (1999). Structure-function studies of an anti-asialo GM1 antibody obtained from a phage display library. *J Neuroimmunol.* 97(1-2):172-81.

Quinn, D.J.; Cunningham, S.; Walker, B. & Scott, C.J. (2008). Activity-based selection of a proteolytic species using ribosome display. *Biochem Biophys Res Commun.* 370(1):77-81.

Rádis-Baptista, G. (2011). Crotamine, a small basic polypeptide myotoxin from rattlesnake venom with cell-penetrating properties. *Curr. Pharm. Design. in press.*

Rádis-Baptista, G.; Kubo, T.; Oguiura, N.; Prieto-da-Silva, A.R.; Hayashi, M.A.; Oliveira, E.B. & Yamane, T. (2004). Identification of crotasin, a crotamine-related gene of *Crotalus durissus terrificus. Toxicon.* 43(7):751-9.

Rádis-Baptista, G.; Oguiura, N.; Hayashi, M.A.; Camargo, M.E.; Grego, K.F.; Oliveira, E.B. & Yamane, T. (1999). Nucleotide sequence of crotamine isoform precursors from a single South American rattlesnake (*Crotalus durissus terrificus*). *Toxicon.* 37(7):973-84.

Rajendra, W.; Hackett, K.J.; Buckley, E. & Hammock, B.D. (2006). Functional expression of lepidopteran-selective neurotoxin in baculovirus: potential for effective pest management. *Biochim Biophys Acta.* 1760(2):158-63

Ratushny, V. & Golemis, E. (2008). Resolving the network of cell signaling pathways using the evolving yeast two-hybrid system. *Biotechniques.* 44(5):655-62

Richards, R.; St Pierre, L.; Trabi, M.; Johnson, L.A.; de Jersey, J.; Masci, P.P. & Lavin, M.F. (2011). Cloning and characterisation of novel cystatins from elapid snake venom glands. *Biochimie.* 93(4):659-68.

Roberts, R.W. & Szostak, J.W. (1997). RNA-peptide fusions for the in vitro selection of peptides and proteins. *Proc Natl Acad Sci U S A.* 94(23):12297-302.

Rokyta, D.R.; Wray, K.P.; Lemmon, A.R.; Lemmon, E.M. & Caudle, S.B. (2011). A high-throughput venom-gland transcriptome for the Eastern Diamondback Rattlesnake (*Crotalus adamanteus*) and evidence for pervasive positive selection across toxin classes. *Toxicon.* 57(5):657-71.

Rottenberg, D.; Bamberger, E.S. & Kochva, E. (1971). Studies on ribonucleic acid synthesis in the venom glands of *Vipera palaestinae* (Ophidiae, Reptilia). *Biochem. J.* 121, 609-612.

Ruiming, Z.; Yibao, M.; Yawen, H.; Zhiyong, D.; Yingliang, W.; Zhijian, C. & Wenxin, L. (2010). Comparative venom gland transcriptome analysis of the scorpion *Lychas mucronatus* reveals intraspecific toxic gene diversity and new venomous components. *BMC Genomics.* 11:452.

Sagerström, C.G.; Sun, B.I. & Sive, H.L. (1997). Subtractive cloning: past, present, and future. *Annu Rev Biochem.* 66:751-83.

Sambrook J. and D.W. Russel. Molecular Cloning - A Laboratroy Manual (2001). CSHL Press, New York.

Sánchez, E.E.; Lucena, S.E.; Reyes, S.; Soto, J.G.; Cantu, E.; Lopez-Johnston, J.C.; Guerrero, B.; Salazar, A.M.; Rodríguez-Acosta, A.; Galán, J.A.; Tao, W.A. & Pérez, J.C. (2010). Cloning, expression, and hemostatic activities of a disintegrin, r-mojastin 1, from the mohave rattlesnake (*Crotalus scutulatus scutulatus*). *Thromb Res.* 126(3):e211-9.

Schaffitzel, C. & Plückthun, A. (2011). Protein-fold evolution in the test tube. *Trends Biochem Sci.* 26(10):577-9.

Schaffitzel, C.; Zahnd, C.; Amstutz, P.; Luginbühl, B. & Plückthun. A. (2002). *In vitro* selection and evolution of protein-ligand interactions by ribosome display. In: E. Golemis. *Protein-Protein Interactions. A Molecular Cloning Manual,* CSHL Press, New York.

Schmidtko, A.; Lötsch, J.; Freynhagen, R. & Geisslinger, G. (2010). Ziconotide for treatment of severe chronic pain. *Lancet.* 375(9725):1569-77.

Schwartz, E.F.; Diego-Garcia, E.; Rodríguez de la Vega, R.C. & Possani, L.D. (2007). Transcriptome analysis of the venom gland of the Mexican scorpion *Hadrurus gertschi* (Arachnida: Scorpiones). *BMC Genomics.* 8:119.

Scott, J.K. & Smith, G.P. (1990). Searching for peptide ligands with an epitope library. *Science.* 249(4967):386-90.

Serebriiski, I. & Joung, J.K. (2002). Yeast and Bacterial Two-hybrid selection systems for studying protein-protein interactions. In: E. Golemis. *Protein-Protein Interactions. A Molecular Cloning Manual*, CSHL Press, New York.

Sergeeva, A.; Kolonin, M.G.; Molldrem, J.J. Pasqualini, R. & Arap W. (2006). Display technologies: application for the discovery of drug and gene delivery agents. *Adv Drug Deliv Rev.* 58(15):1622-54

Singhamatr, P. & Rojnuckarin, P. (2007). Molecular cloning of albolatin, a novel snake venom metalloprotease from green pit viper (*Trimeresurus albolabris*), and expression of its disintegrin domain. *Toxicon.* 50(8):1192-200.

Smith, G.P. (1985). Filamentous fusion phage: novel expression vectors that display cloned antigens on the virion surface. *Science.* 228(4705):1315-7.

Studier, F.W. & Moffatt, B.A. (1986). Use of bacteriophage T7 RNA polymerase to direct selective high-level expression of cloned genes. *J Mol Biol.* 189(1):113-30.

Studier, F.W.; Rosenberg, A.H.; Dunn, J.J. & Dubendorff, J.W. (1990). Use of T7 RNA polymerase to direct expression of cloned genes. *Methods Enzymol.* 185:60-89.

Suter, B.; Kittanakom, S. & Stagljar, I. (2008). Two-hybrid technologies in proteomics research. *Curr Opin Biotechnol.* 19(4):316-23.

Takegawa, K.; Tohda, H.; Sasaki, M.; Idiris, A.; Ohashi, T.; Mukaiyama, H.; Giga-Hama, Y. & Kumagai, H. (2009). Production of heterologous proteins using the fission-yeast (*Schizosaccharomyces pombe*) expression system. *Biotechnol Appl Biochem.* 53(Pt 4):227-35.

Tamiya, T. & Fujimi, T.J. (2006). Molecular evolution of toxin genes in Elapidae snakes. *Mol Divers.* 10(4):529-543.

Torres, F.S.; Silva, C.N.; Lanza, L.F.; Santos, A.V.; Pimenta, A.M.; De Lima, M.E. & Diniz, M.R. (2010). Functional expression of a recombinant toxin - rPnTx2-6 - active in erectile function in rat. *Toxicon.* 56(7):1172-80.

Trueman, L.J. (1995). Heterologous expression in yeast. *Methods Mol Biol.* 49:341-54.

Valadon, P.; Garnett, J.D.; Testa, J.E.; Bauerle, M.; Oh, P. & Schnitzer, J.E. (2006). Screening phage display libraries for organ-specific vascular immunotargeting *in vivo. Proc Natl Acad Sci U S A.* 103(2):407-12.

Wagstaff, S.C. & Harrison, R.A. (2006). Venom gland EST analysis of the saw-scaled viper, *Echis ocellatus*, reveals novel alpha9beta1 integrin-binding motifs in venom metalloproteinases and a new group of putative toxins, renin-like aspartic proteases. *Gene.* 377:21-32.

Walter, G.; Konthur, Z. & Lehrach, H. (2001). High-throughput screening of surface displayed gene products. *Comb Chem High Throughput Screen.* 4(2):193-205

Warren, T.G. & Shields, D. (1984). Expression of preprosomatostatin in heterologous cells: biosynthesis, posttranslational processing, and secretion of mature somatostatin. *Cell.* 39(3 Pt 2):547-55.

Watson J.D.; Baker, T.A.; Bell, S.P.; Gann, A.; Levine, M. & Losick, R. (2008). Molecular Biology of the gene. 6th edition, CSHL Press, New York.

Watson, J.D. & Crick, F.H. (1953). Molecular structure of nucleic acids; a structure for deoxyribose nucleic acid. *Nature.* 171(4356):737-8.

Whittington, C.M.; Koh, J.M.; Warren, W.C.; Papenfuss, A.T.; Torres, A.M.; Kuchel, P.W. & Belov, K. (2009). Understanding and utilising mammalian venom via a platypus venom transcriptome. *J Proteomics.* 72(2):155-64

Whittington, C.M.; Papenfuss, A.T.; Locke, D.P.; Mardis, E.R.; Wilson, R.K.; Abubucker, S.; Mitreva, M.; Wong, E.S.; Hsu, A.L.; Kuchel, P.W.; Belov, K. & Warren, W.C. (2010). Novel venom gene discovery in the platypus. *Genome Biol.* 11(9):R95.

Winter, G.; Griffiths, A.D.; Hawkins, R.E. & Hoogenboom, H.R. (1994). Making antibody by phage display technology. *Annu. Rev. Immunol.* 12:433-455.

Wu, H.; Yang, W.P. & Barbas 3rd, C.F. (1995). Building zinc fingers by selection: toward a therapeutic application. *Proc Natl Acad Sci U S A.* 92(2):344-8.

Yamaguchi, J.; Naimuddin, M.; Biyani, M.; Sasaki, T.; Machida, M.; Kubo, T.; Funatsu, T.; Husimi, Y. & Nemoto, N. (2009). cDNA display: a novel screening method for functional disulfide-rich peptides by solid-phase synthesis and stabilization of mRNA-protein fusions. *Nucleic Acids Res.* 37(16):e108.

Yamamoto, M.; Kominato, Y. and Yamamoto, F. (1999). Phage display cDNA cloning of protein with carbohydrate affinity. *Biochem Biophys Res Commun.* 255(2):194-9.

Yanagida, H.; Matsuura, T. & Yomo, T. (2010). Ribosome display for rapid protein evolution by consecutive rounds of mutation and selection. *Methods Mol Biol.* 634:257-67.

Yang, D.; Peng, M.; Yang, H.; Yang, Q. & Xu, J. (2009). Expression, purification and characterization of *Gloydius shedaoensis* venom gloshedobin as Hsp70 fusion protein in Pichia pastoris. *Protein Expr Purif.* 66(2):138-42.

Yang, Y.; Cun, S.; Xie, X.; Lin, J.; Wei, J.; Yang, W.; Mou, C.; Yu, C.; Ye, L.; Lu, Y.; Fu, Z. & Xu A. (2003). EST analysis of gene expression in the tentacle of *Cyanea capillata.* *FEBS Lett.* 538(1-3):183-91.

Yang, Y.M.; Barankiewicz, T.J.; He, M.; Taussig, M.J. & Chen, S.S. (2007). Selection of antigenic markers on a GFP-Ckappa fusion scaffold with high sensitivity by eukaryotic ribosome display. *Biochem Biophys Res Commun.* 359(2):251-7.

Yau, K.Y.; Groves, M.A.; Li, S.; Sheedy, C.; Lee, H.; Tanha, J.; MacKenzie, C.R.; Jermutus, L. & Hall, J.C. (2003). Selection of hapten-specific single-domain antibodies from a non-immunized llama ribosome display library. *J Immunol Methods.* 281(1-2):161-75.

Yonamine, C.M.; Prieto-da-Silva, A.R.; Magalhães, G.S.; Rádis-Baptista, G.; Morganti, L.; Ambiel, F.C.; Chura-Chambi, R.M.; Yamane, T. & Camillo, M.A. (2009). Cloning of serine protease cDNAs from *Crotalus durissus terrificus* venom gland and expression of a functional Gyroxin homologue in COS-7 cells. *Toxicon.* 54(2):110-20.

Yu, X.; Li, Z.; Xia, X.; Fang, H.; Zhou, C. & Chen, H. (2007). Expression and purification of ancrod, an anticoagulant drug, in *Pichia pastoris.* *Protein Expr Purif.* 55(2):257-61.

Zahid, M.; Phillips, B.E.; Albers, S.M.; Giannoukakis, N.; Watkins, S.C. & Robbins, P.D. (2010). Identification of a cardiac specific protein transduction domain by in vivo biopanning using a M13 phage peptide display library in mice. *PLoS One.* 5(8):e12252.

Zhang, B.; Liu, Q.; Yin, W.; Zhang, X.; Huang, Y.; Luo, Y.; Qiu, P.; Su, X.; Yu, J.; Hu, S. & Yan, G. (2006). Transcriptome analysis of *Deinagkistrodon acutus* venomous gland focusing on cellular structure and functional aspects using expressed sequence tags. *BMC Genomics.* 7:152.

Zhang, W.; Li, H.; Cheng, G.; Hu, S.; Li, Z. & Bi, D. (2008). Avian influenza virus infection induces differential expression of genes in chicken kidney. *Res Vet Sci.* 84(3):374-81

Zheng, Y-M.; Liu, C.; Chen, H.; Locke, D.; Ryan, J.C. & Kahn, M.L. (2001). Expression of the platelet receptor GPVI confers signaling via the Fc receptor g-chain in response to the snake venom convulxin but not to collagen. *J. Biol. Chem.* 276:12999-3006.

Zhijian, C.; Yun, X.; Chao, D.; Shunyi, Z.; Shijin, Y.; Yingliang, W. & Wenxin, L. (2006) Cloning and characterization of a novel calcium channel toxin-like gene BmCa1 from Chinese scorpion *Mesobuthus martensii* Karsch. *Peptides.* 27(6):1235-40.

Zhu, L.; Yuan, C.; Chen, Z.; Wang, W. & Huang, M. (2010). Expression, purification and characterization of recombinant Jerdonitin, a P-II class snake venom metalloproteinase comprising metalloproteinase and disintegrin domains. *Toxicon.* 55(2-3):375-80.

Part 5

Parasitology

12

Genome Based Vaccines Against Parasites

Yasser Shahein[1] and Amira Abouelella[2]
[1]Department of Molecular Biology, National Research Centre,
[2]Department of Biochemistry, National Centre for Radiation
Research and Technology
Egypt

1. Introduction

In its original concept, vaccination aims to mimic the development of naturally acquired immunity by inoculation of nonpathogenic but still immunogenic components of the pathogen in question, or closely related organisms. Vaccination is one of the most effective tools allowing near or complete eradication of fatal diseases. In recent times, vaccination was based on conventional approaches that were successful in several diseases but they require the pathogen to be grown in laboratory conditions, are time-consuming and allow for the identification of only the most abundant antigens, which can be purified in quantities suitable for vaccine testing. The conventional approach to vaccine development uses two methods: first, attenuation of pathogens by serial passages *in vitro* to obtain live-attenuated strains to be used as vaccines, and second, identification of protective antigens to be used in non-living, subunit vaccines.

Using conventional vaccinology, vaccines targeting pathogens with low antigenic variability and those for which protection depends on antibody mediated immunity like polio, MMR, Tetanus, Influenza, and Diptheria, had been licensed. Furthermore, if we are dealing with non-cultivatable microorganisms or have high antigenic variability or need T cell-dependent induced immunity like MenB, TB, HIV, Hepatitis B & C, Malaria and parasite diseases, there is no advance to development a specific vaccine.

With the development of the DNA sequencing technologies in the 1970s till recent years, we observed the completion of the sequencing of the genome of humans, a number of invertebrate species including *Drosophila melanogaster* and *Caenorhabditis elegans* and expanding number of microbial pathogens of medical and economic importance (Dalton et al., 2003). Expansion of these sequence information developed new approaches to decode, analyze and share these massive genetic data. The most widely used term for these approaches is the genomics.

Genomics, studying the genome of organisms as a whole, and postgenomics technologies including investigating RNA (transcriptomics), proteins (proteomics), identification of immunogenic proteins (immunomics) and metabolites (metabolomics), have had a considerable impact in all areas of biological research (Bambini and Rappuoli, 2009), and the field of vaccinology is no exception. There are many examples using these approaches to develop vaccines but this chapter will focus on the possible genome based approaches to confer vaccine developed immunity against two examples of endo- and ecto-parasites such as malaria and ticks, respectively.

2. Reverse vaccinology

Genome sequencing is a powerful tool for understanding and controlling infectious pathogens. Using this technology, researchers can identify target genes for drug discovery and reveal small genetic variations between strains of a specific organism to define its virulence and improve the method of control. The approach of reverse vaccinology uses the genome sequences of viral, bacterial or parasitic pathogens of interest as starting material for the identification of novel antigens, whose activity should be subsequently confirmed by experimental biology (Fig. 1) (Rappuoli, 2001). One of the earlier applications of genomics to vaccinology (reverse vaccinology) had been the identification of vaccine candidates against serogroup B meningococcus by the completion of the whole-genome sequencing (Pizza et al., 2000). They had cloned the open reading frames (ORFs) that encode putative virulence factors and surface-localized proteins of meningococcus. Several hundred ORFs (350 surface-exposed protein coding frames) were cloned into expression vectors, purified and used to immunize mice. The antibodies binding properties to the products of ORF were analyzed using fluorescent activated cell sorter (FACS) analyses and Enzyme linked immunosorbent assay (ELISA). The primary vaccine candidates were then tested *in vitro* and/or animal models to provide an insight on the protective efficacy. Twenty nine of these surface-exposed proteins were found to be bactericidal. The selected candidates were then checked for sequence conservation across a panel of strains representing the genetic diversity of meningococcus allowing further selection of antigens capable of eliciting cross-bactericidal response against the majority of strains included in the panel.

Reverse vaccinology is based on the high throughput analyses of genome sequences. With continuous flow of new genomic sequence and functional annotation data from different taxonomic lineages permits scientists to confine correlations depending on the wide range of data bases, enabling the design of more reliable analytical and predictive tools. One of the most important tools is the alignment of multiple homologous sequences that permitted the identification of large number of structural and functional signatures including ligand-binding sites, sorting signals, protein domain profile, different motifs with catalytic sites and more (Vivona et al., 2008). This has generated several integrated databases containing combined sets of such signatures and of scanning tools capable of inferring possible functions or regulatory mechanisms from the presence of either canonical or degenerate signatures in a sequence of interest. Such 'sequence-to-function' approaches (Oliver, 1996) have enhanced the production of further functional evidence by prediction-driven experiments. Early examples included using the identification of specific signatures in a sequence to suggest functional assays able to identify otherwise long-sought functions (Filippini et al., 1996). They had shown that a plant oncogene *rolB* from *Agrobacterium rhizogenes*, induces differentiation and growth of neoplastic roots (Hairy-roots) in dicotyledonous plants. In addition, the prediction-driven experiments may imply functions for disease gene products (Emes and Ponting, 2001, Vacca et al., 2001). Later, more complex, genome-wide analyses have led to the identification of proteomic complements that underlie regulatory pathways or interaction network organization in model organisms (Carpi et al., 2002, Li et al., 2006).

More recently, bioinformatic approaches are used to uncover functional information and enable researchers to tackle biological and biotechnological problems that require the integration of diverse strategies of both *in silico* (on computer) and experimental evidences. Besides data analyses, a variety of algorithmic approaches have been used to develop novel tools. The functional potential of these *in silico* approaches has found its pattern in reverse vaccinology.

Reverse vaccinology presents a revolution in both immunology and biotechnology and shows how a biological problem like designing a vaccine could be solved by applying integrating tools. However, reverse vaccinology presents a huge advance compared to the conventional vaccine production protocols. It takes advantage of the growing number of genome sequences available for many organisms. The approach uses computer analysis of the genomic sequence to predict suitable candidate vaccine molecules. Unfortunately, the approach does not provide certain evidence that the selected antigens are either immunogenic or protective. On the contrary, the approach permits the identification of novel protein antigens besides the antigens discovered by the traditional protocols.

3. Malaria vaccine research

Malaria is caused by parasites of the genus *Plasmodium* of which there are four species that infects humans, *Plasmodium falciparum, P. vivax, P. ovale* and *P. malariae. P. falciparum* is the most fatal and common causative species of malaria in humans. The parasite originally develops in the gut of an infected *Anopheles* mosquito and is passed into humans in the saliva of infected mosquitoes. Sporozoites; the pre-erythrocytic stage is transported to the liver in the blood or lymph of the human. In the liver, the parasite invades and replicates in hepatocytes and after one to two weeks, the sporozoite become mature into thousands of merozoites which penetrate the circulating red cells. The replication of the merozoites in blood cells causes rapid destruction of blood cells leading to fever and anemia. In the past years, antimalarial drugs were the first line of defense in endemic areas but unfortunately, there is increasing resistance to these drugs (Wykes and Good, 2007).

The 500 million new infections each year and 2.5 million annual deaths indicated that all measures used so far to control the disease have failed (WHO, 1997, Chaudhuri et al., 2008). This world malaria situation had directed researchers to conduct other strategies including the identification of only few dozen of malarial proteins (Duffy et al., 2005) by recombinant vaccine technology and more than 40 clinical trials were carried out but no vaccine has provided a strong and lasting immunity.

The accessibility of complete genome sequences of *Plasmodium falciparum* (Gardner et al., 2002), *P. yoelii* (Carlton et al., 2002) and *P. vivax* has provided new opportunity for applying the principles of reverse vaccinology. Reverse vaccinology uses bioinformatics in the initial steps to identify potential antigens, which are consequently examined for their effectiveness and toxicity. The use of algorithm for prediction of subcellular location improved the power of identifying potential vaccine candidates (Serruto and Rappuoli, 2006). Subsequently, developments have been proposed to reverse vaccinology by suggesting the use of additional algorithms to find probability of being an adhesin, of topology (transmembrane domains) and to find similarity with host protein (Vivona et al., 2006). Recently, integrative approaches are proposed for reverse vaccinology by including prediction of multiple features of proteins such as signal peptides, membrane spanning regions, functional motifs and differences in amino acid composition unique to specific cellular compartments (Vivona et al., 2008). Implementing this approach, the following predictions were included:

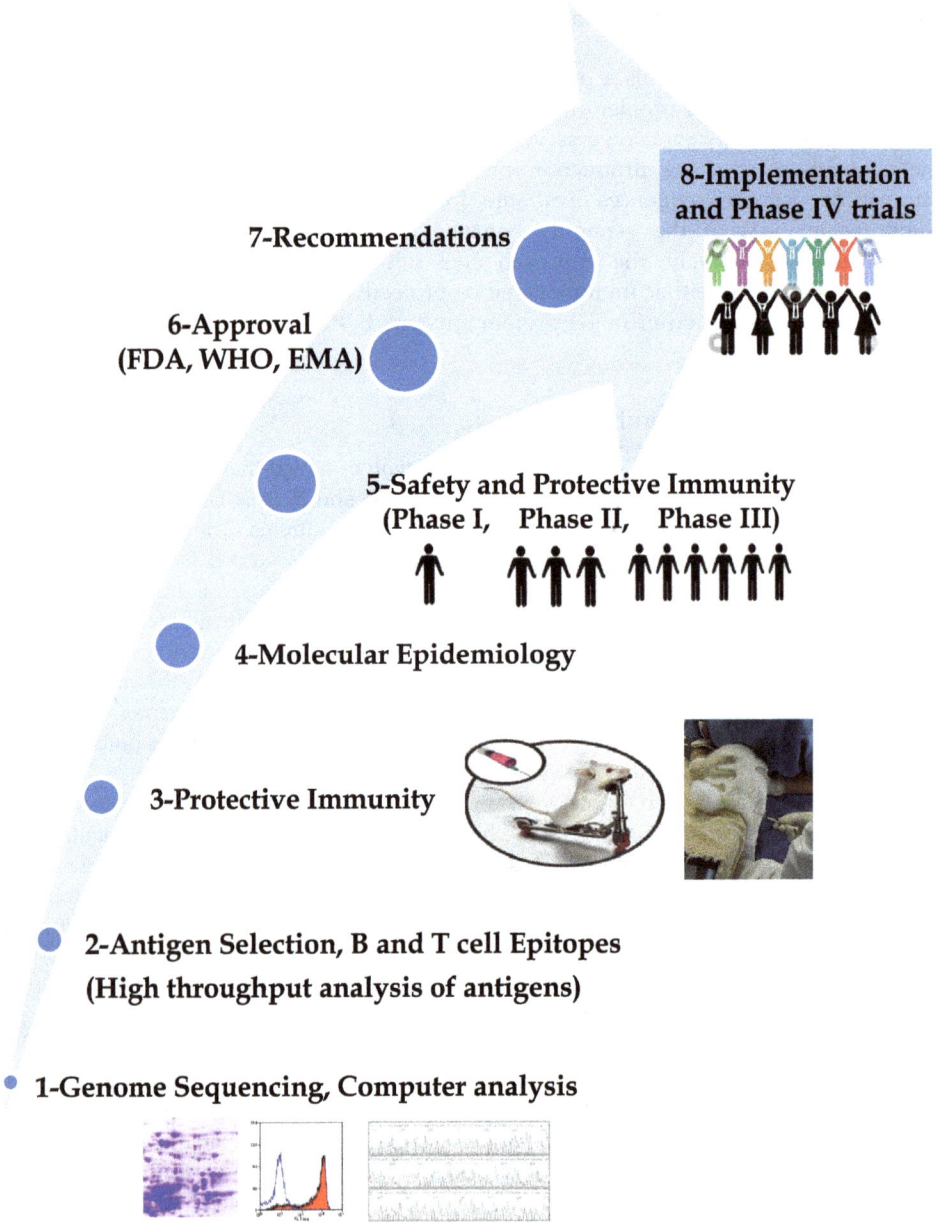

Fig. 1. Diagrammatic Illustration Summarizing the Vaccine Development Pathway Starting from Reverse Vaccinology. (1) First, computer analysis of the entire genome identifies the genes coding for predicted antigens and removes those antigens with homologies to human proteins. (2) Second, these identified antigens are screened for expression by the pathogen and for immunogenicity during infection. (3) Then the selected antigens are used to immunize

experimental animals and examine the protective efficacy of immunization. (4) The presence and conservation of protective antigens in a collection of strains representative of the species (molecular epidemiology), is examined. (5) Finally, large scale production of target antigens are manufactured for clinical trials and candidate vaccines are examined for their safety and protective immunity in humans. (6) Authorized agencies, such as the Food and Drug Administration (FDA), World Health Organization (WHO) or the European Medicinal Agency (EMA) analyze and approve the scientific, clinical, and technical information. (7) Policy-making organizations, such as the Advisory Committee on Immunization Practices (ACIP) and equivalent bodies from other nations, prepare the recommendation on how the vaccine should be used. (8) In this stage, the approved vaccine is then commercialized and used in large scale. At this point, phase IV clinical studies confirm safety.

3.1 Adhesins

One of the preventive approaches targets adhesion of parasites to host cells and tissues. Adhesion of parasites is mediated by proteins called adhesins. Abrogation of adhesion by either immunizing the host with adhesins or inhibiting the interaction using structural analogs of host cell receptors holds the potential to develop novel preventive strategies. The availability of complete genome sequence offers new opportunities for identifying adhesin and adhesin-like proteins. A nonhomology-based approach using 420 compositional properties of amino acid dipeptide and multiplet frequencies was used to develop Malarial Adhesins and Adhesin-like proteins Predictor (MAAP) Web server with Support Vector Machine (SVM) model classifier as its engine for the prediction of malarial adhesins and adhesin-like proteins. Several new predictions were obtained. This list includes hypothetical protein PF14 0040, interspersed repeat antigen, STEVOR, liver stage antigen, SURFIN, RIFIN, stevor (3D7-stevorT3-2), mature parasite-infected erythrocyte surface antigen or *P. falciparum* erythrocyte membrane protein 2, merozoite surface protein 6 in *P. falciparum*, circumsporozoite proteins, microneme protein-1, Vir18, Vir12-like, Vir12, Vir18-like, Vir18-related and Vir4 in *P. vivax*, circumsporozoite protein/thrombospondin related anonymous proteins, 28 kDa ookinete surface protein, yir1, and yir4 of *P. yoelii* (Ansari et al., 2008).

3.2 Paralogs

It means gene duplication, which have a major role in the evolution of new biological functions. Theoretical studies often assume that a duplication *per se* is selectively neutral and that, following a duplication, one of the gene copies is freed from purifying (stabilizing) selection, which creates the potential for evolution of a new function (Kondrashov et al., 2002).

3.3 Transmembrane topologies

Krogh et al. (2001), predicted a membrane protein topology prediction method; transmembrane protein topology with a hidden Markov model (TMHMM). The model is based on a hidden Markov model showing that it correctly predicts 97-98 % of the transmembrane helices. Additionally, TMHMM can discriminate between soluble and membrane proteins with both specificity and sensitivity better than 99 %. The method shows that proteins with N(in)-C(in) topologies are strongly preferred in all examined organisms, except *Caenorhabditis elegans*, where the large number of 7TM receptors increases the counts for N(out)-C(in) topologies.

3.4 β-helix supersecondary structural motifs

A program named BETAWRAP (http://theory.lcs.mit.edu/betawrap) implements the prediction of parallel beta -helices from primary sequences and recognizes each of the seven known parallel beta-helix families, when trained on the known parallel beta-helices from outside that family. BETAWRAP identifies 2,448 sequences among 595,890 screened from the National Center for Biotechnology Information (NCBI; http://www.ncbi.nlm.nih.gov/) nonredundant protein database as likely parallel beta-helices. It identifies surprisingly many bacterial and fungal protein sequences that play a role in human infectious disease; these include toxins, virulence factors, adhesins, and surface proteins of Chlamydia, Helicobacteria, Bordetella, Leishmania, Borrelia, Rickettsia, Neisseria, and *Bacillus anthracis*.

3.5 Subcellular localization

A neural network-based tool, TargetP, for large-scale subcellular location prediction of newly identified proteins has been developed. Using N-terminal sequence information only, it discriminates between proteins destined for the mitochondrion, the chloroplast, the secretory pathway, and "other" localizations with a success rate of 85% (plant) or 90% (nonplant) on redundancy-reduced test sets. TargetP is available as a web-server at http://www.cbs.dtu.dk/services/TargetP/ (Emanuelsson et al., 2000).

3.6 Similarity against human proteins

Using the Basic Local Alignment Search Tool (BLAST) (Altschul et al., 1990).

3.7 Antigenic regions

Analysis of data from experimentally determined antigenic sites on proteins reveal that the hydrophobic residues Cys, Leu and Val, if they occur on the surface of a protein, are more likely to be a part of antigenic sites. The semi-empirical method which makes use of physicochemical properties of amino acid residues and their frequencies of occurrence in experimentally known segmental epitopes was developed to predict antigenic determinants on proteins. Application of this method to a large number of proteins has shown that the method can predict antigenic determinants with about 75% accuracy which is better than most of the known methods (Kolaskar and Tongaonkar, 1990).

3.8 Conserved domains

The conserved domain search results for protein sequences in Entrez are pre-computed to provide links between proteins and domain models, and computational annotation visible upon request. Protein–protein queries submitted to the BLAST search service are scanned for the presence of conserved domains by default (Marchler-Bauer et al., 2005).

3.9 Epitopes

The identification of B-cell and T-cell epitopes is a crucial step in peptide vaccine design. The experimental scanning of B-cell epitope active regions requires the synthesis of overlapping peptides, which span the entire sequence of a protein antigen. This process is time consuming and costly. The *in silico* approaches are the alternative procedures to figure out target regions of a protein with possible immunogenic capacity. In this respect, two new prediction tools; BcePred and ABCpred servers were developed. BcePred can predict continuous B-cell epitopes, and physic-chemical scales used were hydrophilicity, flexibility/mobility, accessibility, polarity, exposed surface, turns and antigenicity. The ABCpred was developed to predict continuous or linear B-cell epitopes obtained from Bcipep database. This server is based on machine learning techniques (Saha and Raghava, 2007).

3.10 Allergens

The prediction of allergenic proteins is becoming very important due to the use of modified proteins in foods (genetically modified foods), therapeutics and bio-pharmaceuticals. The protein is considered allergen if it has one or more IgE epitopes. This kind of prediction can be achieved using the prediction tool created by Saha S, Raghava (2006). The predition tool is based on various approaches. First, a standard method for predicting allergens based on amino acid and dipeptide composition of proteins using support vector machine (SVM). In the second approach, motif-based technique has been used for predicting allergens using the software MEME/MAST. Third, they assigned a protein as an allergen, if it has a segment similar to allergen representative proteins (ARPs).

4. Data bases and predictive tools in malarial diseases

A community resource database; MalVac was created. The database is based on data analysis of available proteins including 161 adhesin proteins from *P. falciparum*, 137 adhesin proteins from *P. vivax* and 34 adhesin proteins from *P. yoelii*. The ORF identification tags (ORF ID) assigned to proteins of malaria parasites as given in PlasmoDB 5.4 which is is the official database of the *Plasmodium falciparum* genome sequencing consortium, were used as primary keys. The database was developed using MySQL version 4.1.20 at back end and operated in Red Hat Enterprise Linux ES release 4. The web interfaces have been developed in HTML and PHP 5.1.4, which dynamically execute the MySQL queries to fetch the stored data and is run through Apache2 server. The frame of MalVac consists of four basic concepts as follows:

4.1 Motif and topology: includes the transmembrane helices and right handed parallel bata helices.

4.2. Location: the position of the signal peptides and subcellular localization of the protein.

4.3 Immunoinformatics: includes T cell epitopes (MHC class I and II), B cell epitopes (conformationaland linear epitopes), Allergens and antigenic regions.

4.4 Homology: includes the paralogs, orthologs, conserved domains and similarity to host proteins.

The first step towards MalVac database creation is the collection of known vaccine candidates and a set of predicted vaccine candidates identified from the whole proteome sequences of Plasmodium species provided by PlasmoDB 5.4 release. These predicted vaccine candidates are the adhesins and adhesin-like proteins from Plasmodium species, *P. falciparum*, *P. vivax* and *P. yoelii* using MAAP server. Subsequently these protein sequences will be analysed with 20 algorithms important from the view of reverse vaccinology (Table 1).

Algorithm	Principle Role in MalVac data base
1. MAAP	Predicts Malarial adhesins and adhesins-like proteins.
2. BLASTCLUST	Clusters protein or DNA sequences based on pair wise matches found using the BLAST algorithm in case of proteins or Mega BLAST algorithm for DNA. Paralogs finding
3. TMHMM Server v. 2.0	Predicts the transmembrane helices in proteins based on Hidden Markov Model. Transmembrane helices prediction
4. BetaWrap (Betawrap finding)	Predicts the right-handed parallel beta-helix supersecondary structural motif in primary amino acid sequences by using beta-strand interactions learned from non-beta-helix structures.
5. TargetP1.1 (Localization Prediction)	Predicts the subcellular location of eukaryotic proteins based on the predicted presence of any of the N-terminal presequences: chloroplast transit peptide (cTP), mitochondrial targeting peptide (mTP) or secretory pathway signal peptide (SP).
6. SignalP 3.0 (Signal Peptide Prediction)	Predicts the presence and location of signal peptide cleavage sites in amino acid sequences from different organisms.
7. BlastP	It uses the BLAST algorithm to compare an amino acid query sequence against a protein sequence database.
8. Antigenic (Antigenic region prediction)	Predicts potentially antigenic regions of a protein sequence, based on frequency occurrence of amino acid residue types in known epitopes..
9. Conserved Domain Database (Conserved Domain Finding)	It is used to identify the conserved domains present in a protein query sequence.
10. ABCPred (Linear B Cell Epitope Prediction)	Predict *B cell epitope(s)* in an antigen sequence, using artificial neural network.
11. BcePred (Linear B Cell Epitope Prediction)	Predicts linear B-cell epitopes, using physico-chemical properties. Linear B Cell Epitope Prediction.
12. Discotope 1.1 (Conformational B Cell Epitope Prediction)	Predicts discontinuous B cell epitopes from protein three dimensional structures utilizing calculation of surface accessibility (estimated in terms of contact numbers) and a novel epitope propensity amino acid score..
13. CEP (Conformational B Cell Epitope Prediction)	The algorithm predicts epitopes of protein antigens with known structures.
14. NetMHC 2.2 (HLA Class I Epitope prediction)	Predicts binding of peptides to a number of different HLA alleles using artificial neural networks (ANNs) and weight matrices.

Algorithm	Principle Role in MalVac data base
15. MHCPred 2.0 (MHC Class I and II epitope prediction)	MHCPred uses the additive method to predict the binding affinity of major histocompatibility complex (MHC) class I and II molecules and also to the Transporter associated with Processing (TAP).
16. Bimas (MHC Class I and II epitope prediction)	Ranks potential 8-mer, 9-mer, or 10-mer peptides based on a predicted half-time of dissociation to HLA class I molecules. The analysis is based on coefficient tables deduced from the published literature by Dr. Kenneth Parker, Children's Hospital Boston.
17. Propred (MHC Class I and II epitope prediction)	Predicts MHC Class-II binding regions in an antigen sequence, using quantitative matrices derived from published literature.
18. AlgPred (Allergen Prediction)	Predicts allergens in query protein based on similarity to known epitopes, searching MEME/MAST allergen motifs using MAST.
19. Allermatch (Allergen Prediction)	Predicts the potential allergenicity of proteins by bioinformatics approaches as recommended by the Codex alimentarius and FAO/WHO Expert consultation on allergenicity of foods derived through modern biotechnology.
20. WebAllergen (Allergen Prediction)	The query protein is compared against a set of pre-built allergenic motifs that have been obtained from 664 known allergen proteins.

Table 1. Different Algorithms used in MalVac predictions tool: Novel malarial candidate vaccines using reverse vaccinology approach.

5. Ticks

Ticks are obligate heamatophagous ectoparasite classified in the subclass Acari, order Parasitiformes, suborder Ixodida, which are distributed worldwide from the Arctic to tropical regions. Ticks include 899 species that parasitize terrestrial vertebrates, including amphibians, reptiles, birds and mammals. The three families of ticks include the Argasidae or soft ticks (185 species) which is divided into two subfamilies, Argasinae and Ornithodorinae, the Ixodidae or hard ticks (713 species) which is divided into the prostriata (Ixodinae; Ixodes) and the Metastriata (subfamilies Amblyomminae, Haemaphysalinae, Hyalomminae and Rhipicephalinae) and the Nuttalliellidae, which only contains one species. The recent classification of the genus *Boophilus* as a subgenus of *Rhipicephalus* is still controversial.

Ticks, as blood-feeding ectoparasites, affect their hosts both directly and as vectors of viral, bacterial and protozoal diseases. Their impact as disease vectors on human wellbeing is second only to that of mosquitoes, and their effect on livestock, wildlife and domestic animals is immeasurably greater. The veterinary importance of ticks compared to other ectoparasites is obvious as they consume large quantities of host blood during their lengthy attachment period (7-14 days), which may be extended depending on the tick species and unique host association.

5.1 Tick genomes

Ticks have large genomes and their estimated sizes vary from 1.04~7.1 x10(9) bp, about one third to over two times the size of the human genome. Karyotype studies have revealed a range in chromosome number and the sex determining system seems to be primarily driven by a XY or a XO format. Estimates for three species are currently available, the smallest being ~10^9 bp (*Amblyomma americanum*) and the largest ~7×10^9 bp [*Rhipicephalus (Boophilus) microplus*]; the *Ixodes scapularis* genome is estimated to be 2×10^9 bp (Jongejan et al., 2007).

Nevertheless, sequencing of the genome of *I. scapularis* is under way. Pilot scale studies on the genome of *R. (Boophilus) microplus* and *R. Annulatus* are progressing and it is expected that this will strengthen a future proposal to sequence the genome more fully. Information on ESTs is available for several tick species, including *R. (Boophilus) microplus, A. americanum* and *Amblyomma variegatum*. Some of these have been used to create a repository of clustered and auto-annotated data in the form of species-specific gene indices.

Ixodes scapularis, the black-legged tick or deer tick, is a hard (ixodid) tick vector of the causative agents of Lyme disease, babesiosis and anaplasmosis in the United States. The Ixodes genome project was initiated in 2004, and aimed to sequence the genome of a medically significant tick. The project represents two important scientific firsts; it is the first sequencing project for a tick and a chelicerate. The project is a multi-phase and multi-investigator undertaking. Current plans call for whole genome shotgun sequencing to approximately six fold coverage of the genome. Trace reads from small (2–4 Kbp), medium (10–12 Kbp) and large (40–50 Kbp) insert genomic clones will be the basis for assembly of the genome sequence. Also included are reads of ~160,000 bacterial artificial chromosome (BAC) clone ends (BAC-end sequencing), the complete end-to-end sequencing of 60 BAC clones and ~240,000 expressed sequence tag (EST) reads. Paired BAC end reads span large segments of the genome and will be used to help assemble whole genome shotgun sequence into scaffolds. Sequenced BACs will provide an early insight into the Ixodes genome and will have utility as probes for physical mapping of assembled sequence to chromosomes. ESTs are arguably one of the most valuable resources generated as part of any genome project. These ESTs will be used to identify expressed genes, confirm gene predictions and to train automated gene finding algorithms. To date, 20 Ixodes BAC clones have been shotgun sequenced and assembled. Also available are approximately 370,000 BAC-end reads and more than 80,000 ESTs have been sequenced from a normalized pooled stage/tissue library. All sequences have been deposited at the National Center for Biotechnology Information (NCBI) (Van Zee et al., 2007).

5.2 Tick proteins and proteomics approaches

The impact of genomics on the knowledge of expressed parasite proteins has barely been experienced, yet it is here that genome and gene sequences could have a major, short-term impact. Detailed molecular information is available for few proteins from tick-borne pathogens and, relatively, for even fewer tick proteins. As an indication of the current limitations, for 20 of the most abundantly expressed proteins from unfed *R. (Boophilus) microplus* larvae, only one could be identified from a mass fingerprint alone, whereas 18 others were tentatively identified through BLAST searches and the limited existing EST database, that is, using genomic resources. The importance of translated gene or genome sequences is likely to be even greater for less abundant proteins. Probably the most studied tick proteins are those involved in the control of host hemostasis, thought to be critical to the success of the feeding tick.

Hemostasis occurs following vascular injury and comprises three distinct events: vascular constriction, platelet aggregation and blood coagulation. Following hemostasis, clot dissolution occurs, enabling the resumption of blood flow after tissue repair. Ticks have developed a diverse array of anti-hemostatic agents that are considered to be essential for successful feeding and tick survival. Ticks circumvent the host defense mechanism through the injection of saliva containing large array of bioactive molecule including immunomodulators, vasodilators, anticoagulants, inhibitorsof platelet adhesion and aggregation and also fibrinolytic and/or fibrinogenolytic agents (Maritz-Olivier et al., 2007).

These anti-hemostatics have been found in salivary glands, saliva, eggs and hemolymph, appearing not only to prevent blood clot formation in the host, as well as the ingested blood meal, but also to regulate hemolymph coagulation in the tick itself. Recently, alternative ways to control ticks have been developed, including the employment of anti-tick vaccines with either concealed or exposed antigens. Tick anti-hemostatics are predominantly exposed antigens because they are secreted from tick salivary glands by regulated exocytosis. The importance of these inhibitors to successful tick feeding is exemplified by the finding of many such compounds in the tick salivary gland transcriptomes of *Ixodes pacificus, Ixodes scapularis* and *Haemaphysalis longicornis*.

A large number of tick anti-hemostatics have been identified, isolated and characterized, however, sequence, kinetic and structural data are lacking for the vast majority. Examples of the possible hemostatic candidates for vaccine production are summarized in (Table 2).

Current information on tick protein sequences is extremely restricted, limiting the application of proteomics. Homology matches have limited usefulness. We urgently need more information on the complement of proteins expressed in a variety of tick tissues, life stages and species, ideally from more than one species. Sequencing of another tick species within the metastriata should be initiated.

Recent advances in vector biology open new possibilities in target identification and vaccine development. The efforts to characterize the genomes of *I. scapularis* and *B. microplus* and also the current studies implying the use of cDNA libraries of *R. annulatus* (Shahein et al., 2008, 2010), *R. microplus* (Guerrero et al., 2005), *A. variegatum* (Nene et al., 2002), *I scapularis* (Ribeiro et al., 2006) in the identification of novel molecule, will impact on the discovery of new tick-protective antigens. The use of the information in conjugation with functional analysis using RNAi, bioinformatics, mutagenesis, transcriptomics, proteomics and other technologies will allow for a rapid, systematic approach to tick vaccine discovery by addressing the sequencing, annotation and functional analysis. Vaccination trials can be designed to evaluate the effect of selected tick antigens in combination with other tick protective and pathogen-specific antigens, for improving the level of tick infestations and reducing transmission of tick-borne diseases in cattle and other domestic animals. The future of research on development of tick vaccines is exciting because of new and emerging technologies for gene discovery that facilitate the efficient and rapid identification of candidate vaccine antigens.

Anti-hemostatic candidate	Source	Target	Molecular Mass (kDa)
Thrombin inhibitors			
R. microplus (Boophilin (G2 and H2))	Unknown	Thrombin	G2=14, H2=14
Amblyomma variegatum	Salivary Glands	Thrombin	NI
R. microplus (Microphilin)	Saliva	Thrombin exosite 1	1.8
H. dromedarii NTI-1	Nymphs	Thrombin and FX	3.2
H. dromedarii NTI-2	Nymphs	Thrombin and FX	15
Ixodes ricinus Ixin	Whole tick Extract	NI	7
FX, FXa and Tissue Factor Pathway Inhibitors (TFPI)			
Ixodes ricinus	Whole tick	FXa	NI
I. scapularis Ixolaris	Salivary glands	FX (Heparin binding exosite)	15.7
I. scapularis Salp14	Saliva	FXa	14
R. appendiculatus	Salivary glands	FXa, not the active site	65
Ornithodoros moubata TAP	Salivary glands	FXa	6
O. savignyi	Salivary glands	FXa	7
Uncharacterized intrinsic (INPI) and Extrinsic (EXPI) pathway			
R. microplus	Eggs, larvae	INPI, EXPI	NI
H. longicornis HLS1	Gut	INPI	41
O. savignyi BSAP1	Salivary glands	EXPI	9.3
O. savignyi BSAP2	Salivary glands	EXPI	9.7
Kallikrein-Kinin inhibitors			
R. microplus	Larvae	NI	18
R. microplus BmTI-D	Larvae	NI	8
R. sanguineus RsTIQ2	Larvae	NI	12
H. longicornis	Salivary glands	FXII, FXIIa and HK	16
Platelet aggregation/Adhesion inhibitors			
D. variabilis Variabilin	Salivary glands	$\alpha_{IIb}\beta_3$	5
O. moubata Moubatin	Whole tick	NI	17
O. moubata TAI	Salivary glands	$\alpha_1\beta_2$	15

Anti-hemostatic candidate	Source	Target	Molecular Mass (kDa)
Fibrin (ogen) olytic agents			
I. scapularis MP1	Salivary glands	Fibrin or Fibrinogen	36.9
I. ricinus Iris	Saliva	Elastase	44

HK: High molecular weight kininogen; NI: Not Identified; TAI: Tick Adhesion Inhibitor; TAP: Tick anticoagulant peptide

Table 2. Properties of anti-hemostatic molecules from different tick species.

6. References

Altschul, S. F., Gish, W., Miller, W., Myers, E. W. & Lipman, D. J. (1990) Basic local alignment search tool. *Journal of Molecular Biology*. 215, 403-410.

Ansari, F. A., Kumar, N., Bala Subramanyam, M., Gnanamani, M. & Ramachandran, S. (2008). MAAP: Malarial adhesins and adhesin-like proteins predictor. *Proteins*, 70, 659-666.

Bambini, S. & Rappuoli R. (2009). The use of genomics in microbial vaccine development. *Drug Discovery Today*, 14(5-6), (2009 Mar), 252-260.

Carlton, J. M., Angiuoli, S. V., Suh, B. B., Kooij, T. W., Pertea, M., Silva, J. C., Ermolaeva, M. D., Allen, J. E., Selengut, J. D., Koo, H. L., Peterson, J. D., Pop, M., Kosack, D. S., Shumway, M. F., Bidwell, S. L., Shallom, S. J., van Aken, S. E., Riedmuller, S. B., Feldblyum, T. V., Cho, J. K., Quackenbush, J., Sedegah, M., Shoaibi, A., Cummings, L. M., Florens, L., Yates, J. R., Raine, J. D., Sinden, R. E., Harris, M. A., Cunningham, D. A., Preiser, P. R., Bergman, L. W., Vaidya, A. B., van Lin, L. H., Janse, C. J., Waters, A. P., Smith, H. O., White, O. R., Salzberg, S. L., Venter, J. C., Fraser, C. M., Hoffman, S. L., Gardner, M. J. & Carucci, D. J. (2002). Genome sequence and comparative analysis of the model rodent malaria parasite *Plasmodium yoelii yoelii*. *Nature*, 419, 512-519.

Carpi, A., Di Maira, G., Vedovato, M., Rossi, V., Naccari, T., Floriduz, M., Terzi, M. & Filippini, F. (2002). Comparative proteome bioinformatics: identification of a whole complement of putative protein tyrosine kinases in the model flowering plant Arabidopsis thaliana. *Proteomics*. 2(11), (2002 Nov), 1494-1503.

Chaudhuri, R., Ahmed, S., Ansari, F. A., Vir Singh, H. & Ramachandran, S. (2008). MalVac: Database of malarial vaccine candidates. *Malaria Journal*, 7, (2008 Sep), 184-190.

Dalton, J. P., Brindley, P. J., Knox, D. P, Brady, C. P., Hotez, P. J., Donnelly, S., O'Neill, S. M., Mulcahy, G. & Loukas, A. (2003). Helminth vaccines: from mining genomic information for vaccine targets to systems used for protein expression. *International Journal for Parasitology*, 33 (5-6), (2003 May), 621-640.

Duffy, P., Krzych, U., Francis, S. & Fried, M. (2005). Malaria vaccines: using models of immunity and functional genomics tools to accelerate the development of vaccines against *Plasmodium falciparum*. *Vaccine*, 23 (17-18), (2005 March), 2235-2242.

Emanuelsson, O., Nielsen, H., Brunak, S. & von Heijne, G. (2000) Predicting subcellular localization of proteins based on their N-terminal amino acid sequence. *Journal of Molecular Biology*. 300, 1005-1016.

Emes, R.D., Ponting, C.P. (2001). A new sequence motif linking lissencephaly, Treacher Collins and oral-facial-digital type 1 syndromes, microtubule dynamics and cell migration. *Human Molecular Genetics*, 10 (24), (2001 Nov), 2813-2820.

Filippini, F., Rossi, V., Marin, O., Trovato, M., Costantino, P., Downey, P.M., Lo Schiavo, F. & Terzi, M. (1996). A plant oncogene as a phosphatase. *Nature*, 379 (6565), (1996 Feb), 499-500.

Gardner, M. J, Hall, N., Fung, E., White, O., Berriman, M., Hyman, R. W., Carlton, J. M., Pain, A., Nelson, K. E., Bowman, S., Paulsen, I. T., James, K., Eisen, J. A., Rutherford, K., Salzberg, S. L., Craig, A., Kyes, S., Chan, M. S., Nene, V., Shallom, SJ, Suh B, Peterson J, Angiuoli S, Pertea M, Allen J, Selengut J, Haft D, Mather M. W., Vaidya, A. B., Martin, D. M., Fairlamb, A. H., Fraunholz, M. J., Roos, D. S., Ralph, S. A., McFadden, G. I., Cummings, L. M., Subramanian, G. M., Mungall, C., Venter, J. C., Carucci, D. J., Hoffman, S. L., Newbold, C., Davis, R. W., Fraser, C. M. & Barrell, B. (2002). Genome sequence of the human malaria parasite *Plasmodium falciparum*. *Nature*, 419, 498-511.

Guerrero, F. D., Miller, R. J., Rousseau, M. E., Sunkara, S., Quackenbush, J., Lee, Y. & Nene, V. (2005). BmiGI: a database of cDNAs expressed in *Boophilus microplus*, the tropical/southern cattle tick. *Insect Biochemistry and Molecular Biology*. 35, 585–595.

Jongejan, F., Nene, V., de la Fuente, J., Pain, A. & Willadsen P. (2007).Advances in the genomics of ticks and tick-borne pathogens. *Trends in Parasitology*, 23 (9), 391-396.

Kolaskar, A. S & Tongaonkar, P. C. (1990) A semi-empirical method for prediction of antigenic determinants on protein antigens. *FEBS Letters*. 276, 172-174.

Kondrashov, F. A., Rogozin, I. B., Wolf, Y. I. & Koonin, E. V. (2002). Selection in the evolution of gene duplications. *Genome Biology*, 3,RESEARCH0008.

Krogh, A., Larsson, B., von Heijne, G. & Sonnhammer, E. L. (2001). Predicting transmembrane protein topology with a hidden Markov model: application to complete genomes. *Journal of Molecular Biology*. 305, 567-580.

Li, X., Zhou, L. & Gorodeski, GI. (2006). Estrogen regulates epithelial cell deformability by modulation of cortical actomyosin through phosphorylation of nonmuscle myosin heavy-chain II-B filaments. *Endocrinology*, 147(11), (2006 Aug), 5236-5248.

Marchler-Bauer, A., Anderson, J. B., Cherukuri, P. F., DeWeese-Scott, C., Geer, L. Y., Gwadz, M., He, S., Hurwitz, D. I., Jackson, J. D., Ke, Z., Lanczycki, C. J., Liebert, C. A., Liu, C., Lu, F., Marchler, G. H., Mullokandov, M., Shoemaker, B. A., Simonyan, V., Song, J. S., Thiessen, P. A., Yamashita, R. A., Yin, J. J., Zhang, D. & Bryant, S. H. (2005). CDD: a Conserved Domain Database for protein classification. *Nucleic Acids Research*. D192-196.

Maritz-Olivier, C., Stutzer, C., Jongejan, F., Neitz, A. & Gaspar, A. (2007). Tick anti-hemostatics: targets for future vaccines and therapeutics. *Trends in Parasitology*, 23 (9), 397-407.

Nene, V., Lee, D., Quackenbush, J., Skilton, R., Mwaura, S., Gardner, M. J. & Bishop, R. (2002). AvGI, an index of genes transcribed in the salivary glands of the ixodid tick *Amblyomma variegatum*. *International Journal for Parasitology*. 32, 1447–1456.

Oliver, S.G. (1996). From DNA sequence to biological function. *Nature*, 379, 597–600.

Rappuoli, R. (2001). Reverse vaccinology, a genome-based approach to vaccine development. *Vaccine*, 19(17-19), (2001 Mar), 2688-26891.

Ribeiro, J. M., Alarcon-Chaidez, F., Francischetti, I. M., Mans, B. J., Mather, T. N., Valenzuela, J. G. & Wikel, S. K. (2006). An annotated catalog of salivary gland transcripts from *Ixodes scapularis* ticks. *Insect Biochemistry and Molecular Biology*. 36, 111–129

Saha, S. & Raghava, G. P. (2007). Prediction methods for B-cell epitopes. *Methods in Molecular Biology*. 409, 387-394.

Serruto, D. & Rappuoli, R. (2006). Post-genomic vaccine development. *FEBS Letters*, 580, 2985-2992.

Shahein, Y. E. (2008). Molecular cloning and expression of an immunogenic larval protein from the cattle tick *Boophilus annulatus*. *Veterinary Immunology and Immunopathology*. 121(3-4), 281-289.

Shahein, Y. E., Abd El- Rahim, M. T., Hussein, N. A., Hamed, R. R., El Hakim, A. E., & Barakat, M. M. (2010). Molecular cloning of a small Heat Shock Protein (sHSPII) from the cattle tick *Rhipicephalus (Boophilus) annulatus* salivary gland. *International Journal of Biological Macromolecules* 47(5), 614-622.

Vacca, M., Filippini, F., Budillon, A., Rossi, V., Della Ragione, F., De Bonis, M.L., Mercadante, G., Manzati, E., Gualandi, F., Bigoni, S., Trabanelli, C., Pini, G., Calzolari, E., Ferlini, A., Meloni, I., Hayek, G., Zappella, M., Renieri, A., D'Urso, M., D'Esposito, M., Macdonald, F., Kerr, A., Dhanjal, S. & Hulten, M. (2001). MECP2 gene mutation analysis in the British and Italian Rett Syndrome patients: hot spot map of the most recurrent mutations and bioinformatic analysis of a new MECP2 conserved region. *Brain Development*., 23 Suppl 1, (2001 Dec), S246-S250.

Van Zee, J. P., Geraci, N. S., Guerrero, F. D., Wikel, S. K., Stuart, J. J., Nene, V. M. & Hill, C. A. (2007). Tick genomics: The Ixodes genome project and beyond. *International Journal for Parasitology* 37, 1297–1305

Vivona, S., Gardy, J.L., Ramachandran, S., Brinkman, F.S., Raghava, G.P., Flower, D.R. & Filippini, F. (2008). Computer-aided biotechnology: from immuno-informatics to reverse vaccinology. *Trends in Biotechnology*, 26(4), (2008 Feb), 190-200.

Vivona, S., Bernante, F. & Filippini, F. (2006). Nerve: New Enhanced Reverse Vaccinology Environment. *BMC Biotechnology*, 6, 35-42.

World Health Organization (1997). World malaria situation in 1994. *The Weekly Epidemiological Record*. 72, 269-274.

Wykes, M. & Good, M.F. (2007). A case for whole-parasite malaria vaccines. International Journal for Parasitology. 37(7), (2007 Feb), 705-712.

13

Cloning the Ribokinase of Kinetoplastidae: *Leishmania Major*

Patrick Ogbunude[1], Joy Ikekpeazu[1], Joseph Ugonabo[2],
Michael Barrett[4] and Patrick Udeogaranya[3]
[1]*University of Nigeria, Enugu-Campus, Enugu*
[2]*Department of Microbiology, University of Nigeria, Nsukka*
[3]*Department of Pharmacy, University of Nigeria, Nsukka,*
[4]*University of Glasgow, Glasgow,*
[1,2,3]*Nigeria*
[4]*Scotland*

1. Introduction

The kinetoplastidae are flagellated protozoans that are widely distributed in nature and cause diseases in both plants and vertebrates. They are distinguished by presence of kinetoplast, that is, the DNA-containing region in their single large mitochondrion. The diseases in crops and lifestocks cause considerable economic loss while serious human suffering and death occur in infections in man. In humans, the diseases include trypanosomiasis, leishmaniasis and chagasis. Trypanosomiasis is caused by infection with two of the three subspecies of *Trypanosoma brucei*; Chagas disease is caused by infection with *Trypanosoma cruzi* and various forms of leishmaniasis are caused by different species of *Leishmania*. These forms of diseases have been classified by the World Health Organization (WHO) as major tropical diseases. According to WHO estimates, about 12 million people suffer from the disease but close to 350 million people in 80 countries world-wide are at risk (Anon, 1990). They are endemic in tropics, subtropics and southern Europe in settings ranging from rain forests in the Americas to deserts in Asia and Middle East. The cellular biology of these kinetoplastids is essentially similar, for example, they are all motile protozoans with a single flagellum that originates close to their large single mitochondrion. They all have glycosomes, that is, micro-bodies that perform glycolysis. All typically grow asexually although sexual recombination has been shown or inferred but is not obligate in any one of them. They divide by binary fission during which their nucleus does not undergo membrane dissolution or chromosome condensation. They are well adapted to their hosts and evade immune elimination by antigenic variation, and alteration of immune responsiveness. There is no effective immune response against human trypanosomiasis which invariably results in fatality. In the case of *T. cruzi* and Leishmania spp., the immune response tends to control rather than eliminate them.

Leishmaniases constitute a broad spectrum of diseases, including the localized cutaneous and the disseminated visceral and mucocutaneous forms. Cutaneous and mucosal leishmaniases cause chronic skin sores and facial disfigurement respectively while untreated visceral leishmaniasis, otherwise known as kala-azar, causes life-threatening systemic infection.   There are a total of about 21 leishmanial species that cause leishmaniases and these are transmitted by about 30 species of phlebotomine sandflies (Desjeux 1996, Ashford 1997).   Globally, leishmaniasis caused approximately 59000 deaths and 2.4 million disability adjusted life years in 2001 (World Health Organization Report, 2001).   The life cycle of Leishmania involves two stages, a stage in the sand fly vectors where they exist mostly as promastigotes and another stage in the mammalian host.   The promastigotes are injected into human host where they invade the macrophages by receptor-mediated endocytosis, transforming into amastigotes that multiply by binary fission.

Leishmaniasis is a treatable disease, however, the antileishmanial therapy is bewildering largely because of the complexities of the disease.   The few effective agents available generally are potentially toxic and mostly are difficult to administer.   Furthermore the treatment of cutaneous leishmaniasis, for example, is often complicated by rapid self healing making it difficult to assess efficacy of trials.   The pentavalent antimony compounds, sodium stibogluconate (Pentostam, Glaxo Wellcome, UK) and meglumine antimonite (Glucantime, Rhône-Poulenc Rorer, France) have been the mainstays of antileishmanial therapy since 1940s (Berman, 1997, Herwaldt & Berman 1999).   These drugs, although, effective suffer from disadvantages of long duration of therapy, parenteral mode of administration, almost always reversible toxic effects among other disadvantages. Other new approaches to management of leishmaniasis have some merits but unfortunately most of the non-parenteral agents that have been assessed at best have modest activity against a limited range of species and strains (Herwaldt & Berman, 1992). New effective, safe and affordable drugs are needed for all kinetoplastids. It will be better to have more than one new drug so that combination therapy can be employed whenever drug resistance arises and also provide back-up when resistance emerges.

Kinetoplastids have been useful for study of fundamental molecular and cellular phenomena like antigenic variation, RNA editing and mRNA trans-splicing (Borst & Rudenko, 1994, Stuart, 1991, Perry & Agabian, 1991).   The comparison of their sequences with other eukaryotes may be useful in identification of ancient conserved motifs considering their early evolutionary divergence.   Also their protein sequence may be a useful source of diversity for protein engineering.

Rational development of anti-leishmanial drugs that will exploit biochemical differences between host and parasite is most desirable.

There are challenges to achieving this goal but with current available technologies, it is possible.   With post-genomic bioinformatics and experimental research, it is possible to identify drug targets, vaccine candidates and pathogenic processes.   It is also possible to identify candidate diagnostics which realistically should be non-invasive, inexpensive and deployable at the poor resource sites. The new diagnostics in addition should discriminate between types of diseases.

Ribose metabolism in Leishmania is of interest because like in other protozoa parasites, it is auxotrophic for purines and expresses multiple pathways for purine uptake (Landfear et al., 2004) and salvage (Hwang & Ulman, 1997) that enable it to acquire and use these vital metabolites from the hosts.   A metabolic pathway of interest is that involved in production of ribose 5-phosphate (R5P) required for the synthesis of 5-phosphoribosyl-1-pyrophosphate used with nucleobases for the synthesis of nucleic acids.   Classically ribose 5-phosphate is generated in cells by one or combination of the following pathways:

- Ribokinase conversion of ribose (reaction 1)
- Condensation of fructose 6-phosphate and glyceraldehyde 3-phosphate mediated by transaldolase and transketolase (reaction 2).
- Conversion of glucose via oxidative pathway of pentose phosphate pathway (reaction 3).
- Hydrolysis of nucleoside to nucleobase and ribose followed by reaction 1

The stoichiometric expressions of these reactions are:

1. Ribose + ATP $\longrightarrow$ R5P + ADP
2. 2F6P + G3P $\longrightarrow$ 3R5P
3. G6P + 2NADP$^+$ + H$_2$O $\longrightarrow$ R5P + 2NADPH 2H$^+$ + H$_2$O
4. Nucleoside $\longrightarrow$ Base + Ribose $\longrightarrow$ R5P

Reaction 4 is unique to trypanosomatids including *Leishmania* .

The consequence of reaction 4, that is the nucleoside hydrolase activity, is the abundance of intracellular ribose available to *Leishmania* parasites. The extracellular ribose has been shown previously to be efficiently incorporated into the nucleic acids of leishmania (Maugeri *et al.*, 2003). Ribose may also serve as a source of energy for organisms (Berens *et al.*, 1980) hence there is every need to study the enzyme (ribokinase) that is involved in the mobilization of ribose. Ribokinase is an ATP-dependent phosphoribosyl kinase (EC 2.1.7.15), which catalyses the conversion of ribose to ribose 5-phosphate, a substrate of 5-phosphoribosyl 1-pyrophosphate synthatase, that uses nucleobases and ribose 5-phosphate to synthesize nucleic acids. X-ray crystallography of *Escherichia coli* ribokinase shows the protein to be a homodimer in solution with each subunit having two domains. Each subunit has a molecular weight of about 33 KDalton (Sigrell *et al.*, 1997, Sigrell *et al.*, 1998, Sigrell *et al.*, 1999). The turnover numbers for *Leishmania* ribokinase for the substrates D-ribose and ATP are respectively10.8 s^{-1} and 10.2 s^{-1} and the catalytic activity is strongly dependent on the presence of monovalent cations(Ogbunude *et al.*, 2003, Chuvikovsky *et al.*, 2006).

Because of these important roles played by ribokinase in *Leishmania* metabolism, we cloned the genes for the enzyme and expressed it in *E coli* as previously published (Ogbunude *et al.*, 2007). Here we summarize the method used in cloning and expression of the enzyme.

2. Materials and methods

2.1 Chemicals

Synthetic oligonucleotides were obtained from MWG-Biotech AG, UK. All restriction and DNA modifying enzymes were obtained from Promega (UK) or Roche, Roche Diagnostics GmbH, Germany. All other chemicals and reagents were of the highest quality commercially available.

2.2 Methods
2.2.1 Culture of *L.major* promastigotes

The *L. major* promastigotes (MHOM/JL/80/Friedlin) were grown in tissue culture flasks in HOMEM medium supplemented with heat-inactivated 10% fetal bovine serum at 27°C. Promastigotes were harvested at approximately 10^7 cells/ml at the end of logarithmic growth phase. Following isolation and washing in wash buffer (50 mM Tris-HCl, pH 8.0), the genomic DNA of the promastigotes was isolated by published procedures with some modifications (Wilson *et al.*, 1991).   Briefly, promastigotes were resuspended to a density of 5 x 10^9 per 450 µl in 50 mM Tris-HCl, pH8, 50 mM EDTA, 100 mM NaCl. 50 µl of 10% SDS was added to this mixture with 50 µl of proteinase type IV (40 mg ml⁻¹ stock) and 2000 U ml⁻¹ RNAse inhibitor. The mixture was incubated at 37°C for 2 hours with intermittent shaking. To the mixture an equal quantity of phenol was added and then the supernatant collected and an equal volume of 3:1 v/v phenol/chloroform added to this supernatant. After mixing and separation of the phases by centrifugation the aqueous phase was again taken and this time mixed with 4 volumes of chloroform. The supernatant was collected, and 0.1 volumes of 3 M sodium acetate (pH 7.0) added followed by 2.5 volumes of ethanol. Nucleic acid was left to precipitate at –20°C overnight and then collected by centrifugation. The precipitate was washed once in 70% ethanol, air-dried and then dissolved in 50 µl of double distilled H_2O.

2.2.2 Design of oligonucleotide primers

The gene coding a putative ribokinase was identified in the *L. major* genome using BLAST searches with ribokinase of diverse organisms (Fig 1).   Oligonucleotide primers were designed against conserved regions of proteins, which represented an open reading frame. The forward primer (5'-AAA<u>CATATG</u>CACCGTGTGCAGAACGTT-3') was designed from the   first   peptide   of   the   protein   while   the   reverse   (5'-AAA<u>CTCGAG</u>CTACGTGACACCAGCC-3') was designed from the second peptide.   The underlined bases represent restriction sites (*Nde1* and *xho1*) that were inserted to facilitate cloning of the PCR amplified products.   The oligonucleotides were synthesized by MWG-Biotech AG, UK.

2.2.3 PCR amplification

Amplification was carried out in a DNA thermal cycler (MJ Research Inc., Western Town, MA, USA).   PCR was routinely performed in 100 µl reaction containing 10 mM Tris-HCl, pH 8.3, 1.5 mM $MgCl_2$, 50 µM KCl, 200 µM each of dNTPs, 40 ng of each of the primer and 1 unit of Pfu DNA polymerase.   The tube containing the reactants were placed in the thermo cycler programmed for 30 cycles, a single cycle at 94°C for 120 s; was followed by 30 cycles at 63°C for 15 s and 72°C for 120 s; with a final cycle at 72°C for 600 s.

2.2.4 Detection of PCR products

Five micro liters of the PCR product was electrophoresed in a 1% agarose gel containing 0.1 µg of ethidium bromide per ml, and bands were visualized by UV transillumination.

2.2.5 Preparation of cloning vector and insertion

The amplified DNA, after gel electrophoresis was isolated using the commercially available QIAquick Gel Extraction Kit (Qiagen, West Sussex).   The gel-purified PCR product (1.2-kb amplicon) was cut with *Nde1* and *xho1* at the primer sequence sites underlined.   This was ligated with the pGEM-T Easy plasmid (Novagen) cut with the same enzymes.   The recombinant plasmid was introduced into *E. coli* strain DH5α (Invitrogen) by heat transformation in Luria broth (LB) medium at 42°C for 50 sec. Agar plates with LB medium supplemented with ampicillin, isopropyl-1-thio-β-D-galactopyranoside (IPTG) and 5-bromo-4-chloro-3-indolyl-β-galactoside (X-gal) were used to test the recombinant strain. Positive strain was used for recombinant protein over-expression experiments.

2.2.6 Protein overproduction

For recombinant protein over expression, the Nde 1 and xho 1 digest of pGEM-T was inserted between the same sites of plasmid pET28a$^+$ (Novagen). The resulting plasmid pET28a$^+$ribo construct harbors the rbk gene under the control of a hybrid promoter-operator region, consisting of sequence of T7 promoter and a *lac* operator. The pET28a$^+$ribo was introduced into the BL21 (strain DE3)(Statagene) for protein over expression. Expression of the His-tagged ribokinase was induced by 0.5 mM IPTG overnight at room temperature. Cells were harvested by centrifugation, washed once and stored at –20°C in 5 ml of 50 mM Hepes buffer, pH 7.0 containing 300 mM NaCl and 10 mM EDTA. The cells were lysed by sonication in buffer A (50 mM NaH$_2$PO$_4$ containing 300 mM NaCl, pH 8.0) and the soluble fraction recovered by centrifugation at 10,000 g for 30 min at 4°C. This was applied to nickel-nitrilotriacetic acid column (bioCAD) pre-equilibrated with the buffer A. The column was washed with 100 ml of the buffer A containing 0.5 mM imidazole and then with the same buffer containing 50 mM imidazole and finally the his-tagged recombinant protein was eluted with 500 mM imidazole in the buffer A. The eluant was dialysed overnight in 50 mM Tris-HCl, pH 7.0 at 4°C and stored at –70°C in buffer/glycerol (1:1).

3. Results

Data base searches of *L. major* ribokinase sequence revealed a gene in chromosome 27 (LmjF27.0420) with an open reading frame of 990 base pairs encoding a 329 amino acid protein orthologous with ribokinase from other organisms . Figure 1 shows the alignment of this protein with ribokinase from phylogenetically diverse organisms. The alignment revealed a high degree of conservation of primary structure among the four aligned proteins.

The SDS-Page analysis of the bacterial extract shown in figure 2 demonstrated protein overproduction at the expected molecular weight of 34 Kdalton. The Coomassie staining indicated that the recombinant protein represented more than 90% of the total protein. After purification, the enzyme was judged to be over 95% homogenous.

The figure 3 graph shows time dependence of conversion of ribose to R5P. In this coupled assay system, ADP is utilized by phosphoenolpyruvate in reaction catalyzed by pyruvate kinase to regenerate ATP and the resulting pyruvate is converted by lactate dehydrogenase to lactate and in the process NADH is oxidized to NAD$^+$. The reaction was followed at 340 nm as a decrease in NADH absorbance when converted to NAD$^+$. The reaction demonstrates that the recombinant protein is indeed a functional enzyme. The enzyme when stored at -70°C in buffer/glycerol (1:1) retained more than 50% of activity after a year. It retained full activity for two months when stored at -70°C with or without glycerol. The enzyme is highly specific for D-ribose in species so far studied although it insignificantly phosphorylates other five and six carbon sugars like D-arabinose, D-xylose, D-galactose and D-fructose. The catalytic efficiency for catalysis of D-ribose ranges from 10.8 s^{-1} mM^{-1} to 40.6 s^{-1} mM^{-1} while it is in the order of 0.05 s^{-1} mM^{-1} to 1.8 s^{-1} mM^{-1} for other D-sugars. The ribokinae activity depends absolutely on the presence of inorganic phosphate. Omission of inorganic phosphate from reaction mixture reduced the enzyme activity to insignificant level.

```
S. cerevisiae (1) ----------------MGITVIGSLNYDLDTFTDRLPNAGETFRANHFETHAGGKGLNQAAAIGKLKNPSS
E. coli    (1) -----------MQNAGSLVVLGSINADHILNLQSFPTPGETVTGNHYQVAFGGKGANQAVAAGRSG----
H. sapiens  (1)
MAASGEPQRQWQEEVAAVVVVGSCMTDLVSLTSRLPKTGETIHGHKFFIGFGGKGANQCVQAARLG----
L. major    (1)
MHRVQNVQSHVGEYAPDILVVGSCFLDYVGYVDHMPQVGETMHSESFHKGFGGKGANQAVAAGRLG----


S. cerevisiae (56)
RYSVRMIGNVGNDTFGKQLKDTLSDCGVDITHVGTYEGINTGTATILIEEKAGGQNRLIVEGANSKTIY
E. coli   (56) -ANIAFIACTGDDSIGESVRQQLATDNIDITPVSVIKGESTGVALIFVNGEGEN--
VIGIHAGANAALSP
H. sapiens  (67) -AMTSMVCKVGKDSFGNDYIENLKQNDISTEFTYQTKDAATGTASIIVNNEGQN--
IIVIVAGANLLLNT
L. major   (67) -AKVAMVSMVGTDGDGSDYIKELERNGVHTAYMLRTGKSSTGLAMILVDTKSSNN-
EIVICPNATNYFTP


S. cerevisiae(126) DPKQLCEIFPEGKEE--EEYVVFQHEIPDPLSIIKWIHANRPNFQIVYN----
PSPFKAMPKKDWELVDL
E. coli  (123) ALVEAQRE---RIAN--ASALLMQLESPLESVMAAAKIAHQNKTIVALN----PAPARELPD--------
H. sapiens  (134) EDLRAAAN---VISR--AKVMVCQLEITPATSLEALTMARRSGVKTLFN----PAPAIADLDP-------
L. major  (135) ELLRAQTSNYEKILHTGLKYLICQNEIPLPTTLDTIKEAHSRGVYTVFNSAPAPKPAEVEQIK------
-


S.cerevisiae (190)
LVVNEIEGLQIVESVFDNELVEEIREKIKDDFLGEYRKICELLYEKLMNRKKRGIVVMTLGSRGVLFCSH
E. coli  (176) ------ELLALVDIITPNETEAEKLTGIRVENDEDAAKAAQVLHEKGIR-----TVLITLGSRGVWASVN
H. sapiens  (188) -------QFYTLSDVFCCNESEAEILTGLTVGSAADAGEAALVLLKRGCQ-----
VVIITLGAEGCVVLSQ
L. major  (198) ------PFLPYVSLFCPNEVEAALITGMKVTDTESAFRAIKALQQLGVR-----DVVITLGAAGFALSEN


S.cerevisiae (260) ESPEVQFLPAIQNVSVVDTTGAGDTFLGGLVTQLYQGET--
LSTAIKFSTLASSLTIQRKGAAESMPLYK
E. coli  (235) G--EGQRVPGFR-VQAVDTIAAGDTFNGALITALLEEK--
PLPEAIRFAHAAAAIAVTRKGAQPSVPWRE
H. sapiens  (247) TEPEPKHIPTEK-
VKAVDTTGAGDSFVGALAFYLAYYPNLSLEDMLNRSNFIAAVSVQAAGTQSSYPYKK
L. major  (257) G-AEPVHVTGKH-VKAVDTTGAGDCFVGSMVYFMSRGR--
NLLEACKRANECAAISVTRKGTQLSYPHPS


S. cerevisiae  (328) DVQKDA----
E. coli   (300) EIDAFLDRQR
H. sapiens  (316) DLPLTLF---
L. major   (323) ELPAGVT---
```

Fig. 1. Alignment of a putative ribokinase gene product from *L. major* with ribokinase sequences from *S. cerevisiae E. coli* and *H. sapiens*.

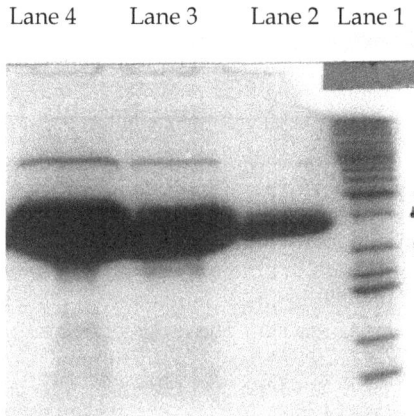

Fig. 2. SDS-Page analysis of bacterial extract

The SDS-Page (12.5% polyacrylamide) analysis of the bacterial extract shows protein overproduction at the expected molecular weight of 34,000 daltons.   The 1kb molecular weight marker which served as a control is in Lane 1. Lanes 2, 3 and 4 are fractions collected at intervals from the nickel-nitrilotriacetic acid column.

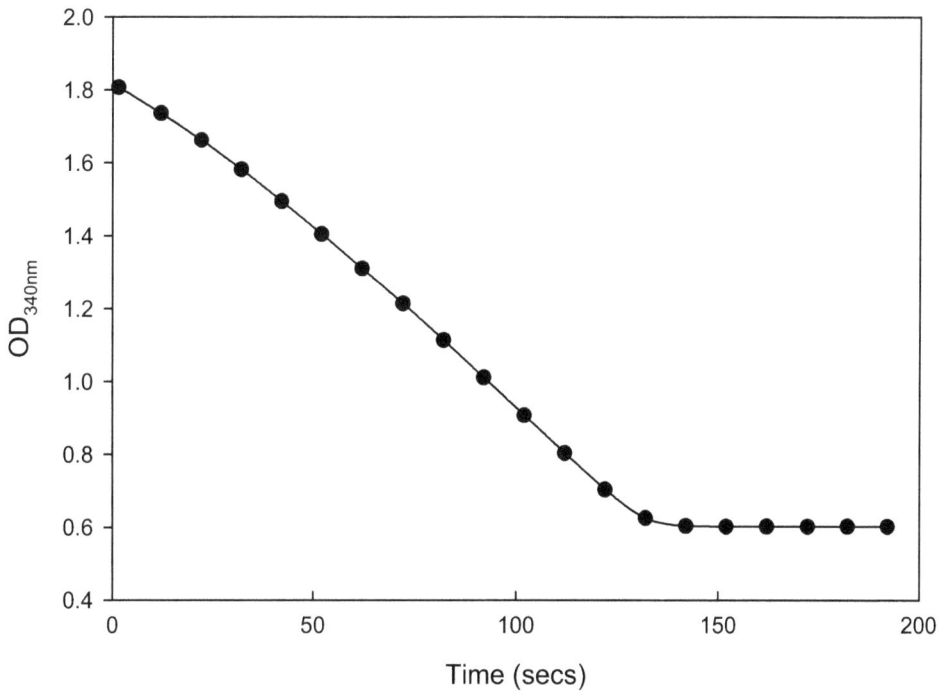

Fig. 3. Time dependence of conversion of ribose to ribose 5-phosphate by ribokinase

4. Conclusion

The interest in the uptake and metabolism of ribose as a potential metabolic supplement has justified the cloning and characterization of the enzyme. Administration of exogenous ribose for instance have been shown to increase both the repletion and maintenance of high level of ATP in various types of cells (Smolenski et al., 1998, St Cyr et al., 1989) and maintenance of higher levels of ATP in rat heart and dog kidney during transplantation experiments (Muller et al, 1998). Thus ribose has potential therapeutic applications involving cardiovascular cells. Studies into the major carbon sources used by intracellular Leishmania amastigotes show that L. mexicana mutant lacking the three high affinity hexose transporters was unable to establish infection in macrophages or susceptible mice (Burchmore et al., 2003), indicating that hexose uptake is essential for intracellular growth. The levels of hexose in the phagolysosome may, however, be limiting as mutant L. major amastigotes that have defect in gluconeogenesis was also found to be poorly virulent in macrophages and susceptible mice (Nadereer et al., 2006). These studies indicate that intracellular amastigotes depend on both salvage and de novo synthesis of hexoses. The phagolysosome of macrophages could however contain a range of other sugars including ribose which could be used directly as an energy source with the pentose phosphate pathway (Barrett 1997, Maugeri et al., 2003) shunting intermediates into the glycolytic pathway.

In parasitic organism such as Leishmania which, lacks purine nucleoside phosphorylase enzyme and instead has high activity of purine nucleoside hydrolase, the organism will generate high concentration of intracellular ribose. As well, it accumulates ribose from the host via a specific carrier-mediated protein (Pastakia & Dwyer, 1987). That ribose is encountered throughout the life cycle of the parasite is well known (Burchmore & Barrett, 2001). In absence of glucose in culture medium, ribose served as a substitute and became essential for the growth of the Leishmania promastigotes (Steiger & Black, 1980). Thus ribose plays an important role in the metabolism of Leishmania. For ribose to be utilized, it must be converted to ribose 5-phosphate by ribokinase, hence the interest in identifying and cloning of this enzyme responsible for this critical first step in ribose metabolism.

The ribokinase gene has been identified from variety of sources including human, E. Coli and Leishmania. There is a high degree of similarity between the ribokinase gene from animal (human) and that from E. Coli (Park et al., 2007). The structure of ribokinase as revealed by X-ray crystal structure of E.coli shows that it is a homodimer in solution with each subunit composed of two domains (Sigrell et al., 1997, Sigrell et al., 1998, Bork et al., 1993). Also the crystallographic studies show that oxygen atoms of the α-D-pentofuranose ring are involved in hydrogen bonding interactions between the enzyme and the substrates, ribose and ADP (Andersson & Mowbray 2002, Maj & Gupta 2001). The amino acids residues, asparagines and glutamic acid, at sequence positions 187 and 190, appear to be conserved in all the species studied and have been shown to be the site for phosphate binding (Park et al. 2007). Site directed mutation of these amino acid residues have led to formation of mutant enzymes with altered $MgATP^{2+}$ and phosphate requirements (Parducci et al., 2006).

In summary, L. major has a gene that expressed functional ribokinase. The enzyme was specific for D-ribose and did not phosphorylate related sugars unlike what was seen with E.coli ribokinase that phosphorylated related sugars to a minor degree. Presently, it is not known whether the enzyme is essential. A lot more information is still required before specific compounds are designed to target the enzyme for chemotherapy.

5. Acknowledgement

The authors appreciate the financial support of The University of Nigeria in defraying part of the cost of publishing this article. The research was supported by a grant from the European Commission INCO-DC Program (ERBIC18CT980357).

6. References

Andersson, CE. & Mowbray, SL. (2002) Activation of ribokinase by monovalent cations. *Journal of Molecular Biology*, 315, 409-419

Anon (1990) World Report on Tropical Diseases, World Health Organization

Ashford, RW. (1997) The leishmaniases as model zoonoses. *Annals of Tropical Medicine Parasitology.* 91: 693-701.

Barrett, MP. (1997) The pentose phosphate pathway and parasitic protozoa . *Parasitology Today* 13, 11-16.

Berens, RL. Deutsch-King, LC. & Marr, JJ. (1980) *Leishmania donovani* and *Leishmania braziliensis*: hexokinase, glucose 6-phosphate dehydrogenase, and pentose phosphate shunt activity. *Experimental Parasitology* 49, 1-8.

Berman, JD. (1997) Human leishmaniasis: clinical, diagnostic and chemotherapeutic developments in the last 10 years. *Clinical Infectious Diease* 24:684-703.

Bork, P. Sander, C. & Valencia, A. (1993) Convergent evolution of similar enzymatic function on different protein folds: The hexokinase, ribokinase, and galactokinase families of sugar kinases. *Protein Science* 2, 31-40

Borst, P. & Rudenko, G. (1994) Antigenic variation in African trypanosomes. *Science* 264, 1872-1873

Burchmore, RJ. & Barrett, MP. (2001) Life in vacuoles-nutrient acquisition by *Leishmania* amastigotes. *International Journal Parasitology* 31, 1311-1320.

Chuvikovsky, DV. Esipov, RS. Skoblov, YS. Chupova, LA. Muravyova, TI. Miroshnikov, AI. Lapinjoki, S. & Mikhailopulo, IA. (2006) Ribokinase from *E. Coli*: Expression, purification, and substrate specificity. *Bioorganic Medicinal Chemistry* 14, 6327-6332.

Desjeux, P. (1996) Leishmaniasis: public health aspects and control. *Clinical Dermatololgy*, 14:417-423.

Herwaldt, BL. Berman, JD. (1992) Recommendations for treating leishmaniasis with sodium stibogluconate (Pentosam) and review of pertinent clinical studies. *American Journal of Tropical Medicine and Hygiene* 46: 296-306.

Hwang, HY. & Ullman, B. (1997) Genetic analysis of purine metabolism in *Leishmania donovani*. *Journal of Biological Chemistry* 272, 19488-19496.

Landfear, SM. Ullman, B. Carter, NS. Sanchez,MA. (2004) Nucleoside and nucleobase transporters in parasitic protozoa. *Eukaryotic Cell* 3, 245-254

Maj, MC. & Gupta, RS. (2001) The effect of inorganic phosphate on the activity of bacterial ribokinase. *Journal of Protein Chemistry* 20, 139-144.

Maugeri, DA. Cazzulo, JJ. Burchmore,JS, Barrett, MP. Ogbunude, POJ. (2003) Pentose phosphate metabolism in *Leishmania Mexicana*. *Molecular and Biochemical Parasitology* 130, 117-125.

Muller, C. Zimmer, H. Gross, M. Gresser, U. Brotsack, I. Wehling, M. & Plim, lW. (1998) Effect of ribose on cardiac adenine nucleotides in a donor model for heart transplantation. *European Journal of Medical Research* 3, 554-558.

Naderer, T. Ellis, MA. Sernee, MF. De Souza DP. Curtis, J. Handman, E. & McConville, MJ. (2006) Virulence of Leishmania major in macrophages and mice requires the gluconeogenic enzyme fructose-1,6-bisphosphatase. Proceedings of National Academy of Science USA 103, 5502-5507.

Ogbunude, POJ. Lamour, N. Barrett, P. (2007) Molecular cloning, expression and characterization of ribokinase of Leishmania major. Acta Biochimica Biophysica (Sinica) 39, 462-466.

Parducci, RE. Cabrera, R. Baez, M. & Guixe, V. (2006) Evidence for catalytic Mg2+ ion and effect of phosphate on the activity of Escherichia coli phosphofructokinase-2: regulatory properties of a ribokinase family member. Biochemistry 45, 9291-9299.

Park, J. Koeverden, P. Singh, B. & Gupta, RS. (2007) Identification and characterization of human ribokinase and comparison of its properties with E. Coli ribokinase and human adenosine kinase. FEBS Letters 581, 3211-3216.

Pastakia, KB. Dwyer, DM. (1987) Identification and characterisation of a ribose transport system in Leishmania donovani promastigotes. Molecular Biochemical Parasitology 26, 175-182.

Perry, K.& Agabian, N. (1991) mRNA processing in the trypanosomatidae. Experientia 47, 118-128

Shi, W. Schramm,VL., Almo, SC. (1999) Nucleoside hydrolase from Leishmania major. Cloning, expression, catalytic properties, transition state inhibitors and 2.5-Å crystal structure. Journal of Biological Chemistry 274, 21114-211120.

Sigrell, JA. Cameron, AD. Jones, TA. & Mowbray, SL. (1997) Purificatioin, characterization and crystallization of Escherichia coli ribokinase. Protein Science 6, 2474-2476

Sigrell, JA. Cameron,AD. Jones, TA.& Mowbray, SL. (1998) Structure of Eschericia coli ribokinase in complex with ribose and dinucleotide determined to 1.8 Å resolution: Insights into a new family of kinase structures. Structure 6, 183-193.

Sigrell, JA. Cameron,AD. & Mowbray, SL. (1999) Induced fit on sugar binding activities ribokinase. Journal of Molecular Biology 290, 1009-1018.

Smolenski, RT. Kalsi, KK. Zych, M. Kochan, Z. & Yacoub, MH. (1989) Adenine/ribose supply increases adenosine production and protects ATP pool in adenosine kinase-inhibited cardiac cells. Journal of Molecular Cell Cardiology, 30,673-683.

St Cyr, JA. Bianco, RW. Schneider, JR. Mahony Jr, JR. Tveter, K. Einzig, S. & Foker, JE. (1989) Enhanced high energy phosphate recovery with ribose infusion after global myocardial ischemia in a canine model. Journal of Surgery Research 46, 157-162.

Steiger, RF. & Black, CD. (1980) Simplified defined media for cultivating Leishmania donovani promastigotes. Acta Tropica. 37, 195-198

Stuart K (1991) RNA editing in trypanosomatid mitochondria Annual Review of Microbiology 45, 327-344.

Wilson, K. Collart, FR. Huberman, E. Stringer, JL. & Ullman, B. (1991) Amplification and molecular cloning of the IMP dehydrogenase gene of Leishmania donovani. Journal of Biological Chemistry 266, 1665-1671.

World Health Organization, Annex 3: Burden of disease in DALYs by cause, sex and mortality stratum in WHO regions, estimates for 2001. In: The world health report, Geneva: WHO, 2002: 192-7 (www.who.int/whr/2002/whr 20002_annex3.pdf).

14

Phosphagen Kinase System of the Trematode *Paragonimus westermani*: Cloning and Expression of a Novel Chemotherapeutic Target

Blanca R. Jarilla[1] and Takeshi Agatsuma[2]
[1]Department of Immunology, Research Institute for Tropical Medicine (RITM),
[2]Department of Environmental Health Sciences, Kochi University, Kochi,
[1]Philippines
[2]Japan

1. Introduction

Paragonimiasis is a food-borne trematode infection that affects 22 million people in at least 20 countries (World Health Organization [WHO], 2002) with 293 million more at risk of infection (Keiser & Utzinger, 2007). In East, Southeast, and South Asia, pulmonary paragonimiasis is commonly caused by the trematode *Paragonimus westermani* (Blair et al., 2007). Humans get infected by this lung fluke by ingesting metacercariae present in raw freshwater crabs or by eating raw meat of paratenic hosts such as omnivorous mammals (Miyazaki & Habe, 1976). Human paragonimiasis has been reported to be re-emerging in previously endemic areas in Japan (Mukae et al., 2001; Nakano et al., 2001, and Kirino et al., 2009) and increasing in some regions of China (Lieu et al., 2008). New foci of transmission have also been reported in Lao PDR (Odermatt et al., 2009).

Currently, either praziquantel or triclabendazole is effective for the treatment of pulmonary paragonimiasis. However, since these are the only viable drugs against this infection, there is a need to develop back up drugs while drug resistance is not yet emerging (Keiser & Utzinger, 2007). In addition, a more specific and sensitive detection tool is needed for diagnosing pulmonary paragonimiasis since a considerable number of cases are misdiagnosed as tuberculosis or vice versa due to similarity of some signs and symptoms (WHO, 2002).

The rapid availability of parasite genomic sequences coupled with development of robust bioinformatics tools have resulted in the identification of numerous potential drug targets. These include the phosphagen kinases (PKs) that catalyze the reversible transfer of a phosphate between ATP and naturally occurring guanidino substrates. These enzymes play a key role in maintaining cellular energy homeostasis through temporal energy buffering that stabilizes the cellular ATP/ADP hydrolysis (Ellington, 2001). Studies on PKs of parasitic protozoans and nematodes have shown that PKs are important in energy metabolism and adaptation to stress conditions (Platzer et al, 1999; Alonso et al., 2001; Miranda et al., 2006; Pereira et al., 2003; Pereira et al., 2002). Enzyme activity of PK in *Trypanosoma cruzi*, the causative agent of Chagas disease, has been shown to be inhibited by various compounds such as catechin gallate (Paveto et al., 2004) and arginine analogs

(Pereira et al., 2003). In this chapter the cloning, expression and molecular characterization of a novel phosphagen kinase from the lung fluke *P. westermani* will be described. Mutation studies for the elucidation of substrate binding mechanism and phylogenetic analyses will also be presented.

2. Cloning and molecular characterization of *P. westermani* PK

At present, eight PKs and their corresponding phosphagens have been identified (Ellington, 2001). Creatine kinase (CK) is the sole PK in vertebrates. In addition to CK, the following PKs are found in various invertebrate species: arginine kinase (AK) (Uda et al. 2006), hypotaurocyamine kinase (HTK), glycocyamine kinase (GK), thalessemine kinase (ThK); opheline kinase (OK), lombricine kinase (LK), and taurocyamine kinase (TK) (Robin, 1974; Thoai, 1968; Morrison, 1973). Among the PKs, AK is the most widely distributed being present in deuterostomes, protostomes, basal metazoans, some protozoans (Uda et al. 2006) and in the prokaryote *Desulfotalea psychrophila* (Andrews et al., 2008).

Besides *T. cruzi*, phosphagen kinases have also been identified in other important animal and human parasites. AKs were cloned from the nematodes *Ascaris suum* and *Toxocara canis* which can cause visceral larva migrans (VLM) in humans (Nagataki et al., 2008; Wickramasinghe et al., 2007) and from *T. brucei* which causes human sleeping sickness and Nagana in livestock (Pereira et al., 2002). The PK from the trematode *Schistosoma mansoni* was also recently described by Awama et al. (2008).

To determine the cDNA sequence of the PK in *P. westermani*, total RNA was first isolated from samples collected from definitive hosts in Bogil Island, South Korea using the acid guanidinium thiocyanate–phenol–chloroform extraction method (Chomczynski & Sacchi, 1987). Messenger RNA (mRNA) was purified with a poly (A)+ isolation kit (Nippon Gene, Tokyo, Japan). First-strand cDNA was synthesized from 20 ng to 2 µg of mRNA using the Ready-To-Go You-Prime First-Strand Beads (Amersham Pharmacia Biotech, NJ, USA) which utilizes Moloney Murine Leukemia Virus (M-MuLV) as reverse transcriptase. One microliter of 10 pmol lock-docking oligo(dT) primer was used for the first-strand synthesis. The 5′ half of *P. westermani* PK cDNA was first amplified by RT-PCR using the universal PK primers SmTcPKptnF1 and SmTcPKptnR1 and ExTaq DNA polymerase (Takara, Kyoto, Japan) as the amplifying enzyme. The PCR reaction components are listed in Table 1 while the amplification conditions are in Table 2.

Components	Concentration/25 µL reaction volume
cDNA	5 µL
Forward primer	10 pmol
Reverse primer	10 pmol
Ex Taq™ dNTPs	0.2 mM each
10 × Ex Taq™ buffer	1X
ExTaq™ polymerase	2.5 U

Table 1. Components of the PCR reaction mixture for the amplification of *P. westermani* PK cDNA.

Step	Temperature	Time	Cycles
Initial denaturation	94°C	2 min	1
Denaturation	94°C	30 sec	35
Annealing	55°C	35 sec	35
Extension	72°C	2 min	35
Final Extension	72°C	4 min	1

Table 2. Thermal cycling conditions for the amplification of *P. westermani* PK cDNA.

The amplified products were purified using GeneClean® II Kit (QBIOgene, USA) and subcloned into the pGEM® T-vector (Promega, USA) and transformed into *E. coli* JM109 cells (Takara, Japan). After transformation, the bacteria are plated on Luria-Bertani plate containing 5-bromo-4-chloro-3-indoyl-β-D-galactoside (X-gal), isopropyl β-D-1 galactopyranoside (IPTG), and ampicillin. Positive clones were obtained and cultured overnight in liquid 2 ml Luria Bertani (LB) medium with ampicillin. Plasmid DNA was extracted using the alkaline SDS method and nucleotide sequencing was done with an ABI PRISM 3100-Avant DNA sequencer using a Big Dye Terminators v3.1 Cycle Sequencing Kit (Applied Biosystems, CA, USA). From the obtained partial sequence, the specific primer PwKoreaPKF1 was designed and used, together with the lock-docking oligo (dT) primer, to amplify and determine the remaining sequence of the 5′ half. A poly (G)+ tail was added to the 3′ end of the *P. westermani* cDNA pool with terminal deoxynucleotidyl transferase (Promega, WI, USA). The 3′ half of the PK cDNA was then amplified using the oligo(dC) primer and a specific primer PwKoreaPKR3 designed from the sequence of the 5′ half. The amplified products were purified, subcloned, and sequenced as described above. The sequences of the primers used for cDNA amplification are listed on Table 3.

Primer name	Sequence (5′-3′)
Oligo(dT)	GACTCGAGTCGACATCGA(T)$_{17}$
SmTcPKptnF1	CTNMCNAARAARTAYCT
SmTcPKptnR1	AGNCCNAGNCGNCGYTRTT
PwKoreaPKF1	TCTGTGAGGAGGATCATAT
Oligo(dC)	GAATT(C)$_{18}$
PwKoreaPKR3	TTTTTGTTGTGGAAGATCCC

Table 3. Primers used for the amplification of *P. westermani* PK cDNA.

P. westermani PK cDNA comprises 2, 305 bp with 163 bp of 3′ UTR; the 5′UTR was not successfully amplified. The ORF consisting of 2, 142 bp codes for 713 amino acid residues and the translated protein has a calculated mass of 80, 216 Da and an estimated pI of 7.86. Alignment of *P. westermani* PK amino acid sequence with other PKs indicated that this enzyme has a contiguous two-domain structure potentially encoding for two distinct PK enzymes. Domain 1 (D1) (Fig. 1) consists of 360 amino acids with a calculated mass of 40, 422 Da and an estimated pI of 8.47. Domain 2 (D2) consists of 353 amino acids with a calculated mass of 39, 583 Da and an estimated pI of 7.63.

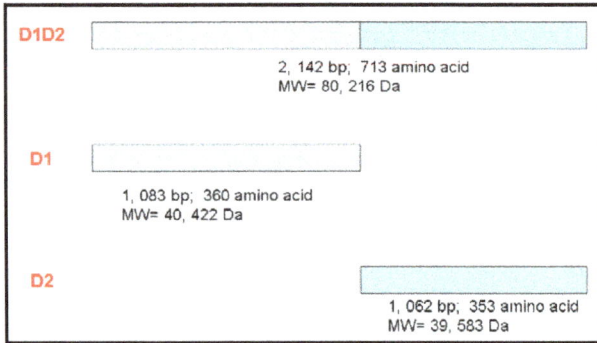

Fig. 1. Schematic representation of *P. westermani* PK

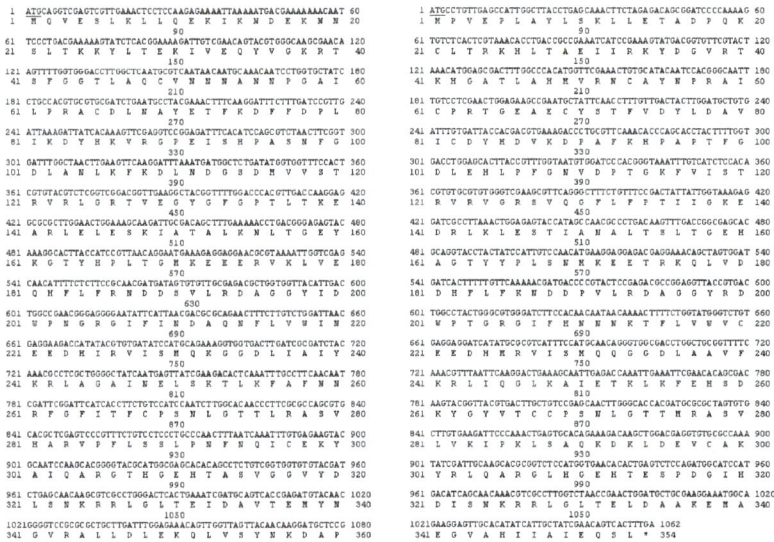

a) Domain 1   b) Domain 2

Fig. 2. Nucleotide and amino acid sequence of ORF of *P. westermani* PK (start codon is underlined and stop codon is marked with *).

Two-domain AKs were also reported for the sea anemone *Anthopleura japonicus* (Suzuki et al., 1997), the clams *Pseudocardium sachalinensis* (Suzuki et al., 1998), *Corbicula japonica, Solen strictus* (Suzuki et al., 2002), *Ensis directus* (Compaan & Ellington, 2003), and *Calyptogena kaikoi* (Uda et al., 2008). The PK found in *S. mansoni* also has a contiguous two-domain structure (Awama et al., 2008). These multiple domain AKs are products of gene duplication and subsequent fusion as suggested by the presence of a bridge intron separating the two domains of *A. japonicus* and *P. sachalinensis* AKs (Suzuki & Yamamoto, 2000).

Members of the phosphagen kinase family share high sequence identity suggesting that these enzymes have evolved from a common ancestor (Suzuki et al., 1997). Phylogenetic analyses have shown that PKs can be divided into two distinct lineages, an AK lineage and a CK lineage (Uda et al., 2005). It is probable that the ancestral PK was monomeric and an early gene duplication event resulted in the formation of these lineages. Moreover, the genes coding for the ancestral AK and CK could have been present early in the evolution of metazoans since AKs and CKs are found in both deuterostomes and protostomes (Ellington & Suzuki, 2006). The phylogenetic tree constructed based on the amino acid sequence of *P. westermani* PK and other phosphagen kinases using the Neighbor-joining method in MEGA version 4 (Tamura et al., 2007) shows the presence of two major clusters: a CK cluster and an AK cluster. The CK cluster includes the CKs from vertebrates and other PKs from annelids. The AK cluster, on the other hand, is divided into two subclusters. The first subcluster contains the nematode, arthropod, and protozoan AKs. *P. westermani* PK falls in the second subcluster together with *S. mansoni* PK, molluscan AKs and sipunculid HTK.

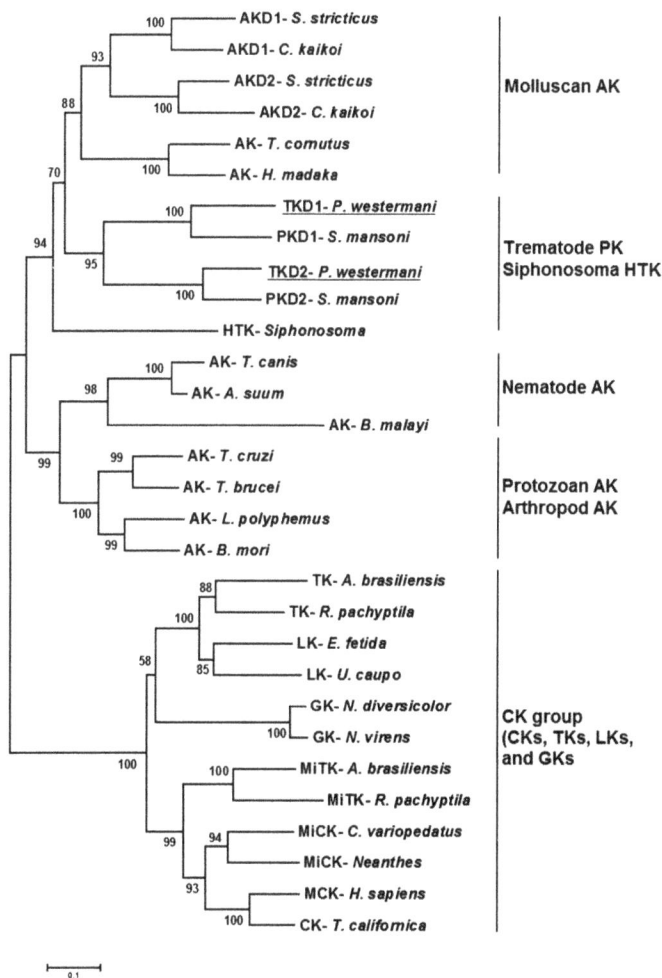

Fig. 3. Neighbor-joining tree for the amino acid sequences of phosphagen kinases (Jarilla et al., 2009).

3. Expression of recombinant *P. westermani* PK and enzyme kinetics

The full-length and truncated domains of *P. westermani* PK were expressed in *E. coli* as MBP fusion proteins. Specific primers with EcoRI and PstI restriction sites were designed to amplify the open reading frames (ORFs) of *P. westermani* PK. The forward primer PwKorPK5D1EcoRI (5'TT<u>GAATTC</u>GCCGATGTCATGCCTGTTGAG-3') and reverse primer PwKorPK3D1PstI (5'-TT<u>CTGCAG</u>TTACGGAGCATCCTTGTTGTAA-3') were used to amplify D1 while the primers PwKorPK5D2EcoRI (5'-TT<u>GAATTC</u>G CCGATG TCATGCCTGTTGAG-3') and PwKorPK3D2PstI (5'-TT<u>CTGCAG</u>TCAAA GTGACTGTTCGATAGC-3') were used for D2. PwKorPK5D1EcoRI and PwKorPK3D2PstI were used to amplify the ORF of the full-length construct. PCR was done in a total volume of 50 μL using KOD DNA polymerase for high fidelity amplification. The thermal cycling condition used was similar as described above. The components of the PCR reaction mixture are shown on Table 4.

Components	Concentration/25 μL reaction volume
cDNA	5 μL
Forward primer	10 pmol
Reverse primer	10 pmol
KOD dNTPs	0.2 mM each
MgSO$_4$	2 mM
10 × KOD Plus buffer	1X
KOD Plus DNA polymerase	1 U

Table 4. Components of the PCR reaction mixture for amplification of *P. westermani* PK ORF.

Since KOD DNA polymerase produces blunt-ended DNA products, an A-tail was added to the 3' end of the PCR product. Amplified products were first purified using the QIA quick PCR purification columns (QIAGEN, GmbH, Hilden, Germany). The 30 μL A-tailing reaction mixture contained purified KOD PCR product, 15 U of Gene Taq DNA Polymerase (Wako Nippon Gene, Japan), 3 μL of 10X Gene Taq Buffer, and 1.2 μL of 5 mM dATP. The mixture was incubated at 70°C for 30 minutes. A-tailing product was purified, subcloned, and sequenced as described above.

To isolate the ORFs of *P. westermani* PK D1, D2, and D1D2, plasmids with verified sequence were restriction digested and cloned into the EcoRI/PstI site of pMAL-c2 (New England Biolabs, MA, USA). Plasmids from selected clones were isolated and sequenced as above for final verification of orientation and sequence of the inserts. The maltose binding protein (MBP)- phosphagen kinase fusion protein was expressed in *E. coli* TB1 cells by induction with 1 mM IPTG at 25°C for 24 h. The cells were resuspended in 5X TE Buffer, sonicated, and the soluble protein was extracted. The recombinant enzymes were obtained as soluble fractions, and successfully purified by affinity chromatography using amylose resin (New England Biolabs, MA, USA). SDS-PAGE was used to determine the purity of the expressed protein. A single 120 kDa band (PwTK D1D2+MBP) was obtained from SDS-PAGE of the full-length recombinant protein and 80 kDa band (truncated domain+MBP) for each of the truncated domain.

The purified enzymes were placed on ice until enzyme activity assay within 12 h. The
enzyme activity of the recombinant proteins was measured with an NADH-linked assay
spectrophotometric assay at 25°C (Morrison & James, 1965) and determined for the forward
reaction or phosphagen synthesis (Fujimoto et al., 2005). The following available substrates
were used to determine specificity: L-arginine, D-arginine, creatine, glycocyamine, and
taurocyamine. The MBP tag was not removed from the recombinant enzymes due to
possible enzyme inactivation if the tag was to be digested. Tada et al. (2008), based on
results of previous studies, concluded that the presence of MBP tag had no significant effect
on the substrate binding properties of AK or CK activity. The full-length and truncated
domains 1 and 2 showed significant activity for the substrate taurocyamine (0.715 - 32.857
umol/min*mg protein) (Table 1). Therefore, it was concluded that the PK of *P. westermani* is
a taurocyamine kinase (Jarilla et al., 2009).

Substrate	PK activity (µmol/min*mg protein)		
	D1	D2	D1D2
Blank (control)	0.028	0.010	0.198
Taurocyamine	32.857	0.715	14.360
L-arginine	0.024	0.014	0.184
D-arginine	0.031	0.006	0.100
Creatine	0.019	0.006	0.085
Glycocyamine	0.014	0.013	0.156

Table 5. Enzyme activity of *P. westermani* phosphagen kinase for various guanidine
compounds (Jarilla et al., 2009).

Taurocyamine kinase was previously suggested to be exclusively found in marine annelids
(Uda et al., 2005). This enzyme was first purified from the lugworm *Arenicola marina* by
sequential ammonium sulfate precipitation and gel-sieve chromatography (Thoai et al., 1963
as cited in Surholt, 1979) and inferred to be localized in the cytosol and mitochondria
(Surholt, 1979). Indeed, cytoplasmic and mitochondrial isoforms of TK were found in *A.
brasiliensis* (Uda et al., 2005) and the tubeworm *Riftia pachyptila* (Uda et al., 2005). It has
been observed that cytoplasmic isoforms are more specific to taurocyamine compared to the
mitochondrial isoforms (Uda et al., 2005). On the basis that *P. westermani* TK showed
activity to taurocyamine only, it can be surmised that this TK is cytoplasmic. But this cannot
be confirmed at this point, since the 5′ UTR of the cDNA was not succesfully amplified;
thus, the absence or presence of the mitochondrial targeting sequence cannot be determined.
The kinetic parameters (K_m, K_d, and k_{cat}) and V_{max} of the MBP-tagged *P. westermani* TK were
also obtained for the forward reaction with various concentrations of the substrate
taurocyamine and ATP. Shown in Table 6 are the components of the reaction mixture (total
1.0 ml). The reaction was started by adding 0.05 ml of an appropriate concentration of
guanidine substrate made up in 100 mM Tris–HCl (pH 8). The initial velocity values were
obtained by varying the concentration of guanidine substrate (taurocyamine) versus fixed
concentrations of the ATP. The $K_m{}^{Tc}$ value was determined from the enzyme reaction using
nine different substrate concentrations of taurocyamine around the rough $K_m{}^{Tc}$ value. To
determine the K_d value, the above reactions were done at four different concentrations of
ATP (10 mM, 7 mM, 5 mM, and 3 mM). For the estimation of kinetic constants (K_m and K_{cat}),
a Lineweaver–Burk plot was made and fitted by the least-square method in Microsoft Excel.

The kinetics of phosphagen kinase can be explained as a random-order, rapid-equilibrium kinetic mechanism (Morrison and James, 1965), and the K_d, the dissociation constant, was obtained graphically as described by Suzuki et al. (1997) or by fitting data directly according to the method of Cleland (Cleland, 1979), using the software written by Dr. R. Viola (Enzyme Kinetics Programs, ver. 2.0). Protein concentration was estimated from the absorbance at 280 nm (0.77 AU at 280 nm in a 1 cm cuvette corresponds to 1 mg protein/ml).

Components	Volume µL
100 mM Tris-HCl (pH 8)	650.00
750 mM KCl	50.00
250 mM Mg-acetate	50.00
25 mM phoephoenolpyruvate (made up in 100 mM imidazole/HCl, pH 7.0)	50.00
5 mM NADH (made up in Tris-HCl, pH 8.0)	50.00
Pyruvate kinase/lactate dehydrogenase mixture (made up in 100 mM imidazole/HCl, pH 7.0)	50.00
ATP (appropriate concentration; made up in 100 mM imidazole/HCl, pH 7)	50.00
Recombinant enzyme	50.00

Table 6. Components of the reaction mixture used for enzyme kinetics assays.

X-ray crystal structures of substrate-free as well as transition state forms of both AKs and CKs showed that these enzymes can be divided into two structural domains, a smaller amino-terminal (N-terminal) domain and a carboxyl-terminal (C-terminal) domain (Zhou et al., 1998, Lahiri et al., 2002, Gattis et al., 2004). During substrate binding, a flexible loop from each domain folds over the substrate at the active site resulting to large conformational changes (Zhou et al., 1998) which appear to be necessary in aligning the two substrates for catalysis, configuring the active site only when productive phosphoryl transfer is possible, and excluding water from the active site to avoid wasteful ATP hydrolysis (Zhou et al., 2000). These conformational changes are elicited by the combination of Mg^{2+} + ADP or ATP which are substrates common to all PKs (Forstner et al., 1998).

Recent studies by Yousef et al. (2003) and Fernandez et al. (2007) on crystal structures of AKs from *Limulus polyphemus* and *Trypanosoma cruzi*, respectively, suggested that instead of the movement of two domains, the differences in substrate binding can be attributed to the motion of three domains relative to a fixed one. Dynamic domain 1 comprises the amino-terminal globular domain, as well as other elements of the active site that are critical to substrate specificity and catalysis. It also contains the active-site cysteine conserved in phosphagen kinases that is proposed to mediate the synergism in substrate binding (Yousef et al., 2003) that appears to be a common feature in PKs (Wu et al., 2008). The substrate synergism may be associated with substrate-induced conformational changes within the tertiary complex (Maggio et al., 1977; Zhou et al., 1998). Gattis et al. (2004) suggested that the active-site cysteine is relevant to catalysis and that one of its roles is enhancing the catalytic rate through electrostatic stabilization of the transition state. Also included in the dynamic domain 1 is the highly conserved segment "NEEDH" regarding which interactions link

Phosphagen Kinase System of the Trematode Paragonimus westermani: Cloning and Expression of a
Novel Chemotherapeutic Target

255

conformational changes to phosphagen binding. The other two dynamic domains, together with the dynamic domain 1, close the active site with separate hinge rotations relative to the fixed domain (Yousef et al. 2003).

The complex conformational changes during substrate binding may be affected by the presence of two or more catalytic domains on a single polypeptide chain (Compaan & Ellington, 2003). Contiguous dimeric AKs from Ensis and C. japonica have high sequence conservation in both domains of the protein but only their second domain showed activity (Compaan & Ellington, 2003; Suzuki et al., 2003). However, this was not the case for P. westernai TK since both truncated domains exhibited activity for taurocyamine (Jarilla et al, 2009).

Table 7 shows the kinetic parameters of P. westermani TK and annelid TKs. The second domain and the contiguous domain of P. westermani TK have stronger affinity for taurocyamine than D1 as indicated by their lower K_m^{Tc} values. However, D1 has stronger affinity for ATP. P. westermani TK also has lower K_m^{Tc} compared to the Arenicola cytoplasmic and mitochondrial TKs and Riftia mitochondrial TK. All P. westermani TK constructs exhibit synergism during substrate binding since the K_d^{Tc}/K_m^{Tc} and K_d^{ATP}/K_m^{ATP} values obtained for the three recombinant enzymes were greater than one.

It appears that the full-length P. westermani TK is catalytically more efficient than the truncated domains since the k_{cat} value (a measure of the number of substrate molecules converted to product per enzyme molecule per unit time) for D1D2 accounts for the k_{cat} values of the truncated domains. This is further corroborated by the values obtained for the V_{max} and k_{cat}/K_m^{Tc}.

Source	K_m^{Tc} (mM)	K_d^{Tc} (mM)	K_d^{Tc}/K_m^c	K_m^{ATP} (mM)	K_d^{ATP} (mM)	K_d^{ATP}/K_m^{ATP}	k_{cat} (S^{-1})	k_{cat}/K_m^{Tc}	V_{max} (umol/min* mg protein
P. westermani TK D1	0.75 ± 0.07	4.22 ± 1.12	5.63	0.66 ± 0.11	3.58 ± 0.27	5.42	24.16 ± 1.54	32.21	40.31 ± 2.51
P. westermani TK D2	0.51 ± 0.04	1.49 ± 0.29	2.92	1.43 ± 0.36	4.03 ± 0.76	2.82	11.56 ± 0.45	22.67	21.43 ± 1.75
P. westermani TK D1D2	0.57 ± 0.10	1.95 ± 0.43	3.42	0.98 ± 0.16	3.37 ± 0.70	3.44	33.44 ± 1.01	58.67	60.01 ± 3.01
Arenicola TK	4.01 ± 0.41	NA	NA	NA	NA	NA	9.43 ± 0.45	2.35	28.71 ± 1.06
Arenicola MiTK	0.88 ± 0.08	NA	NA	NA	NA	NA	14.3 ± 1.01	16.23	17.82 ± 1.24
Riftia MiTK	2.12 ± 0.45	NA	NA	NA	NA	NA	12.5 ± 1.52	5.9	10.4 ± 0.59

Table 7. Kinetic parameters of P. westermani and annelid TKs.

4. Role of the amino acids on the guanidino specificity (GS) region

The guanidine specificity (GS) region has been proposed by Suzuki et al. (1997) as a potential candidate for the guanidine substrate recognition site based on amino acid sequence analysis results. This is the substrate specifity loop (residues 61-68), included in the part of the N-terminal domain, which moves substantially closer to the phosphagen substrate-binding site (Yousef et al., 2003). It has been suggested that there is a proportional relationship between the size of the deletion in the GS region and the mass of the guanidine substrate used. For instance, CK and GK, which use the smallest substrate, have no deletion while LK, AK, and TK, which recognize relatively large guanidine substrates, have a five-amino acid deletion (Suzuki et al., 1997; Uda et al., 2005). Previous studies on AKs of *Nautilus* and *Stichopus*, CK of *Danio* and LK of *Eisenia* showed that introduction of mutations on the GS region significantly reduced the activity of the said enzymes (Suzuki et al., 2000; Suzuki et al., 2000; Uda & Suzuki, 2004; Suzuki & Yamamoto, 2000; Tanaka and Suzuki, 2004). Unlike cytoplasmic TKs from *A. brasiliensis* and *R. Pachyptila* which have five amino acid deletions on the GS region (Uda et al., 2005; Uda et al., 2005), the two domains of *P. westermani* TK have six amino acid deletions in the GS region (Fig. 4) (Jarilla et al., 2009).

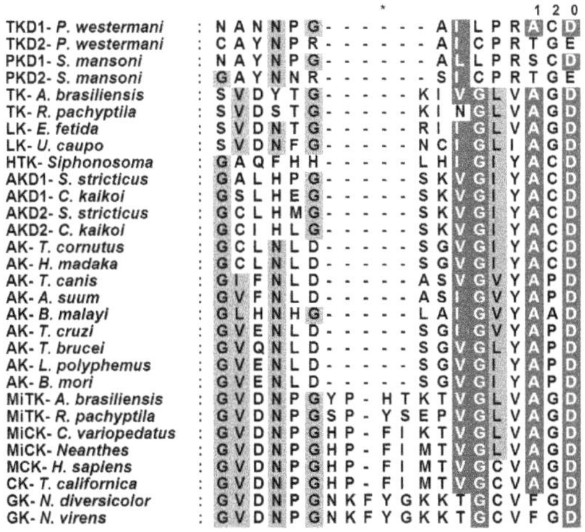

Fig. 4. Alignment of the guanidino specificity region of *P. westermani* PK with other selected PK.

Elucidation of the amino acids involved in substrate binding and maintenance of substrate-bound structure in *P. westermani* TK are important for the exploration of inhibitors against this enzyme. Currently, the functional properties and substrate binding mechanisms in TK are not well known. To characterize the substrate recognition system in *P. westermani* TK we introduced mutations (D1: Gly58Arg, Ala59Gly, Ile60Val, Tyr84His, and Tyr84 Ile; D2: Arg61Leu, Ala62Gly, Ile63Val, Tyr87His, and Tyr87 Ile) on the amino acids on and near the GS region of pMAL/*P. westermani* TK template by PCR-based mutagenesis. The PCR products were digested with DpnI and the target DNA was purified using the QIA quick purification column (Qiagen, USA). The mutated cDNA was self-ligated after blunting and phosphorylation and sequenced as described above to check for mutations. After the mutation has been confirmed, enzyme expression and determination of kinetic properties were done as above and the obtained values are shown on Table 8.

The conformational change or the change from open to close structure during substrate binding in PKs is reflected by the kinetic parameters K_m^P (the value comparable to the dissociation constant of the phosphagen substrate in the absence of ATP) and K_d^P (the dissociation constant of the phosphagen substrate in the absence of ATP) (Suzuki et al., 2003). For Domain 1, mutations on the GS region caused a decrease in the affinity for taurocyamine as seen in the increase in K_m^{Tc} values. The most significant decrease was observed for the Ala[59] to Gly mutant which had a K_m^{Tc} value 4 times greater than the wild type. The K_d^{Tc}/K_m^{Tc} of this mutant was almost equal to 1 indicating the absence of synergism upon ATP substrate binding. Moreover, the V_{max} value for the Ala[59] to Gly mutant was reduced to only 50% of the wild type. Mutation of the equivalent position in Domain 2 (Ala[62]Gly mutant) resulted in the loss of detectable enzyme activity. These suggest that the replacement of Ala on the GS region may have affected the stabilization of the closed structure implying that this amino acid may play an important role in taurocyamine binding.

Source	K_m^{Tc} (mM)	K_d^{Tc} (mM)	K_d^{Tc}/K_m^{Tc}	K_m^{ATP} (mM)	K_d^{ATP} (mM)	k_{cat} (S⁻¹)	k_{cat}/K_m^{Tc}	V_{max} (umol/min* mg protein)
Domain 1								
WT	0.75 ± 0.07	4.22 ± 1.12	5.63	0.66 ± 0.11	3.58 ± 0.27	24.16 ± 1.54	32.21	40.31 ± 2.51
Gly[58]Arg	1.02 ± 0.03	4.84 ± 0.88	4.74	0.64 ± 0.10	3.00 ± 0.09	34.10 ± 1.46	54.30	57.66 ± 3.47
Ala[59]Gly	3.54 ± 0.41	4.41 ± 1.77	1.25	1.89 ± 0.45	2.20 ± 0.23	7.98 ± 2.16	2.25	17.06 ± 3.52
Ile[60]Val	1.33 ± 0.33	4.61 ± 1.37	3.47	1.05 ± 0.32	3.56 ± 0.89	31.00 ± 13.50	26.29	54.30 ± 22.21
Tyr[84]His	1.76 ± 0.05	6.20 ± 0.83	3.52	0.70 ± 0.11	2.43 ± 0.25	18.10 ± 4.01	10.28	29.25 ± 6.93
Tyr[84]Ile	1.91 ± 0.29	8.85 ± 3.15	4.63	0.97 ± 0.15	4.41 ± 0.66	11.38 ± 3.05	6.13	20.07 ± 5.17
Domain 2								
WT	0.51 ± 0.04	1.49 ± 0.29	2.92	1.43 ± 0.36	4.03 ± 0.76	11.56 ± 0.45	22.67	21.43 ± 1.75
Arg[61]Leu	0.51 ± 0.17	2.44 ± 1.17	4.78	1.83 ± 0.99	8.757 ± 2.45	6.88 ± 0.48	13.49	16.17 + 2.53
Ala[62]Gly	ND	ND	ND	ND	ND	ND	ND	ND
Ile[63]Val	1.19 ± 0.14	2.42 ± 1.22	2.03	2.12 ± 0.63	4.11 ± 0.74	6.99 ± 1.51	6.00 ± 1.78	14.10 ± 2.51
Tyr[87]His	ND	ND	ND	ND	ND	ND	ND	ND
Tyr[87]Ile	ND	ND	ND	ND	ND	ND	ND	ND

Table 8. Kinetic parameter of wild-type and mutant P. westermani TK.

In *Danio rerio* CK, Arg-96 which is located close to the substrate-binding site has been proposed to have a key role in substrate recognition and in organizing a hydrogen-bond network for active center configuration (Uda et al., 2009). For annelid TKs, the equivalent amino acid is His (Uda et al., 2005; Uda et al., 2005); however, it has been replaced with tyrosine in *P. westermani* TK similar to that in AKs (Jarilla et al., 2009). Replacement of this amino acid with isoleucine (Tyr[84] Ile and Tyr[87] Ile mutants) did not change the substrate specificity from taurocyamine to glycocyamine but caused a decrease in affinity and V_{max} in domain 1 and loss of activity in domain 2 (Table 8). Similar results were observed when Tyr was replaced with His. These imply that, in the case of *P. westermani* TK, Tyr may play a role in substrate binding but it may not be the key residue for substrate recognition.

5. Gene arrangement of *P. westemani* TK

We have determined the exon/intron organization of *P. westermani* TK (Fig. 5) to further elucidate the phylogentic relationship with other PKs. Genomic DNA was extracted from an adult worm of *P. westermani* using Easy-DNA™ Kit (Invitrogen, USA). PCR was performed with Ex *Taq*™ (Takara, Tokyo, Japan) and primers designed from the cDNA. PCR conditions were as follows: initial denaturation at 94 °C for 2min, followed by 35 cycles of 94 °C for 30 sec, annealing at 50 °C for 30 sec and extension at 72 °C for 3 min and a final extension at 72 °C for 4min. The PCR products were purified, cloned in T-vector, and sequenced as described above. Thus far, for Domain 1, we have identified four exons and three introns with intron sizes ranging from 88 bp to 3,627 bp. Domain 2 has five exons and four introns with sizes ranging from 99 bt to 3,145 bp. We have also identified a bridge intron (2,806 bp) between the two domains. All of these introns typically began with gt and ended with ag.

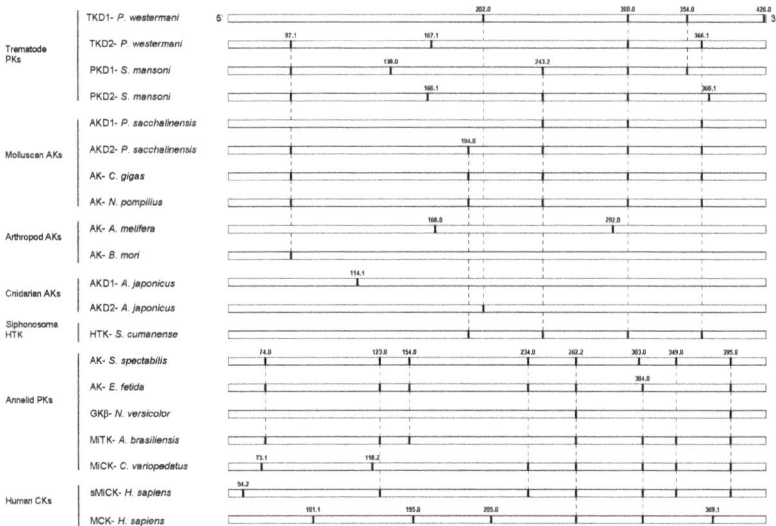

Fig. 5. Comparison of the gene strucrures (exon/inton organization) of *P. westermani* TK with PKs from other sources. The intron postions are based on aligned amino acid sequences. Intron phases are indicated by ".0", ".1", or ".2" followed by the amino acid sequence position. The conserved introns are shown by vertical lines

Interestingly, no intron positions are conserved between *P. westermani* TK and TK from the annelid *A. brasiliensis*. Instead, *P. westermani* TK share more intron positions with the molluscan AKs (D1: 300.0, D2: 97.1, 300.0, 366.1) suggesting that *P. westermani* TK evolved from an AK gene, consistent with the phylogenetic analysis of amino acid sequence. On the contrary, several intron positions are conserved among annelid TKs, AKs, GKs and vertebrate CKs. Annelid TKs and other annelid-specific PKs are hypothesized to have evolved from a MiCK-like ancestor early in the divergence of the protostome metazoans (Tanaka et al., 2007) with cytoplasmic TKs diverging earliest together with cytoplasmic AKs and LKs (Suzuki et al., 2009).

Three intron positions were also shared with the sipunculid HTK which was suggested to have evolved from an AK gene (Uda et al., 2005). It should also be noted that less intron positions are conserved in the first domain which could imply that it has recently diverged. However, the 5' half of *P. westermani* TK gene still needs to be further amplified to confirm for the presence of intron.

6. Conclusion

The phosphagen kinase system found in the lung fluke *P. westermani*, a taurocyamine kinase, appears to be different from those identified in annelids. This TK, which probably evolved from an AK gene, consists of two enzymatically active domains that may have a unique substrate recognition and binding mechanisms.

Since TK is not present in mammalian hosts, *P. westermani* TK could be a potential novel chemotherapeutic target for the effective control and eradication of paragonimiasis. This enzyme could also be utilized in the development of diagnostic tools specific for pulmonary paragonimiasis to avoid misdiagnoses especially in regions where tuberculosis is also enedemic. However, for the validation of this enzyme as a therapeutic target, there is still a need for further studies to determine the specific role of TK in the metabolic routes of the lung fluke. Further elucidation of the structure and substrate binding mechanisms are also necessary for the subsequent search of specific TK inhibitors.

7. Acknowledgments

The authors would like to thank all the members of the Department of Environmental Health Sciences in Kochi Medical School for their encouragement and help.

8. References

Alonso, G. D.; Pereira, C.A.; Remedi, M. S.; Paveto, M.C.; Cochella, L.; Ivaldi, M. S.; Gerez de Burgos, N. M.; Torres, H. N. & Flawia', M. M. (2001). Arginine kinase of the flagellate protozoa *Trypanosoma cruzi*; regulation of its expression and catalytic activity. *Federation of European Biochemical Society Letters*, Vol.498, No.1 (June 2001), pp. 22-25, ISSN 1873-3468

Andrews, L.D.; Graham, J.; Snider, M.J. & Fraga D. (2008). Characterization of a novel bacterial arginine kinase from *Desulfotalea psychrophila*. *Comparative Biochemistry and Physiology Part B: Biochemistry and Molecular Biology*, Vol.150, No.3 (July 2008), pp. 312-319, ISSN 1096-4959

Awama, A.M.; Paracuellos, P.; Laurent, S.; Dissous, C.; Marcillat, O. & Gouet, P. (2008). Crystallization and X-ray analysis of the *Schistosoma mansoni* guanidine kinase. *Acta Crystallographica Section F: Structural Biology and Crystallization Communications*, Vol.F64, No.9 (September 2008), pp. 854-857, ISSN 1744-3091

Blair, D.; Agatsuma, T. & Wang, W. (2007). Paragonimiasis. In: *Food-Borne Parasitic Zoonoses: Fish and Plant-Borne Parasites*, K.D. Murrel & B. Fried (Eds.), 117-150, Springer Science, ISBN 978-0-387-71357-1, New York, USA

Chomczynski, P. & Sacchi N. (1987). Single-step method of RNA isolation by acid guanidinium thiocyanate-phenol-chloroform extraction. *Analytical Biochemistry*, Vol.162, No.1 (April 1987), pp. 156-159, ISSN 0003-2697

Cleland, W.W. (1979). Statistical analysis of enzyme kinetic data. *Methods in Enzymology*, Vol.63, pp. 103-108, ISSN 0076-6879

Compaan, D.M. & Ellington ,W.R. (2003). Functional consequences of a gene duplication and fusion event in an arginine kinase. *Journal of Experimental Biology*, Vol.206 (May 2003), pp. 1545-1556, ISSN 1477-9145

Ellington, W. R. & Suzuki, T. (2006). Evolution and divergence of creatine kinase genes, In: *Molecular anatomy and physiology of proteins: creatine kinase*, C. Vial, (Ed.), Nova 1-27, Science, ISBN 1594547157, New York, USA

Ellington, W. R. (2001). Evolution and physiological roles of phosphagen systems. *Annual Review of Physiology*, Vol.63 (March 2001), pp. 289-325, ISSN 0066-4278

Fernandez, P.; Haouz, A.; Pereira, C.A.; Aguilar, C. & Alzari, P.M. (2007). The crystal structure of *Trypanosoma cruzi* arginine kinase. *Proteins: Structure, Function and Bioinformatics*, Vol.69, No.1 (October 2007), pp. 209-212, ISSN 1097-0134

Forstner, M.; Kriechbaum, M.; Laggner, P. & Wallimann, T. (1998). Structural changes of creatine kinase upon substrate binding. *Biophysical Journal*, Vol.75, No.2 (August 1998), pp. 1016-1023, ISSN 1542-0086

Fujimoto, N.; Tanaka, K. & Suzuki, T. (2005). Amino acid residues 62 and 193 play the key role in regulating the synergism of substrate binding in oyster arginine kinase. *Federation of European Biochemical Society Letters*, Vol.579, No.7 (March 2005), pp. 1688-1692, ISSN 0014-5793

Gattis, J.L.; Ruben, E.; Fenley, M.O.; Ellington, W.R. & Chapman, M.S. (2004). The active site cysteine of arginine kinase: structural and functional analysis of partially active mutants. *Biochemistry*, Vol.43, No.27 (July 2004), pp. 8680-8689, ISSN 1520-4995

Jarilla, B.R.; Tokuhiro, S.; Nagataki, M.; Hung, S.; Uda, K.; Suzuki, T. & Agatsuma, T. (2009). Molecular characterization and kinetic properties of a novel two-domain taurocyamine kinase from the lung fluke *Paragonimus westermani. The Federation of European Biochemical Society Letters*, Vol.583, No.13 (June 2009), pp. 2218-24, ISSN 0014-5793

Keiser, J. & Utzinger, J. (2007). Food-borne trematodiasis: current chemotherapy and advances with artemisinins and synthetic trioxolanes. *Trends in Parasitology*, Vol.23, No.11 (November 2007), pp. 555-562, ISSN 1471-5007

Kirino, Y.; Nakano, N.; Doanh, P. H.; Nawa, Y. & Horii, Y. (2009). A seroepidemiological survey for paragonimosis among boar-hunting dogs in central and southern Kyushu, Japan. *Veterinary Parasitology*, Vol.161, No.3-4 (May 2009), pp. 335-338, ISSN 0304-4017

Lahiri, S.D.; Wang, P.; Babbitt, P.C.; McLeish, M.J.; Kenyon, G.L. & Allen, K.N. (2002). The
    2.1 Å structure of *Torpedo californica* creatine kinase complexed with the ADP-Mg^{2+}-
    NO$_3^-$-creatine transition-state analogue complex. *Biochemistry*, Vol.41, No.47
    (November 2002), pp. 13861-13867, ISSN 1520-4995
Liu, Q.; Wei, F.; Liu, W.; Yang, S. & Zhang, X. (2008). Paragonimiasis: an important food-
    borne zoonosis in China. *Trends in Parasitology*, Vol.24, No.7 (July 2008), pp. 318-
    323, ISSN 1471-5007
Maggio, E.T. & Kenyon, G.L. (1977). Properties of a CH$_3$-blocked creatine kinase with
    altered catalytic activity. Kinetic consequences of the presence of the blocking
    group. *The Journal of Biological Chemistry*, Vol.252, No.4 (February 1977), pp. 1202-
    1207, ISSN 1083-351X
Miranda, M.R.; Canepa, G.E.; Bouvier, L.A. & Pereira, C.A. (2006). *Trypanosoma cruzi*:
    oxidative stress induces arginine kinase expression. *Experimental Parasitology*,
    Vol.114, No.4 (December 2006), pp. 341-344, ISSN 1090-2449
Miyazaki, I & Habe, S. (1976). A newly recognized mode of human infection with the lung
    fluke, *Paragonimus westermani* (Kerbert 1878). *Journal of Parasitology*, Vol.62, No.4
    (August 1976), pp. 646-648, ISSN 00223395
Morrison, J. F. & James, E. (1965). The mechanism of the reaction catalyzed by adenosine
    triphosphate-creatine phosphor-transferase. *Biochemistry Journal*, Vol.97, pp. 37-52,
    ISSN 0264-6021
Morrison, J. F. (1973). Arginine kinase and other invertebrate guanidine kinases. In: *The
    Enzymes*, P.C. Boyer (Ed.), 457-486, Academic Press, ISBN 0-12-122708, New York,
    USA
Mukae, H.; Taniguchi, H.; Matsumoto, N.; Iiboshi, H.; Ashitani, J.; Matsukura, S. & Nawa, Y.
    (2001). Clinicoradiologic features of pleuropulmonary *Paragonimus westermani* on
    Kyusyu Island, Japan. *Chest*, Vol. 120, No.2 (August 2001), pp. 514-520, ISSN 1931-
    3543
Nakano, N.; Kirino, Y.; Uchida, K.; Nakamura-Uchiyama, F.; Nawa, Y. & Horii, Y. (2009).
    Large-group infection of boar-hunting dogs with *Paragonimus westermani* in
    Miyazaki Prefecture, Japan, with special reference to a case of sudden death due to
    bilateral pneumothorax. *Journal of Veterinary Medical Science*, Vol.71, No.5 (May
    2009), pp. 657-660, ISSN1347-7439
Nagataki, M.; Wickramasinghe, S.; Uda, K.; Suzuki, T.; Yano, H.; Watanabe, Y. & Agatsuma,
    T. (2008). Cloning and enzyme activity of a recombinant phosphagen kinase from
    nematodes (in Japanese). *Japanese Journal of Medical Technology*, Vol.57, No.1, pp. 41-
    45
Odermatt, P.; Veasna, D.; Zhang, W.; Vannavong, N.; Phrommala, S.; Habe, S.;
    Barennes, H. & Strobel, M. (2009). Rapid identification of paragonimiasis foci
    by lay informants in Lao People's Democratic Republic. *Public Library of Science
    (PLoS) Neglected Tropical Diseases*, Vol.3, No.9 (September 2009), pp. 1-5, ISSN
    1935-2735
Paveto, C.; Guida, M. C.; Esteva, M. I.; Martino, V.; Coussio, J.; Flawia, M. M. & Torres, H. N.
    (2004). Anti-*Trypanosoma cruzi* activity of green tea (*Camellia sinensis*) catechins.
    Antimicrobial Agents and Chemotherapy, Vol.48, No.1 (January 2004), pp. 69-74, ISSN
    0066-4804

Pereira, C. A.; Alonso, G. D.; Ivaldi, M. S.; Silber, A. M.; Alves, M. J. M.; Torres, H. N. & Flawia', M. M. (2003). Arginine kinase overexpression improves *Trypanosoma cruzi* survival capability. *Federation of European Biochemical Society Letters*, Vol.554, No.1-2 (November 2003), pp. 201-205, ISSN 1873-3468

Pereira, C. A.; Alonso, G. D; Ivaldi, S.; Bouvier, L. A.; Torres, H.N. & Flavia, M.M. (2003). Screening of substrate analogs as potential enzyme inhibitors for the arginine kinase of *Trypanosoma cruzi*. *Journal of Eukaryotic Microbiology*, Vol.50, No.2 (March-April, 2003), pp. 132-134, ISSN 1550-7408

Pereira, C.A.; Alonso, G. D.; Ivaldi, M. S.; Silber, A. M.; Alves, M. J. M.; Bouvier, L. A.; Flavia, M.M. & Torres, H. N. (2002). Arginine metabolism in *Trypanosoma cruzi* is coupled to parasite stage and replication. *Federation of European Biochemical Society Letters*, Vol.526, No.1-3 (August 2002), pp. 111-114, ISSN 1873-3468

Platzer, E.G.; Wang, W.; Thompson, S.N. & Borchardt, D.B. (1999). Arginine kinase and phosphoarginine, a functional phosphagen, in the rhabditoid nematode *Steinernema carpocapsae*. *Journal of Parasitology*, Vol.85, No.4 (August 1999), pp. 603-607, ISSN 00223395

Robin, Y. (1974). Phosphagens and molecular evolution in worms. *BioSystems*, Vol.6, No.1 (July 1974), pp. 49-56, ISSN 0303-2647

Surholt, B. (1979). Taurocyamine kinase from the body-wall musculature of the lugworm *Arenicola marina*. *European Journal of Biochemistry*, Vol.93, No.15 (January 1979), pp. 279-285, ISSN 1432-1033

Suzuki, T. & Yamamoto, Y. (2000). Gene structure of two-domain arginine kinase from *Anthopleura japonicus* and *Pseudocardium sachalinensis*. *Comparative Biochemistry and Physiology Part B*, Vol.127, No.4 (December 2000), pp. 513-518, ISSN 1879-1107

Suzuki, T., Uda, K., Adachi, M., Sanada, H., Tanaka, K., Mizuta, C., Ishida, K. & Ellington, W.R. (2009). Evolution of the diverse array of phosphagen systems present in annelids. *Comparative Biochemistry and Physiology Part B*, Vol.152, No.1 (January 2009), pp. 60-66, ISSN 1879-1107

Suzuki, T.; Fukuta, H.; Nagato, H. & Umekawa, M. (2000). Arginine kinase from *Nautilus pompilius*, a living fossil; Site-directed mutagenesis studies on the role of amino acid residues in the guanidino specificity region. *The Journal of Biological Chemistry*, Vol.275, No.31 (August 2000), pp. 23884-23890, ISSN 1083-351X

Suzuki, T.; Kawasaki, Y. & Furukohri T. (1997). Evolution of phosphagen kinase. Isolation, characterization and cDNA-derived amino acid sequence of two-domain arginine kinase from the sea anemone *Anthopleura japonicus*. *Biochemical Journal*, Vol.328 (November 1997), pp. 301-306, ISSN 1470-8728

Suzuki, T.; Kawasaki, Y.; Unemi, Y.; Nishimura, Y.; Soga, T.; Kamidochi, M.; Yazawa, Y. & Furukohri, T. (1998). Gene duplication and fusion have occurred frequently in the evolution of phosphagen kinases—a two-domain arginine kinase from the clam *Pseudocardium sachalinensis*. *Biochimica et Biophysica Acta*, Vol.1388, No.1 (October 1998), pp. 253-259, ISSN 0006-3002

Suzuki, T.; Sugimura, N.; Taniguchi, T.; Unemi, Y.; Murata, T.; Hayashida, M.; Yokouchi, K.; Uda, K. & Furukohri, T. (2002). Two-domain arginine kinases from the clams *Solen strictus* and *Corbicula japonica*: exceptional amino acid replacement of the functionally important D62 by G. *The International Journal of Biochemistry and Cell Biology*,Vol.34, No.10 (October 2002), pp. 1221-1229, ISSN 1878-5875

Suzuki, T.; Tomoyuki, T. & Uda, K. (2003). Kinetic properties and structural characteristics of an unusual two-domain arginine kinase from the clam *Corbicula japonica*. *Federation of European Biochemical Society Letters*, Vol.533, No.1-3 (January 2003), pp. 95-98, ISSN 1873-3468

Suzuki, T.; Yamamoto, Y. & Umekawa, M. (2000). *Stichopus japonicus* arginine kinase: gene structure and unique substrate recognition system. *The Biochemical Journal*, Vol.351 (November 2000), pp. 579-585, ISSN 1470-8728

Tada, H.; Nishimura, Y & Suzuki, T. (2008). Cooperativity in the two-domain arginine kinase from the sea anemone *Anthopleura japonicus*. *International Journal of Biological Macromolecules*, Vol.42, No.1(January 2008), pp. 46-51, ISSN 1879-0003

Tamura, K.; Dudley, J.; Nei, M. & Kumar, S. (2007). MEGA4: Molecular Evolutionary Genetics Analysis (MEGA) software version 4.0. *Molecular Biology and Evolution*, Vol.24, No.8 (August 2007), pp. 1596-1599, ISSN 1537-1719

Tanaka, K. & Suzuki, T. (2004). Role of amino-acid residue 95 in substrate specificity of phosphagen kinases. *The Federation of European Biochemical Society Letters*, Vol.573, No.1-3 (August 2004), pp. 78-82, ISSN 1873-3468

Tanaka, K.; Uda, K.; Shimada, M.; Takahasi, K.; Gamou, S.; Ellington, W.R. & Suzuki, T. (2007). Evolution of the cytoplasmic and mitochondrial phosphagen kinases unique to annelid groups. *Journal of Molecular Evolution*, Vol.65, No.5 (November 2007), pp. 616-625, ISSN 1432-1432

Thoai, V.N. (1968). Homologous phosphagen phosphokinases. In: *Homologous Enzymes and Biochemical Evolution*, N. van Thoai & J. Roche (Eds.), 199-229, Gordon and Breach, New York, USA

Uda, K. & Suzuki, T. (2004). Role of the amino acid residues on the GS region of *Stichopus* arginine kinase and *Danio* creatine kinase. *The Protein Journal*, Vol.23, No.1 (January 2004), pp. 53-64, ISSN 1875-8355

Uda, K.; Fujimoto, N.; Akiyama, Y.; Mizuta, K.; Tanaka, K.; Ellington, W. R. & Suzuki, T. (2006). Evolution of the arginine kinase gene family. *Comparative Biochemistry and Physiology Part D*, Vol.1, No.2 (June 2006), pp. 209-218, ISSN 1744-117X

Uda, K.; Iwai, A. & Suzuki, T. (2005). Hypotaurocyamine kinase evolved from a gene for arginine kinase. *The Federation of European Biochemical Society Letters*, Vol.579, No.30 (December 2005), pp. 6756-6762, ISSN 0014-5793

Uda, K.; Kuwasaki, A; Shima, K; Matsumoto, T. & Suzuki, T. (2009). The role of Arg-96 in *Danio rerio* creatine kinase in substrate recognition and active center configuration. *International Journal of Biological Macromolecules*, Vol.44, No.5 (June 2009), pp. 413-8, ISSN 1879-0003

Uda, K.; Saishoji, N.; Ichinari, S.; Ellington, W. R. & Suzuki, T. (2005). Origin and properties of cytoplasmic and mitochondrial isoforms of taurocyamine kinase. *The Federation of European Biochemical Society Journal*, Vol.272, No.14 (July 2005), pp. 3521-3530, ISSN 1742-4658

Uda, K.; Tanaka, K.; Bailly, X.; Zal, F. & Suzuki, T. (2005). Phosphagen kinase of the giant tubeworm *Riftia pachyptila*: Cloning and expression of cytoplasmic and mitochondrial isoforms of taurocyamine kinase. *International Journal of Biological Macromolecules*, Vol.37 (September 2005), pp. 54-60, ISSN 0141-8130

Uda, K.; Yamamoto, K.; Iwasaki, N.; Iwai, M.; Fujikura, K.; Ellington, W.R. & Suzuki, T. (2008). Two-domain arginine kinase from the deep-sea clam *Calyptogena kaikoi* —

Evidence of two active domains. *Comparative Biochemistry and Physiology Part B,* Vol.151, No.2 (October 2008), pp. 176-182, ISSN 1879-1107

Wickramasinghe, S.; Uda, K.; Nagataki, M.; Yatawara, L.; Rajapakse, R.P.V.J.; Watanabe, Y.; Suzuki, T. & Agatsuma, T. (2007). *Toxocara canis:* Molecular cloning, characterization, expression and comparison of the kinetics of cDNA-derived arginine kinase. *Experimental Parasitology,* Vol.117, No.2 (April 2007), pp. 124-132, ISSN 0014-4894

World Health Organization. (2002). Report: Joint WHO/FAO workshop on foodborne trematode infections in Asia. Hanoi, Vietnam. Retrieved from <http://whqlibdoc.who.int/wpro/2004/RS_2002_GE_40 (VTN).pdf>

Wu, Q.Y.; Li, F. & Wang X.Y. (2008). Evidence that amino-acid residues are responsible for substrate synergism of locust arginine kinase. *Insect Biochemistry and Molecular Biology* Vol.38, No.1 (January 2008), pp. 59-65, ISSN 1879-0240

Yousef, M.; Clark, S.A.; Pruett, P.K; Somasundaram, T.; Ellington, W.R. & Chapman, M.S. (2003). Induced fit in guanidine kinases- comparison of substrate-free and transition state analog structures of arginine kinase. *Protein Science,* Vo.12, No.1 (January 2003), pp. 103-111, ISSN 1469-896X

Zhou, G.; Ellington, W.R. & Chapman, M.S. (2000). Induced fit in arginine kinase. *Biophysical Journal,* Vol.78, No.3 (March 2000), pp. 1541-1550, ISSN 1542-0086

Zhou, G.; Somasundaram, T.; Blanc, E.; Parthasarathy, G.; Ellington, W.R. & Chapman, M.S. (1998). Transition state structure of arginine kinase: Implications for catalysis of bimolecular reactions. *Proceedings of the National Academy of Science USA,* Vol.95, No.15 (July 1998), pp. 8449-8454, ISSN 1091-6490

Part 6

Evolutionary Biology

15

Molecular Cloning and Characterization of Fe-Superoxide Dismutase (Fe-SOD) from the Fern *Ceratopteris thalictroides*

Chen Chen and Quanxi Wang
Shanghai Normal University, Shanghai, China

1. Introduction

Ferns are, evolutionarily, in a pivotal position between bryophytes and seed plants (Pryer et al., 2001). Fern gametophytes, like bryophytes, have no vascular system and live on substrate surfaces as small individual plants. However, fern sporophytes do have a vascular system enabling more vertical growth than the gametophytes, and resulting in a larger herbaceous plant form. The origins of plant vascular systems must have arisen during the evolution of primitive ferns (Kenrick , 2000). Ferns are historic plants and provide many facets of interest for researchers (Dyer, 1979; Raghavan, 1989). An especially important reason for choosing to study ferns is to gain insight into the evolution of higher plants .

Homosporous ferns，such as *Ceratopteris*，is a genus of homosporous ferns found in most tropical and subtropical area of the world (Lloyd, 1974, 1993; Masuyama, 1992). *Ceratopteris* are vascular plants that exhibit a biphasic life cycle with independent autotrophic haploid and diploid generations. Thus, they offer unique opportunities for studying a wide variety of experimental approaches and a large body for literature has been produced（Miller, 1986; Dyer, 1979）. In contrast to most other ferns, *Ceratopteris* possesses a fast life cycle time of less than 120 d, can be cultured easily, and is readily amenable to genetic analyses (Hickok et al., 1987) *Ceratopteris* has been used as a model plant for many years in the study of genetics, biochemistry, cell biology, and molecular biology (Hickok et al., 1995; Chatterjee, 2000).

Plants are continually exposed to environmental fluctuations that lead to oxidative stress. Part of the damage caused by conditions such as intense light, drought, temperature stress, air pollutants etc. is associated with oxidative stress is an increase in the production of reactive oxygen species (ROS) (Levine A., 1999). Reactive oxygen species (ROS), such as hydrogen peroxide (H_2O_2), superoxide anion (O_2^-) and hydroxyl radical (OH-) are generated from normal metabolic process in all aerobic organisms. The damages from ROS include lipid peroxidation, cross-linking and inactivation of proteins, breaks in DNA and RNA, and cell death (Bestwick & Maffulli, 2004; Fridovich, 1995). Aquatic organisms are often subjected to enhanced "oxidative stress" by ROS due to chronic exposure to pollutants in their environments (Marikovsky et al., 2003; Geret et al., 2004). To limit the harmful effect of ROS production and prevent damage from oxidative stress, cells have evolved to use antioxidant systems as part of the innate immune defense to maintain reactive oxygen species at low basal levels and protect themselves from the constant oxidative challenge (Geret et al., 2004; Manduzio et al., 2004).

Superoxide dismutases (SODs) are a family of metalloenzymes that catalyze the disproportionation reactions of two superoxide anions to H_2O_2 and O_2. SODs are classified into three groups according to their metal cofactors; copper-zinc (Cu/Zn-SOD), iron (Fe-SOD) and manganese (Mn-SOD) (Bowler et al., 1992; Scandalios, 1993). A fourth class of SODs with a nickel atom cofactor (Ni-SOD) was also identified, but so far, it has only been found in Streptomyces genera (Youn et al., 1996). The Fe-SOD is found in both prokaryotes and eukaryotes. It has not been found in animals or fungi, but is present in a limited number of seed plants (e.g., *Arabidopsis thaliana*, tobacco (van Camp et al., 1990), soybean (Crowell & Amasino, 1991 b), and rice (Kaminaka et al., 1999). Its absence in animal species has led researchers to propose that the Fe-SOD gene originated in the plastid before moving to the nuclear genome. Nonetheless, many seed plants, including maize, exhibit no Fe-SOD activity. The Fe- SOD activity in pea leaves was induced only by a deficiency of copper (Ayala & Sandmann, 1988). An increase in the expression of Fe-SOD genes caused by copper deficiency was reported in tobacco leaves (Kurepa et al., 1997 a) and moss cells (Shiono et al., 2003) .However, such activity and transcription of the FeSOD gene has been detected in response to various stimuli and at certain developmental stages in barley (Casano et al., 1994), tobacco (Tsang et al., 1991; Kurepa et al., 1997), and rice (Kaminaka et al., 1999).

Here we report the Fe-SOD levels from *Ceratopteris thalictroides* . The cloning of Ct Fe-SOD may provide information to help further fern research in the area of functional genes.

2. Materials and methods

2.1 Plant material and culture

Spores of *Ceratopteris thalictroides* were collected from the Tianmu Mountains (Zhejiang, Chain) , sterilized with 2% sodium hypochlorite, and sown on to Murashige and Skoog (MS) solid medium (Sigma) with the addition of 1% pure agar (Sigma). All plants were cultured in an environmentally controlled chamber with 16 h light 26°Cand 8 h dark 20°C periods. After 21 days the gametophytes were used for RNA and protein extraction.

2.2 Total RNA isolation

0.5 g gametophyte material was ground in liquid nitrogen with 0.25g pvpp (polyvinylpolypyrrolidone) to a powder. The total RNA was extracted from this powder by using Trizol reagent (Invitrogen) following the manufacturer's instructions.

2.3 Touchdown PCR and RT-PCR

Two degenerate primers P1: 5'-GARTTYCACTGGGGIAARCAYC-3'and P2: 5'-GTARGCRTGCTCCCARACRTC-3'were designed based on highly conserved sequences to clone the mid-fragment of the SOD gene from Ceratopteris thalictroides. Touchdown PCR was performed using the following program: 5 min at 94°C (1 cycle), followed by 30 s at 94°C, 30 s at 59–67°C, and 1 min at 72°C (35 cycles), and a final 20 min 72°C extension step. RNA used for RT-PCR was treated with RQ1 RNase-free DNase I (Promega, Madison, WI) to remove any possible contaminating DNA. RT-PCR was performed by using 500 ng oligo(dT)12-18 primer for first-strand synthesis under standard conditions. Negative controls with water in place of reverse transcriptase were prepared for all samples in order to control for possible genomic DNA contamination of the RNA samples.

2.4 5'- and 3'-rapid amplification of cDNA ends

To extend the Fe-SOD sequence in the 5'- and 3'-directions, we performed rapid amplification of cDNA ends (RACE) using a GeneRacer kit according to the manufacturer's instructions (Invitrogen). The 5'- and 3'- PCR was performed using touchdown PCR with the following respective primers; primer pair one (F5: 5'- CGACTGGAGCA CGAGGACACTG A-3'; R5:5'-TACGCAGTTTACATCCAGGT CG-3') and primer pair two (F3: 5'- GCTGTCAACGATACGTACGTAACG-3'; R3:5'-ACGCTACGT AACG GCATG -3'). Finally, one pair of gene-specific primers of full length Fe-SOD (FS:5'-CGGGATC CGATGGCCACGGCGACTTGCAGCTCTA-3';RS:5'-GCGTCGACCTATTTGTATTTATAT TGATCATCG-3') was designed based on the sequenced 5'- and 3'- fragments. The full-length cDNA of the Fe-SOD gene was amplified by PCR with these primers.

2.5 Bioinformatic analysis

Homologous sequences were identified by searching within the DDBJ/EMBL/ GenBank database using BLAST. Alignments were performed using the CLUSTAL W multiple sequence alignment program.

2.6 Expression and purification of recombinant Fe-SOD

The coding region of the Fe-SOD gene was amplified by PCR with primers. The amplified product was purified, digested with BamHI and SalI, and cloned into the pET32a vector which was predigested with the same restriction enzymes. The resulting plasmid was transformed into E. coli BL21 cells (Invitrogen), and positive clones were selected. Expression of the recombinant protein was induced by adding 1 mM isopropyl β-D-thiogalactopyranoside (IPTG) at 37°C for 6 h and centrifuged at 6,000 rpm for 10 min to collect cells. The cell pellets were washed in 1 ml of Tris-HCl (pH 7.4) and resuspended in 30 ml of binding buffer (10 mM NaH_2PO_4, 10 mM Na_2HPO_4, 500 mM NaCl, 30 mM imidazole, pH 7.4) and sonicated for lysis. The suspension was centrifuged at 10,000 rpm for 10 min to clarify the enzyme solution. The recombinant protein was analyzed by 12 % SDS-PAGE assay. The results show that the expressed proteins were in an insoluble form. The inclusion body was used to purify this recombinant protein. The 'Methods of Purification' were performed according to "Molecular Cloning: A Laboratory Manual Third Edition".

2.7 Western blotting

Five μg of the crude soluble protein samples was separated by using SDS polyacrylamide gel electrophoresis (SDS-PAGE) with 12% seperating gels and 5% collecting gels according to the standard protocol. Proteins were transferred to the polyvinylidene fluoride (PVDF; Millipore, U.S.A.) membrane and Fe-SOD protein was detected by polyclonal anti-FeSOD antibodies. Chemiluminescence was detected using the Amersham Enhanced Chemiluminescent (ECL) Plus Western Blotting Detection System (GE Healthcare Co. Ltd., U.K.) on X-ray film. The optical density from Western blot was conducted by using the Tannon Gel Image System (Tannon, Shanghai, China).

2.8 Stress treatment and crude protein extraction

In order to determine the function of Fe-SOD, the gametophytes were stressed under different illuminations and at low temperatures. The gametophytes were cultured in incubator at 26°C for one month. After that, half of them were put into a low temperature stress at 4°C for 72 hours. The other half were exposed to different illumination stresses at 26°C. It was critical to maintain the same temperature during these stress events.

The gametophytes of *Ceratopteris thalictroides* were ground in liquid nitrogen. Total soluble proteins were extracted by sonication of the ground samples in Medium A (5 mM sodium phosphate, pH 7.5, 10 mM $MgCl_2$, 10 mM NaCl, 25% glycerol, 10 mM HEPES) at 4°C for 15 min. Each extract was centrifuged at 13,000 xg for 15 min and the supernatant was collected. These proteins ware quantified spectrophotometrically via Bradford methods (Bradford, 1976) with bovine serum albumin (BSA) as the standard.

3. Results

3.1 Cloning and characterization of Fe-SOD cDNA

Following PCR with degenerate primers and sequencing analysis of selected clones, we obtained a partial gene sequence of about 458bp that putatively encoded for Fe-SOD. Sequencing analysis and Blast search of NCBI performed on the gene fragment revealed that it contained a partial sequence which was well conserved in Fe-SOD. The full-length gene sequence of Fe-SOD was about 1212 bp and was finally obtained by RACE procedures. DNA sequencing analysis revealed that the amplified full-length sequence showed 77% identity to the Fe-SOD sequence from Matteuccia struthiopteris, and 71% identity to that of Pinus pinaster. The nucleotide sequence of the Ct Fe-SOD gene was deposited to the GenBank database under Accession No. HQ439554 Lane 1，7: DNA Marker(DL2000，HindIII)，Lane 2: PCR product of Fe-SOD，Lane 3: pET32a-FeSOD digested by BamHI/SalI，Lane 4: pET32a digested by BamHI/SalI，Lane 5: pET32a-FeSOD digested by BamHI，Lane 6: pET32a-FeSOD digested by SalI.

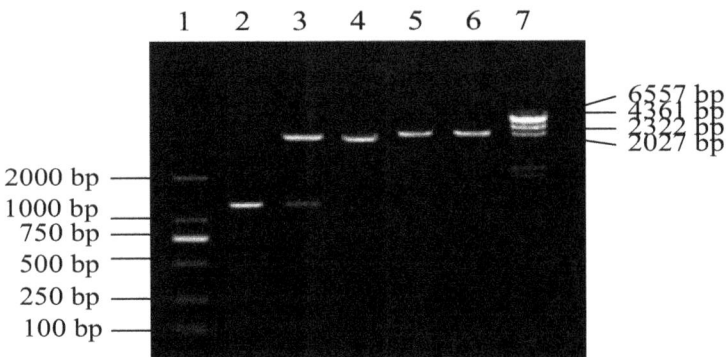

Fig. 1. Gel electrophoresis of pET32a-FeSOD after enzyme digestion.

3.2 Expression and characterization of recombinant enzyme

The gene encoding Fe-SOD harbored an open reading frame consisting of 798 bp that encoded 266 amino acids. The estimated molecular weight of the protein was 43 kDa .The deduced amino acid sequence of Ct FeSOD was compared with the homologous enzymes from *Matteuccia struthiopteris* (MSFeSOD), *Pinus pinaster* (PpFeSOD) and *Solanum lycopersicum* (SlFeSOD). CtFeSOD showed as high as 82% identity with MSFeSOD, and 69% identiy with PpFeSOD.

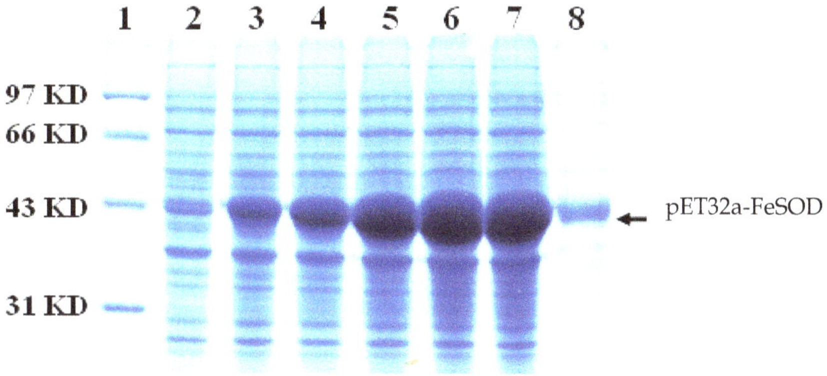

Fig. 2. Expression and purification of the recombinant Fe-SOD proteins in E. coli strain BL21.The arrow indicates recombinant Fe-SOD. The expression of the recombinant pET32a-fesod proteins in E. coli strain BL21. Lane 1, molecular weight standards; Lane 2,total protein of E. coli without induction; Lane 3,4,5,6,7 soluble protein of E.coli induced by 1mM IPTG at 37°C for 0.5,1,2,4 and 6h respectively. Lane 8: purified proteins. The separation gel of SDS-PAGE was with 12%polyacrylamide and stained with Coomassie Brilliant Blue.

```
Chlamy          MALAMKAQASSLVAGQRR---AVRPASGRRAVITR----------AALELKSPPYALDAL
Marchantia      MAMAAVTGARCLLPVPSLRFSKARPSNANFLKLKLPLG----APVAKFDLKDPPYPLDAL
Barbula         --MAAVP-----------------------LPLTT------------VDLRPPPYALDAL
Pinus           MALLHRPISLPLLPLAAPASFISHPKHIHLRKVAVPKFRSAPPTVALLNLNPPPYDFDAL
Arabidopsis     --MAASS-------------------------AVTAN-----------YVLKPPPFALDAL
Ceratopteris    MATATCSSILRATPWPSQILSTSPSLRKGNTRSSYR------PIVMEFHLRPLPFEMDAL
                      .                                          *.  *: :***

Chlamy          EPHMSKQTLEFHWGKHHRAYVDNMNKQVAGTPLDGKSLEEIVLASWNN-GQPTPVFNNAA
Marchantia      EPHMSKETLEFHWGKHHRAYVNNLNGQIENSDLERNTLEEIVQSSYNN-GNPTPAFNNAG
Barbula         EPHMSKETLEYHWGKHHRAYVDNLKKQIEGTELASQTLEDIVRATYNN-GEPTAPFNNAA
Pinus           EPHMSKQTLSFHWGKHHRAYVDNCNNQIKDTELEECTLEDIIKKTYNN-GDPRPVFNNAA
Arabidopsis     EPHMSKQTLEFHWGKHHRAYVDNLKKQVLGTELEGKPLEHIIHSTYNNNGDLLPAFNNAA
Ceratopteris    EPYMSKQTLEFHWGKHHRTYVDNLNKQIKGTELEGMQLDELLKHGYNN-GNPTAYFNNAG
                **:***:**.:*******:**:*  :  *:  .: *     *:.::    :** *:   . ****.

Chlamy          QVWNHTFFWESMKPNGGGAPTGALAEAITRDFG-SLDKFKEEFKQAGMTQFGSGWAWLN-
Marchantia      QIWNHDFFWQSMKPGGGGPPTGEVAELINRDFG-SYDNFVKEFKAAGATQFGSGWAWL--
Barbula         QAWNHEFFWLSMSPHGGKQPDGELMSLLKRDFG-SYDNFVKEFKQAGATQFGSGWAWL--
Pinus           QAWNHEFFWECMKPGGGKEPKGEILELIKRDFG-AYDEFLKEFKQASATQFGSGWAWL--
Arabidopsis     QAWNHEFFWESMKPGGGGKPSGELLALLERDFTTSYEKFYEEFNAAAATQFGAGWAWL--
Ceratopteris    QAWNHDFFWESIKPGGGKKPTGEVLHLIERDFG-SFEAFVDEFKQAAATQFGSGWAWLVG
                * *** *** .:.* **   * * :    : ***   : : * .**: *. ****:*****

Chlamy          -------ADKTGKLSISKSPNAVNPVVEGKTPILTVDVWEHAYYIDVQNRRPDYITTFME
Marchantia      -------VLKNGKLAIEKTPNAVTPILWGSIPLLVVDVWEHAYYLDFQNRRPDYLSTFVN
Barbula         -------TVADGKLMVEKSPNAINPLVFGHVPILVADVWEHAYYLDYQNRRPDYLTTFMN
Pinus           -------VLKDNKLVIEKTPNALNPLIWGHKPLLTIDVWEHAYYLDYQNRRPDYISTFMN
Arabidopsis     -------AYSNEKLKVVKTPNAVNPLVLGSFPLLTIDVWEHAYYLDFQNRRPDYIKTFMT
Ceratopteris    MPTDPLVENKFGKLAIEKTPNALNPLILGHVPLLTIDVWEHAYYLDYQNRRPDYISTVIN
                            ** : *:***:.*:: *   *:*. ********:* *******:.*.:

Chlamy          KLINWDAVAQRYAAATK-----------------
Marchantia      DLISWDAVNERLARAKASVNLSTPSIPES-----
Barbula         ELVSWDAVAKRLQLALQES---------------
Pinus           ELVSWDAVNARLDRAKASAT--------------
Arabidopsis     NLVSMEAVSARLEAAKAASA--------------
Ceratopteris    ELISWEAVTARLERAKATINFYDVVVPDDQYKYK
                .*:. :**  *   *
```

Fig. 3. Multiple sequence alignment of the deduced amino acid sequence of Fe -SOD . The identical residues among these Fe-SODs are marker by asterisks. *Chlamydomonas reinhardtii*(AABO4944.1), *Marchantia polymorpha*(BAC66948.1) , *Barbula unguiculata* (BAC66946.1), *Pinus pinaster*(AY 536055.1), *Arabidopsis thaliana*(NP 199923.1).

Fe-SOD complete mRNA sequences from five species(Chlamydomonas reinhardtii, Marchantia polymorpha, Barbula unguiculata, Ceratopteris thalictroides, Pinus pinaster, Arabidopsis thaliana) were aligned in this study. Fe-SOD complete mRNA was highly homologous, having 65% homology of nucleotides within the coding regions. Amino acid sequence alignment from the same five species displayed five highly conserved domains in the Fe-SOD proteins: FNNA, FGSGW, WEHAYY, WNHHFF and HWGKH. In addition, Fe-SOD sequences of Ceratopteris thalictroides, Pinus pinaster and Arabidopsis thaliana encoded for a unique tripeptide ARL close to the carbox I terminus of the enzyme. ARL is the location signal of peroxisomes in cells. Although this sequence has been shown to direct the proteins to peroxisomes in other proteins, it has yet to be determined whether this is a functional sequence or not. The conserved ARL (or SRL) sequence is not present in the prokaryotic Fe-SOD proteins showing that it is not obligatory for the enzyme function (Van Camp et al., 1994). Experimental studies have shown that different types of SOD could exert their respective antioxidative functions, for example, Mn-SOD providing effective protection to DNA. Fe-SOD was primarily shown to protect the soluble proteins that were most sensitive to oxidation.

3.3 Demonstration of Fe-SOD

The specific Fe-SOD antibody (purchased from Agrisera) was used to demonstrate that the recombined protein was in fact Fe-SOD. Western blotting results showed that hybridization signals of recombined proteins and Fe-SOD from Ceratopteris thalictroides were placed in the same position. In addition, the same trend was found in the two proteins. This proved that the Fe-SOD gene in Ceratopteris thalictroides was expressed within transcribed levels and thus the recombined protein had to be Fe-SOD- based.

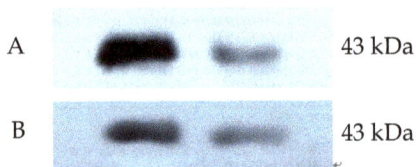

Fig. 4. 10µg and 5µg crude protein was used for the western blotting to detect Fe-SOD expression level by twoantibodies respectively. Experiments were done at least in triplicates. Row A. recombinant pET32Fe-SOD antibody was used in this picture. Row B. Arabidopsis thaliana Fe-SOD antibody (Agrosera) was used in this picture.

3.4 Fe-SOD expression during low temperature stress

Gametophytes exposed to low temperature stress began to wither after 24 h. According to this picture of western blotting, as stress time increased, the enzyme activity increased. Enzyme activity attained a maximum when the time was 8 hour, and then decreased. This result indicates that the Fe-SOD in gametophytes played an important role on resistance to adverse circumstances (low temperature). After 24 hours of cold treatment, the expressionof Fe-SOD was significantly reduced. We hypothesized that the clearance mechanism was restrained due to excessive ROS. The tendency of the gametophytes in Ceratopteris thalictroides was to first increase, then decrease. We saw the same tendency in pea plants under low temperature stress.

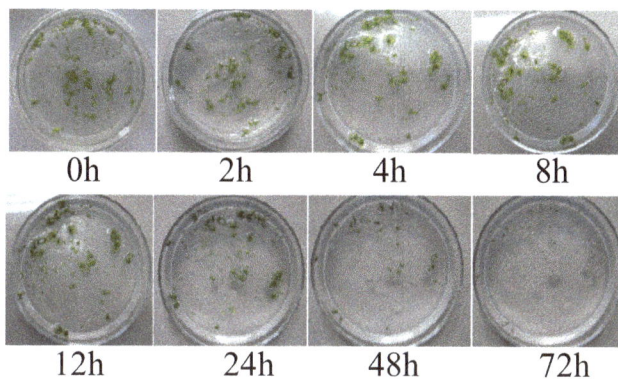

A.

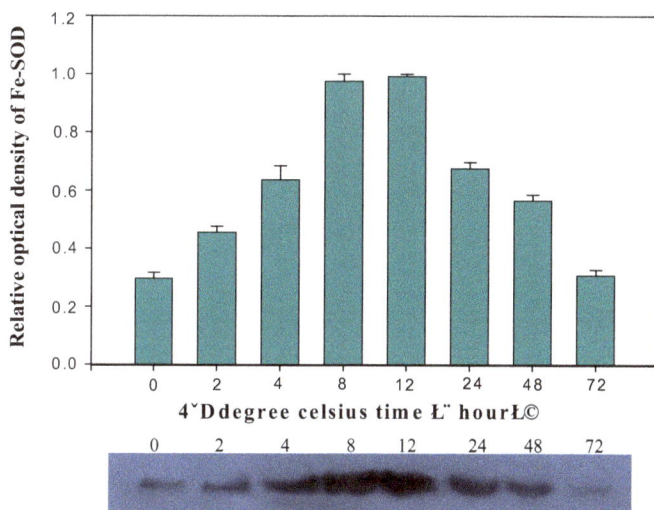

B.

Fig. 5. A. the gametophytes of *Ceratopteris thalictroides* in Chilling stress (4℃). B.Expression
of Fe-SOD in Ceratopteris thalictroides during chilling stress. 5µg crude protein was used
for the western blottingto detect Fe-SOD expression level by using specific Fe-SOD
antibody. Experiments were done at least in triplicates.

3.5 Fe-SOD expression during illumination stress

There was no difference in appearance between these gametophytes, but the results from
western-blotting tests show that, when illuminated with 25×10^2 lux, the expression of Fe-
SOD was the lowest. 25×10^2lux was the light intensity in the incubator. The amounts of
Fe-SOD in the other light illuminations were all greater than that of 25×10^2 lux. Whether
the light source was strong or weak, expression of Fe-SOD was induced. This is because
weak light interferes with plant photosynthesis, and strong light induces photo-inhibition
of plants. The oxygen radicals in the plant cells increase in both cases. Fe-SOD expression

levels were expected to increase in attempt to protect the plants against this adverse environment. We found that the amount of Fe-SOD in 600×10² lux was less than the amount found in 300×10² lux. We speculated that this was due to strong light damaging the scavenging systems of the active oxygen.

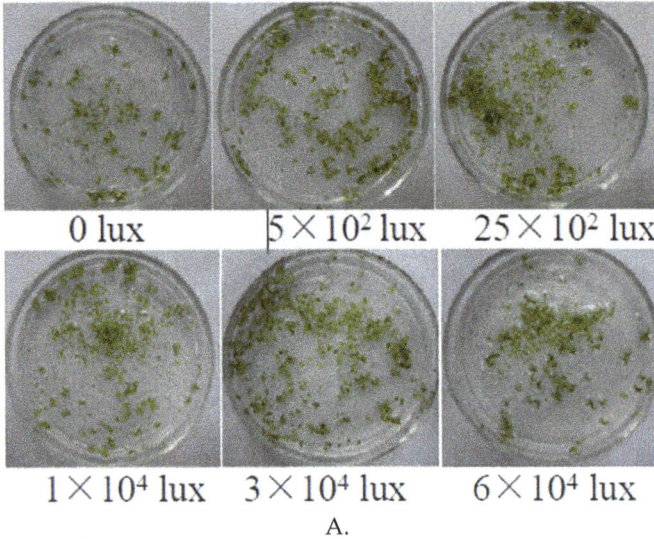

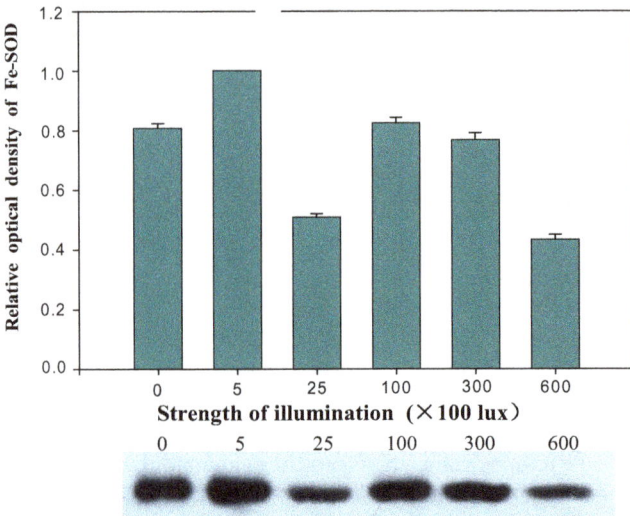

Fig. 6. Expression of Fe-SOD in *Ceratopteris thalictroides* during Illumination stress. 5µg crude protein was used for the western blotting to detect Fe-SOD expression level by using specific Fe-SOD antibody. Experiments were done at least in triplicates.

4. Discussion

The absence of Fe-SOD in animals has given rise to the proposal that the Fe-SOD gene originated in the plastid and moved to the nuclear genome during evolution. Support for this theory comes from the existence of several conserved regions that are present in plant and cyanobacterial Fe-SOD sequence, but absent in non-photosynthetic bacteria (Bowler et al., 1994). Previous studies reported that ، the majority of vascular plants examined contained both the Cu-Zn and Mn enzymes but not the Fe enzyme. But Susan and Marvin considered that the gene for the Fe enzyme is present in all eukaryotic plants but not expressed. Environmental pressures could have resulted in the selection of a modified controlling region arising by mutation and allowing once more for expression of the enzyme (Susan & Marvin, 1981) . Kenichi Murao reported that Fe-SOD activity was not detected in extracts from the leaves of ferns, *Equisetum arvense* and *Matteuccia struthiopteris*. He thought the fern Fe-SOD gene was transcriptionally regulated by Cu (Kenichi Murao et al., 2004). But in this study we found Fe-SOD activity was detected in fern *Ceratopteris thalictroides*.

Photon energy is the only source of energy for plants but it can have harmful effects on plants if irradiance is lower or higher than the physiological requirement for plant growth and development (Long et al., 1996; Lawlor, 2001; Loomis & Connor, 2003; Mandal & Sinhá, 2004). High leaf irradiance reduces photosynthetic efficiency resulting in photodynamic degradation of the photosynthetic apparatus. Pigment-protein complexes present in photosystem II are highly sensitive to photo-damage triggered by the formation of reactive oxygen species (ROS) (Barber & Anderson, 1992). Plants have developed enzymatic and non-enzymatic protection mechanisms against irradiance stress. Superoxide dismutases was part of the enzymatic antioxidative response system (Asada, 1996; Niyogi, 1999). SODs act as the first line defense against ROS, dismutating superoxide-radicals to H_2O_2 (Bowler et al., 1994; Kanematsu & Asada 1994).These enzymes are in different cellular compartments and are controlled by a ROS gene network (Mittler et al., 2004). In the red alga *Eucheuma denticulatum* the photosynthetic production of O_2 (by photosynthesis) under excessive light leads to an increase in the levels of ROS-inducing SOD activity (Mtolera et al. 1995).

However The SOD enzyme not only consumes superoxide and thereby provides tolerance to oxidative stress, but also produces H_2O_2. It is tempting to speculate that an increased steady-state level of H_2O_2 or an increased flux through the H_2O_2 pool enhanced an acclimation process that enabled the plants to tolerate or repair freezing injury more effectively. H_2O_2 has potential toxicity in plants, but it may also have a number of regulatory roles. Recent reports suggest that H_2O_2 mediates some responses to pathogens (Chen et al.,1993), produces a transient Ca^{2+} surge, which is a known signaling component (Price et al., 1994), and initiates the production of other antioxidant enzymes during acclimation (Prasad, 1997). H_2O_2 is metabolized by a number of peroxidases using reducing equivalents to form water. Mcord and Fridawich believe that, the removing process and generation process of ROS were exist simultaneously in plant cells. The ROS would generate and membrane lipid peroxidation increased while the plant in adversity stress (Mocord & Ries, 1997 ; Pryor WA, 1977). The increase of ROS in cells would damage DNA and membranes, thus affecting the protein synthesis. This could lead to metabolic rate reduction and cell death. Certain enzymes (SOD, for example) are able to remove ROS to protect cells from injury. This is referred to as the "protectase system".

In conclusion, our study described the cloning, expression and characterization of the Fe-SOD gene from the fern *Ceratopteris thalictroides*. Sequence analysis of CtFe-SOD predicted that this gene encodes a protein of about 43 kDa and shows high similarity to most known Fe-SOD genes, sharing five highly conserved domains that are most likely essential for enzyme activity. Certainly, further investigation is warranted to determine the functional and biological significance of CtFe-SOD, in particular its role in stress resistance.

5. References

Asada, K. (1996). Radical production and scavenging in the chloroplasts, in Baker, NR. (ed.), Phytosynthesis and the Environment, Kluwer Academic Publ., Dordrecht –Boston – London, pp.123-150

Asano, CS., Okamoto, OK., Hollnagel, HC., Stringher, CG., Oliveira, MC. & Colepicolo, P. (1996). The activity of superoxide dismutase oscillates in the marine dinoflagellate Gonyaulax Polyedra, Ciência e Cultura 48: 64–67

Ayala, MB. & Sandmann, G. (1988). Activities of Cu-containing protein in Cu-depleted pea leaves, Physiol Plant 72: 801–806

Barber, J. & Andersson, B. (1992). Too much of a good thing: light can be bad for photosynthesis, Trends biochem. Sci. 17: 61-66

Bestwick, CS. & Maffulli, N. (2004). Reactive oxygen species and tendinopathy: do they matter? Br J Sports Med 38:672-4.

Bowler, C., Van Montagu, M. & Inzd, D. (1992). Superoxide dismutase and stress tolerance, Ann. Rev. Plant Physiol 43:83-116

Seandalios, J. (1993). Oxygen stress and superoxide dismutase. Plant Physiol 101:7-12.

Bowler, C., Van Camp, W., Van Montagu, M. & Inzé, D. (1994). Superoxide dismutase in plants, CRC crit. Rev. Plant Sci. 13: 199-218

Bradford, MM. (1976). A rapid and sensitive method for the quantitation of microgram quantities of protein utilizing the principle of protein-dye binding, Anal. Biochem 72:248-254

Casano, LM., Martin, M. & Sabater, B. (1994). Sensitivity of superoxide dismutase transcript levels and activities to oxidative stress is lower in mature-senescent than in young barley leaves, Plant Physiol 106:1033-1039

Chatterjee, A. & Roux, S.J. (2000). Ceratopteris richardii: a productive model for revealing secrets of signaling and development, J.plant Groeth regul 19:284-289

Chen, Z., Silva. H., & Klessig, DF. (1993). Active oxygen species in the induction of plant systemic acquired resistance by salicylic acid, Science 262: 1883–1886

Colepicolo, P., Camarero, VCPC. & Hastings, JW. (1992). A circadian rhythm in the activity of SOD in the photosynthetic alga Gonyaulax polyedra, Chronobiol. Int. 9:266–268

Crowell, D.N. & Amasino, R.M. (1991). Induction of specific mRNAs in cultured soybean cells during cytokinin or auxin starcation, Plant Physiology 95:711-715

Crowell, D.N. & Amasino, R.M. (1991 b). l'-lucleotide sequence of an iron superoxide dislllutase complementary DNA from soy-bean, Plant Physiol 96: 1393-1394

Dyer, AF. (1979). The experimental biology of ferns, Academic Press, London, pp. 657

Fahrendorf, T., Ni, WT., Shorrosh, BS. & Dixon, RA. (1995). Stress responses in alfalfa (Medicago sativa L.)19: transcriptional activation of oxidative pentose phosphate pathway genes at the onset of the isoflavonoid phytoalexin response, Plant Mol Biol 28: 885–900

Fridovich, I. (1995) Superoxide radical and superoxide dismutases, Annu Rev Biochem 64:97-112

Geret, F., Manduzio, H., Company, R., Leboulenger, F., Bebianno MJ. & Danger, JM. (2004)
    Molecular cloning of superoxide dismutase (Cu/Zn-SOD) from aquatic mollusks,
    Mar Environ Res 58:619-23

Hickok, LG., Warne, TR. & Slocum, MK. (1987). Ceratopteris richadii: applications for
    experimental plant biology, Am J Bot 74:1304-1316

Hickok, LG., Wame, TR. & Fribourg, RS. (1995). The biology of the fern ceratopteris and its
    use as a model system, Int.J.Plant.Sci 156:332-345

Hollnagel, HC., Di Mascio, P., Asano, CS., Okamoto, OK., Stringher, CG., Oliveira, MC. &
    Colepicolo, P. (1996). The effect of light on the biosynthesis of b-carotene and
    superoxide dismutase activity in the photosynthetic alga Gonyaulax polyedra,
    Braz. J. med. biol. Res. 29:105–110

Hunt, JS. & Sipes, SD. (2001). Horsetails and ferns are a monophyletic group and the closest
    relatives to seed plants, Nature 409:618–622

Kaminaka, H., Morita, S., Tokumoto, M., Yokoyama, H., Masumura, T. & Tanaka, K. (1999).
    Molecular cloning and characterization of a cDNA for an iron-superoxide
    dismutase in rice (Oryza sativa L.), Biosci Biotechnol Biochem 63: 302-308

Kanematsu, S. & Asada, K. (1994). Superoxide dismutase, in: Fuku, T., Soda, K. (ed.),
    Molecular Aspects of Enzyme Catalysis, Kodansha, Tokyo, pp. 191-210

Kenichi Murao, Masayuki Takamiya, & Kanji Ono. (2004). Copper eficiency induced
    expression of Fe-superoxide dismutase gene in Matteuccia struthiopteris, Plant
    Physiology and Biochemistry 42 143–148

Kenrick, P. (2000). The relationships of vascular plants.Phil Trans R Soc Lond B 355:847–855.

Kurepa, J., Van Montagu, M. & Inzé,D. (1997 a). Expression of sodCp and sodB genes in
    Nicotiana tabacum: effects of light copper excess, J. Exp. Bot 48:2007–2014.

Kurepa, J., Herouart, D., Van Montagu, M. & Inze, D. (1997 b). Differential expression of
    CuZn-and Fe superoxide dismutase genes of tobacco during development,
    oxidative stress, and hormonal treatments, Plant Cell Physiol 38:463-470

Lawlor, DW. (2001). Photosynthesis: Molecular, Physiological and Environmental Processes,
    Springer-Verlag, New York

Levine, A. (1999). Oxidative stress as a regulator of environmental responses in plants, in
    Lerner HR.(ed.), Plant responses to environmental stress, New York: Marcel
    Dekker Inc, pp. 247-264

Lloyd, RM. (1974). Systematics of the genus Ceratopteris Brongn. (Parkeriaceae). II
    Taxonomy.Brittonia 26:139-160

Lloyd, RM. (1993). Parkeriaceae Hooker, water fern family. Pages North of Mexico.Vol 2.
    Pteridophytes and Gymnosperms. Oxford University Press, New York

Long, SP., Farage, PK. & Garcia, RL. (1996). Measurement of leaf and canopy photosynthetic
    CO2 exchange in the field, J. exp.Bot 47:1629-1642

Loomis, RS. & Connor, DJ. (2003). Crop Ecology: Productivity and Management in
    Agricultural Systems, Cambridge University Press, Wiltshire

Mandal, KG. & Sinhá, AC. (2004). Nutrient management effects on light interception,
    photosynthesis, growth, dry-matter production and yield of Indian mustard
    (Brassica juncea), J.Agron. Crop Sci. 190:119-129

Manduzio, H., Monsinjon, T., Galap, C., Leboulenger, F. & Rocher, B. (2004). Seasonal
    variations in antioxidant defences in blue mussels Mytilus edulis collected from a
    polluted area: major contributions in gills of an inducible isoform of Cu/Zn-
    superoxide dismutase and of glutathione S-transferase, Aquat Toxicol 70:83-93

Marikovsky, M., Ziv, V., Nevo, N., Harris-Cerruti, C. & Mahler, O. (2003). Cu/Zn superoxide dismutase plays important role in immune response, J Immunol 170:2993-3001

Masuyama, S. (1992). Clinal variation of frond morphology and its adaptive implication in the fern Ceratopteris thalictroides in Japan, Plant Species Biol 7:87-96

Miller, Jh. (1968). Fern gametophytes as experimental material, Bot Rev 34:316-426

Mittler, R., Vanderauwera, S., Gollery, M. & Breusegem, FV. (2004). Reactive oxygen gene network of plants, Trends Plant Sci. 9: 490-498

Mocord, Ries sk. (1997). Pwrification and quantiative relationship with eater - soluble pratein in seedlings , plant physiol. 59 : 315 - 318

Mtolera, MSP., Collén, J., Pedersen, M. & Semesi, AK. (1995). Destructive hydrogen peroxide production in Eucheuma denticulatum (Rhodophyta) during stress caused by elevated pH, high light intensities and competition with other species, Eur. J.Phycol. 30:289–297

Niyogi, KK. (1999). Photoprotection revisited: Genetic and molecular approaches, Annu. Rev. Plant Physiol. Plant mol. Biol. 50: 333-359

Prasad, TK. (1997). Role of catalase in inducing chilling tolerance in pre- emergent maize seedlings, Plant Physiol 114: 1369–1376

Price, AH., Taylor, A., Ripley, SJ., Griffiths, A., Trewavas, AJ. & Knight, MR. (1994). Oxidative signals in tobacco increase cytosolic calcium, Plant Cell 6:1301–1310

Pryer, KM., Schneider, H., Smith, AR., Cranfill, R., Wolf, PG. & Raghavan, V. (1989). Developmental biology of ferns. Cambridge University Press, New York

Pryor, WA. (1977). Free Redical in Biology Volume III, Academic Press Incorporated,New York, NY

Robertson, D. Davies, DR., Gerrish, C., Jupe, SC. & Bolwell, GP. (1995). Rapid changes in oxidative metabolism as a consequence of elicitor treatment of suspension-cultured cells of French bean (Phaseolus vulgaris L), Plant Mol Biol 27: 59–67

Salin, ML. & Bridges, SM. (1981). Absence of the iron-containing superoxide dismutase in mitochondria from muatard(Brassica campestris), Biochemical Journal 195: 229-233.

Shiono, T., Nakata, M., Yamahara, T., Matsuzaki, M., Deguchi, H. & Satoh, T. (2003). Repression by Cu of the expression of Fe-superoxide dismutase of chloroplasts in the moss Barbula unguiculata but not in the liverwort Marchantia paleacea var. diptera, J. Hattori Bot. La. 93:141–153

Susan, MB & Marvin, LS. (1981)Distribution of iron−containing Superoxide dismuasein vascular plant, Plant Physiol 68 ： 275-282

Tsang, EWT., Bowler, C., Herouart, D., van Camp, W., Villaroel, R., Genetello, C., Van Montagu, M. & Inze, D. (1991). Differential regulation of superoxide dismutase in plants exposed to environmental stress, Plant Cell 3:783-792

Van Camp, W., Bowler, C., Viliarroel, R., Tsang, EWT., Montagu, MY. & Inze, D. (1990a). Characterization of iron superoxide dismutase cDNAs from plants obtained by genetic complementation in Escherichia coli, Proc Natl Acad Sci USA 84: 9903-9907

Van Camp, W., Bowler, C. & Villarroel, R. (1990 b). Characterization of iron superoxide dismutase cDNAs from plants obtained by genetic complementation in Escherichia coli, Plant Physiology 112:1703-1714

Van Camp, W., Willekens, H. & Bowler, C. (1994). Elevated levels of suoeroxide dismutase protect transgenic plants against ozone damage, Bio Technology 12:165-168

Youn, HD., Kiln, EJ., Roe, JH., Hah, YC. & Kang, SO. (1996). A novel nickel-containing superoxide dismutase from Streptomyces spp, J. Biochem 318:889-896

16

Molecular Cloning, Expression Pattern, and Phylogenetic Analysis of the *Lysyl-tRNA Synthetase* Gene from the Chinese Oak Silkworm *Antheraea pernyi*

Yan-Qun Liu and Li Qin
Department of Sericulture, College of Bioscience and Biotechnology,
Shenyang Agricultural University
China

1. Introduction

Aminoacyl-tRNA synthetases (AARS) are a class of enzymes that charge tRNAs with their cognate amino acids. There are two classes of tRNA synthetases, classes I and classes II, that are distinguished by the architectures of their active-site catalytic cores (Guo et al., 2008). Lysyl-tRNA synthetase (LysRS) is an AARS, a group of ancient proteins known for their critical role in translation. LysRS is a homodimer localized to the cytoplasm which belongs to the class II family of tRNA synthetases. Its assignment to class II AARS is based upon its structure and the presence of three characteristic sequence motifs in the core domain. The catalytic core domain is primarily responsible for the ATP-dependent formation of the enzyme bound aminoacyl-adenylate.

The *LysRS* gene has been cloned and characterized from various kinds of organisms. In *Escherichia coli*, there are two distinct LysRSs encoded by two widely separated genes, *LysRS* and *LysRU* (VanBogelen et al., 1983; Emmerich & Hirshfield, 1987), while in *Campylobacter jejuni* only one *LysRS* is present (Chan & Bingham, 1992). In *E. coli*, the *LysRS* gene is expressed constitutively while *LysRU* can be induced by growth at high temperature (Hirshfield et al., 1981). In human, only one *LysRS* belonging to the class II family of AARSs is present (Guo et al., 2008). The crystal structure of tetrameric form of human LysRS has been recently determined (Guo et al., 2008). Human LysRS has been shown to be secreted and to trigger a proinflammatory response as a target of autoantibodies in the human autoimmune diseases, polymyositis or dermatomyositis (Park et al., 2005). And, LysRS is required for the translocation of calreticulin to the cell surface in immunogenic death (Kepp et al., 2010). However, none of insect *LysRS* gene has been characterized to date.

The Chinese oak silkworm, *Antheraea pernyi* (Lepidoptera: Saturniidae), is one of the most well-known economic insect species used for silk production and insect food source. This insect is known to be sucessfully domesticated in China around the 16th century (Liu et al. 2010a), and it is commercially cultivated in China, India, and Korea. To isolate the functional genes of *A. pernyi*, we have constructed a full-length cDNA library (Li et al. 2009). By EST sequencing, several *A. pernyi* genes encoding important enzymes have been characterized, such as two enolase genes and a lysophospholipase gene (Liu et al. 2010b, 2010c).

In this chapter, we describe the cloning and characterization of the *A. pernyi LysRS* gene from the full-length pupal cDNA library by random EST sequencing. The expression patterns at various developmental stages and different tissues were investigated. Finally, the deduced protein sequence of the *LysRS* gene from *A. pernyi* and other organisms were used to examine the relationship among these species, and to test the potential use of LysRS protein in phylogenetic study.

2. Materials and methods

2.1 Silkworms and tissues

The *A. pernyi* strain *Shenhuang No. 1* was used in this study. Larvae were reared routinely on oak trees (*Quercus liaotungensis*) in the field. Blood, fat body, midgut, silk glands, body wall, Malpighian tubules, spermaries, ovaries, brain and muscle were taken from silkworm larvae at day 10 of fifth instar. Eggs at day 5, larvae at day 10 of fifth instar, pupae and moths were also sampled. All the samples were immediately frozen in liquid nitrogen and stored at – 80°C for later use.

2.2 Cloning of the *A. pernyi LysRS* gene

A full-length pupal cDNA library of *A. pernyi* has been constructed in our lab (Li et al. 2009). An EST encoding LysRS homolog (GenBank accession no. GH335029) was isolated by random EST sequencing. So, the cDNA clone was used to complete the full-length cDNA sequence of the *A. pernyi LysRS* gene.

2.3 Sequence analysis

DNASTAR software (DNASTAR Inc., Madison, WI) was used to identify open reading frame (ORF), deduce amino acid sequence, and predict the isoelectric point and molecular weight of the deduced amino acid sequence. Blast search was performed at http://www.ncbi.nlm.nih.gov/blast/. Conserved Domains was predicted at http://www.ncbi.nlm.nih.gov/Structure/cdd/wrpsb.cgi/. The *in silico* gene expression analysis based on the available EST resources was employed at http://www.ncbi.nlm.nih.gov/Unigene/ESTprofileViewer/.

2.4 Total RNA extraction and first strand cDNA synthesis

Total RNA was extracted by using RNAsimple Total RNA Extraction Kit (TIANGEN Biotech Co. Ltd., Beijing) according to the manufacturer's instruction. DNase I was used to remove contaminating genomic DNA. The purity and quantity of the extracted RNA was quantified by the ratio of OD260/OD280 by ultraviolet spectrometer. First strand cDNA was generated by using 2 µg of total RNA per sample with TIANScript cDNA Synthesize Kit (TIANGEN Biotech Co. Ltd., Beijing).

2.5 RT-PCR analyses

The cDNA samples were amplified by semi-quantitative PCR method using the gene-specific primer pair LYQ146 (5'-TCCGA GTGGG GAAGA AGTTG-3') and LYQ147 (5'-TTCAG TCAGT CCTGG TATGT-3') for the *A. pernyi LysRS* gene, which generated a 322 bp

fragment. An *actin* gene (GU073316) was used as an internal control, and a 468 bp fragment
was amplified using the primer pair LYQ85 (5'-CCAAA GGCCA ACAGA GAGAA GA 3')
and LYQ86 (5'CAAGA ATGAG GGCTG GAAGA GA 3') (Wu et al, 2010). PCRs were
performed with the following cycles: initial denaturation at 95°C for 5 min; followed by 30
cycles of 1 min at 95°C, 30 s annealing at 55°C, 30 s extension at 72°C; and a final extension
at 72°C for 10 min. To avoid sample DNA contamination, the negative RT-PCRs control
reactions were performed with every total RNA as templates. The amplification products
were analyzed on 1.0% agarose gels, purified from the gel, and directly sequenced to
confirm the specificity.

2.6 Phylogenetic analysis
The amino acid sequences of LysRS homologs from different organisms were retrieved from
GenBank database, and SilkDB database (Duan et al., 2010). Multiple sequence alignments
were performed using Clustal X software (Thompson et al. 1997). A phylogenetic tree was
constructed by MEGA version 4 (Tamura et al. 2007) using Neighbour-Joining (NJ) method
(Saitou and Nei 1987) with bootstrap test of 1000 replications.

3. Results and discussion

3.1 Sequence analysis of the *A. pernyi LysRS* gene
We identified the *A. pernyi LysRS* gene from a pupal cDNA library constructed in our lab by
random EST sequencing (Li et al., 2009). Based on the cDNA clone Appu0107, we isolated
and sequenced a full-length cDNA of the *A. pernyi LysRS* gene. The cDNA sequence and
deduced amino acid sequence of the *A. pernyi LysRS* gene were shown in **Figure 1**. The
obtained 2136 bp cDNA sequence contains a 5'-untranslated region (UTR) of 70 bp, a 3' UTR
of 292 bp with a polyadenylation signal sequence AATAAA at position 2082 and a poly (A)
tail, and an ORF of 1740 bp encoding a polypeptide of 579 amino acids. The LysRS protein
has a predicted molecular weight of 65.62 kDa and isolectric point of 6.1. Blast search
revealed that the deduced amino acid sequence of the *A. pernyi LysRS* gene had 66%
identities and 79% positives with that of *Homo sapiens* LysRS (NP_005539), which belongs to
the class II family of AARSs (Guo et al., 2008). Conserved Domains prediction showed that
the *A. pernyi* LysRS protein contained the LysRS class II core domain (Wolf et al., 1999),
including the conserved active sites and three characteristic sequence motifs (Desogus et al.,
2000). Moreover, the characteristic signature HIGH sequence of LysRS class I was not
present. We therefore referred to the protein as LysRS of *A. pernyi*. This cDNA sequence has
been deposited in GenBank under accession no. JF773568.

3.2 Homologous alignment
By searching in database, the *A. pernyi* LysRS protein homologues were found in various
kinds of life organisms, including bacteria, fungi, plants, invertebrates and vertebrates. To
assess the relatedness of *A. pernyi* LysRS to LysRS proteins from other organisms, identities
were calculated based on a Clustal alignment including 43 representative LysRS protein
sequences (**Figures 2 and 3**). The other 42 LysRS protein or homologue sequences used in
this study were downloaded from GenBank database, with one exception of *B. mori* LsyRS
from SilkDB database (Duan et al., 2010). These protein sequences were from 3 bacteria, 1
fungi, 6 plants, 22 invertebrates and 11 vertebrates. By sequence alignment, the *A. pernyi*

```
1            AGAT TAGT CTGT GACG TTTA GGTT AGTT GCGT TTCA TATT AATT GAGAGTATT ATT TTTGATTT AGAAGTC
71    ATGGCAGAAACCTCCAGTGAAAAGTTTCAAAGAATGAGTTAAAGCGCCGGTTAAAAGCTGAACAGAAATTAAAGGAGAAA
      M  A  E  T  S  S  E  K  V  S  K  N  E  L  K  R  R  L  K  A  E  Q  K  L  K  E  K
152   GCTGAAAAGGTTGCACAACAACCTGTTCAACCTGCGACCGAGAAAAAATATCAAAACAAGAGGGAAGAAATTAGTCCTAAT
      A  E  K  V  A  Q  Q  P  V  Q  P  A  T  E  K  K  I  S  K  Q  E  E  E  I  S  P  N
233   GAGTATTATAAGCTTCGATCTGCCGCTGTGGCAGCACTCAAAACAGGAGCGAAAGAGGAACATCCTTATCCTCATAAATTC
      E  Y  Y  K  L  R  S  A  A  V  A  A  L  K  T  G  A  K  E  E  H  P  Y  P  H  K  F
314   ACTGTGACAATTTCTTTAGAAGAATTCATTAATAAGTACAATAATTTAAATAGTGGAGAAGTGCTAGAAGAATACAACTGTA
      T  V  T  I  S  L  E  E  F  I  N  K  Y  N  N  L  N  S  G  E  V  L  E  N  T  T  V
395   TCACTCGCTGGGCGAGTGCATTCTATTAGGGAGTCCGGAGCTAAACTAATATTCTATGACTTGAGAGCTGAAGGAGTGAAA
      S  L  A  G  R  V  H  S  I  R  E  S  G  A  K  L  I  F  Y  D  L  R  A  E  G  V  K
476   ATTCAGGTTATGGCTAATGCAAAGCTGTATGAATCGGAATTTGAGACTGATACTGATAAGCTGAGCGTGGTGACATCATA
      I  Q  V  M  A  N  A  K  L  Y  E  S  E  F  E  T  D  T  D  K  L  R  R  G  D  I  I
557   GGTTGCATTGGACACCCAGGAAAAACAAAGAAAGGGGAGTTGTCCATTGTACCACAATCAATTAAGTTACTCTCTCCATGT
      G  C  I  G  H  P  G  K  T  K  K  G  E  L  S  I  V  P  Q  S  I  K  L  L  S  P  C
638   CTTCATATGCTGCCTCATTTACACTTTGGTCTCAAAGATAAAGAAACCCGGTTTAGGAAGAGATATTTAGATCTCATTTTA
      L  H  M  L  P  H  L  H  F  G  L  K  D  K  E  T  R  F  R  K  R  Y  L  D  L  I  L
719   AATGATCAGGTAAGGCAAACATTCTATACAAGAGCCAAAATTATAGCATATGTAAGAAGATTTTTAGATAATATGGGCTTC
      N  D  Q  V  R  Q  T  F  Y  T  R  A  K  I  I  A  Y  V  R  R  F  L  D  N  M  G  F
800   TTAGAGATAGAAACCCCAATGATGAACATGATCCCAGGAGGAGCTACAGCAAAACCCTTTATAACACATCACAATGATCTT
      L  E  I  E  T  P  M  M  N  M  I  P  G  G  A  T  A  K  P  F  I  T  H  H  N  D  L
881   AATATGGACCTGTTTATGAGAATTGCTCCAGAGTTGTATCACAGATGTTAGTTGTTGGTGGCCTTGACCGTGTTTATGAA
      N  M  D  L  F  M  R  I  A  P  E  L  Y  H  K  V  L  V  V  G  G  L  D  R  V  Y  E
962   ATTGGAAGGCAGTTCCGTATGAGGGTATAGATTTAACACACAACCCTGAATTTACAACATGTGAATTTTATATGGCATAC
      I  G  R  Q  F  R  N  E  G  I  D  L  T  H  P  E  F  T  T  C  E  F  Y  M  A  Y
1043  GCTGATTACAATGATCTAATTACTATAACAGAAACAATGTTGTCAGGAATGGTGAAAAGCATTCATGGTACTTATAAGGTA
      A  D  Y  N  D  L  I  T  I  T  E  T  M  L  S  G  M  V  K  S  I  H  G  T  Y  K  V
1124  AAATACCATCCAGATGGTCCGAGTGGGGAAGAAGTTGAAATTGATTTTACTCCGCCATTTGCTAGAGTTCCTATGATTGCG
      K  Y  H  P  D  G  P  S  G  E  E  V  E  I  D  F  T  P  P  F  A  R  V  P  M  I  A
1205  ACTTTGGAAAAGGTTCTGAATGTGAAATTGCCTTCTCCAGATAAACTGGACACGGCTGAAGCAAACTCTCTCCTTAGCCAA
      T  L  E  K  V  L  N  V  K  L  P  S  P  D  K  L  D  T  A  E  A  N  S  L  L  S  Q
1286  TTATGTGAAAAACATGAGGTGGAATGCCCACCACCGCGGACAACAGCTAGACTACTTGATAAATTAGTTGGTGAATTCCTT
      L  C  E  K  H  E  V  E  C  P  P  P  R  T  T  A  R  L  L  D  K  L  V  G  E  F  L
1367  GAAGACAAATGCATCAACCCAACATTTATTCTGGATCATCCACAAATTATGAGCCCATTGTCTAAGTATCACAGGGACATA
      E  D  K  C  I  N  P  T  F  I  L  D  H  P  Q  I  M  S  P  L  S  K  Y  H  R  D  I
1448  CCAGGACTGACTGAAAGATTTGAACTATTCGTAATGAAGAAAGAAATCTGTAATGCGTATACTGAGTTAAATGACCCTGCT
      P  G  L  T  E  R  F  E  L  F  V  M  K  K  E  I  C  N  A  Y  T  E  L  N  D  P  A
1529  ACTCAGAGAGAAAGGTTTGAGCAACAAGCTAAAGATCGAGCAGCTGGTGATGATGAGACTCCACCGACAGATGAAGCATTC
      T  Q  R  E  R  F  E  Q  Q  A  K  D  R  A  A  G  D  D  E  T  P  P  T  D  E  A  F
1610  TGTACTGCCCTCGAATATGGACTTCCACCCACTGCTGGTTGGGGACTTGGCGTCGATCGCCTCACTATGTTTTTAACAGAT
      C  T  A  L  E  Y  G  L  P  P  T  A  G  W  G  L  G  V  D  R  L  T  M  F  L  T  D
1691  GCAAACACACATTAAGGAAGTATTGCTTTTCCCTGCAATGAAACCAGATGATCCGAACAAGCATAACAATGAAGAGGGAAAT
      A  N  N  I  K  E  V  L  L  F  P  A  M  K  P  D  D  P  N  K  H  N  N  E  E  G  N
1772  GCAGCCGGATAGCACTCCCTTATTACAAAATGGAGCTTAGTCAATGAGAACAGTGTTCACAATTAACACTGTTAAAAGTTTT
      A  A  D  S  T  P  L  L  Q  N  G  A  *
1853  TCACATAGAAAGTTTTACTAACAAATTTCATAACTATTCACCTCCTACTCATACTTAACCCTTTGCACTCGTAACTTTTT
1934  TCATTTGTATTACCAGCAGCTAGAAACTACGATTGTTTGCTGTTATCGAGAGCATAACAAGCTGTGACAGTCATTTGTCCC
2015  CAACATTTGGAATAACTGCTCCGGAGCCCTTGAGTGCAAAGAGTTAAATAATATTTTCAGACAGTGAAATAAATATATTGG
2096  GAAATAGAAAAAAAAAAAAAAAAAAAAAAAAAAAAAAAA
```

B

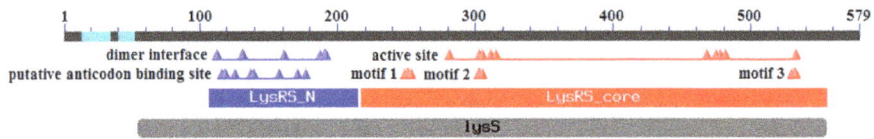

Fig. 1. The complete nucleotide and deduced amino acid sequence of the *A. pernyi LysRS* gene. (A) cDNA sequence and deduced amino acid sequence of the *LysRS* gene. The amino acid residues are represented by one-letter symbols. The initiation codon ATG is bolded and the termination codon TAG is bolded and marked with an asterisk. The polyadenylation signals AATAAA are double-underlined. The gene specific primer sequences used in the semi-quantitative RT-PCR experiment are underlined. The cDNA sequence was deposited in GenBank under accession no. JF773568. (B) Conserved domains of *A. pernyi* LysRS determined by http://www.ncbi.nlm.nih.gov/Structure/cdd/wrpsb.cgi/.

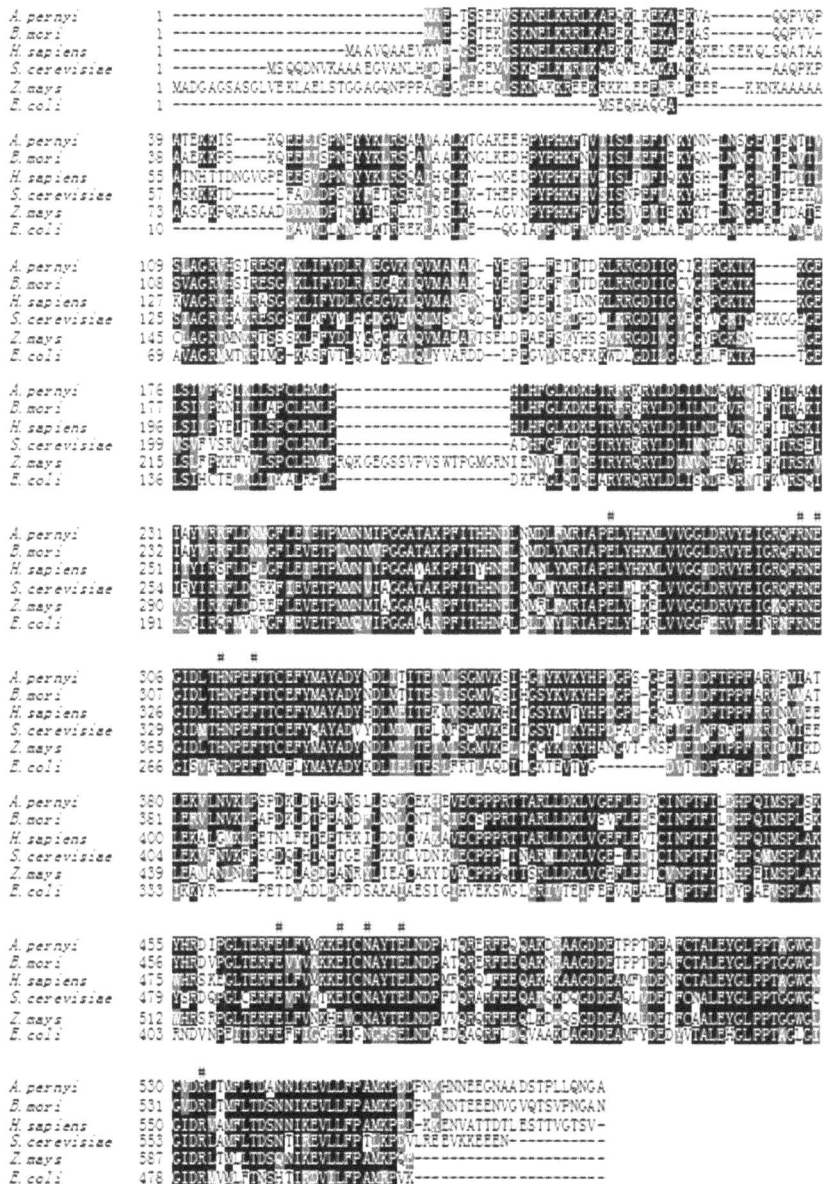

Fig. 2. Sequence alignment of LysRS proteins from *A. pernyi* and other organisms. These LysRs proteins were included from *A. pernyi*, *Bombyx mori* (BGIBMGA002984-PA in SilkDB), *Drosophila melanogaster* (NP_572573), *Homo sapiens* (NP_005539), *Zea mays* (NP_001146902), *Saccharomyces cerevisiae* (AAA6691), *Escherichia coli* (NP_417366). The number sign (#) show the residues which form the active site. Identical amino acids are highlighted in black, and positive amino acids are highlighted in gray.

LysRS revealed 84% sequence identity to LysRS of *Bombyx mori* (BGIBMGA002984-PA in SilkDB), 68-74% to LysRSs of other insects. The *A. pernyi* LysRS also revealed 57-72% identity to other invertebrates, 64–71% identity to vertebrates, 56–59% identity to plants, and 57% identity to fungi (*Saccharomyces cerevisiae*). Note that the *A. pernyi* LysRS protein showed 37-42% sequence identity to LysRS of bacteria. In **Figure 2**, the sequence is shown aligned with six LysRS proteins from *A. pernyi*, *B. mori* (BGIBMGA002984-PA in SilkDB), *Drosophila melanogaster* (NP_572573), *H. sapiens* (NP_005539), *Zea mays* (NP_001146902), *S. cerevisiae* (AAA6691), *E. coli* (NP_417366). High level of conservation of the amino acid sequence among these LysRS proteins indicates that the LysRS proteins are highly conserved during the evolution of life organisms.

Moreover, by sequence alignment, we found that, the active site residues of LysRS (Desogus et al., 2000), responsible for the ATP-dependent formation of the enzyme bound aminoacyl-adenylate, are identical among various kinds of organisms including *A. pernyi* collected in this study (**Figure 2** and data not shown). This finding suggested that the *A. pernyi* LysRS is sufficient to have the catalytic activity.The high level of conservation and identical active sites are suggestive of a critical function that these proteins must play in all the organisms where they are found.

3.3 Phylogenetic analysis

A total of 43 representative LysRS protein sequences from various organisms including *A. pernyi* were used to reconstruct the phylogenetic relationship. The final alignment resulted in 678 amino acid sites including gaps. Of these sites, 101 were conserved, 562 were variable, and 477 were informative for parsimony. A neighbor-joining tree was constructed using amino acid sequences and a poisson-corrected distance with bootstrap test of 1000 replications (**Figure 3**). The *A. pernyi* LysRS protein sequence was found to be closely related to that of *B. mori* with 100% confidence support. The used LysRS sequences were well divided into five groups corresponding to the known bacteria, fungi, plants, invertebrates and vertebrates.

Within the plant group, three subgroups were well defined corresponding to algae (*Chlamydomonas reinhardtii*), magnoliopsida (*Vitis vinifera*) and liliopsida (*Z. mays*).

Within the vertebrate group, four subgroups were well clustered corresponding to mammalia (*H. sapiens*), archosauria (*Gallus gallus*), amphibia (*Xenopus laevis*) and actinopterygii (*Danio rerio*). Within the invertebrate group, four subgroups were also well defined corresponding to insects and hard tick (*Branchiostoma floridae*), cnidaria (*Hydra magnipapillata*), nematoda (*Caenorhabditis elegans*). Moreover, within the insect subgroup, all the lepidopterans, dipterans and hymenopterans species were further clearly separated. These results agreed with the topology tree on the classical systematics and other molecular data, such as the *will die slowly* gene (Li et al., 2011), suggesting the potential value of LysRS protein in phylogenetic inference of life organisms.

3.4 Expression patterns

We performed semi-quantitative RT-PCR to detect the quantify the *A. pernyi* LysRS gene expression levels, by using an *actin* gene as an internal control that was a constitutively expressed gene (Wu et al, 2010). The results showed that the *A. pernyi* LysRS gene was expressed during four developmental stages (egg, larva, pupa and adult) (**Figure 4**), suggesting that the product of the *LysRS* gene plays an essential role throughout the entire life cycle of *A. pernyi*.

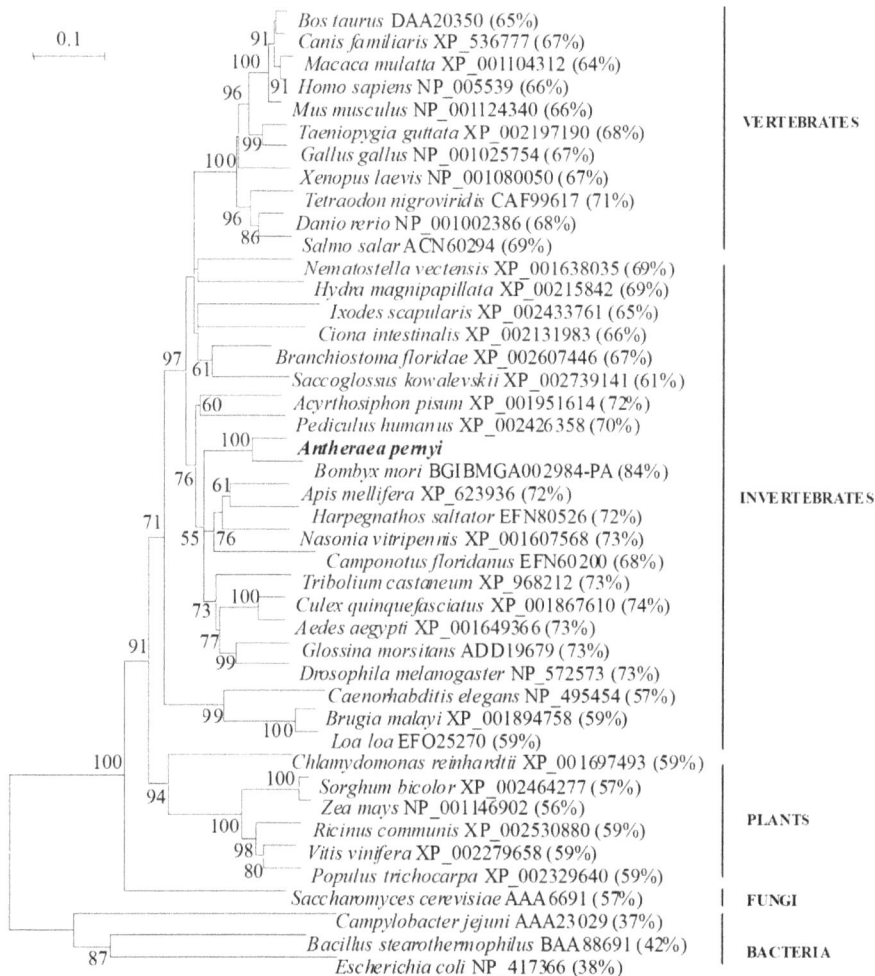

Fig. 3. Phylogenetic tree based on the amino acid sequence comparisons of LysRS proteins from various organisms including *A. pernyi*. Numbers at nodes represent bootstrap P-values (>50%). Public database accession numbers of LysRS proteins are shown following the names of organisms. Identity (%) in parentheses following accession number is obtained by pairwise alignment of amino acid sequence of *A. pernyi* LysRS with indicated LysRSs from other organisms.

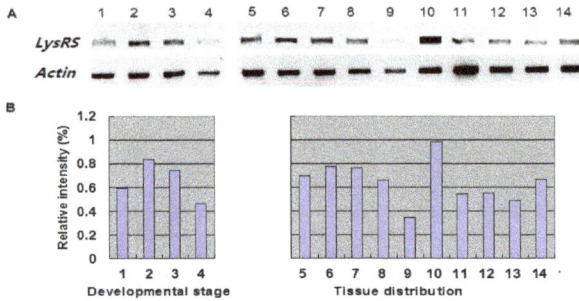

Fig. 4. Expression patterns of the *A. pernyi LysRS* mRNA in different developmental stages and different tissues of fifth instar larvae preformed by semi-quantitative RT-PCR. RT-PCR was amplified after 30 cycles with specific primer pair for the *A. pernyi LysRS* gene. The *actin* gene was used as an internal standard to normalize the templates. Relative expression profiles of *A. pernyi LysRS* were normalized with *actin* level. Lanes: 1, eggs at day 5; 2, larvae of fifth instar; 3, pupae; 4, moths; 5, blood; 6, fat body; 7, midgut; 8, silk glands; 9, body wall; 10, Malpighian tublues; 11, spermaries; 12, ovaries; 13, brain; 14, muscle. (A) The electrophoretic results. (B) The relatively intensity.

Tissue distributions in fifth instar larvae of the *A. pernyi LysRS* gene were also analyzed. The results showed that the *A. pernyi LysRS* gene mRNA was present in all tissues tested including blood, fat body, midgut, silk glands, body wall, Malpighian tubules, spermaries, ovaries, brain and muscle (**Figure 4**). Large-scale EST resource for *B. mori*, a lepidopteran model insect, are available at GenBank database. The *in silico* gene expression analysis based on the available EST resources showed the *B. mori LysRS* gene was also expressed in imaginal disks and pheromone gland not analyzed in this study.

The results also showed that the mRNA levels of the *A. pernyi LysRS* gene were most abundant in Malpighian tubules. Malpighian tubules system is a type of excretory and osmoregulatory system in insect, including *A. pernyi*. The system consists of branching tubules extending from the alimentary canal that absorbs solutes, water, and wastes from the surrounding hemolymph. The wastes then are released from the organism in the form of solid nitrogenous compounds. Therefore, the high expression of the gene in the Malpighian tubules corresponds to its role in the development of *A. pernyi*.

4. Conclusion

In the present study, the full length cDNA of the *A. pernyi LysRS* gene was isolated and characterized from a pupal cDNA library by random EST sequencing. The obtained cDNA sequence consists of 2136 bp nucleotides encoding a polypeptide of 579 amino acids which contains the LysRS class II core domain including the conserved active sites and three characteristic sequence motifs. RT-PCR analysis showed that the *A. pernyi LysRS* gene was transcribed during four developmental stages (egg, larva, pupa, and moth) and in all the tissues tested (blood, midgut, silk glands, Malpighian tublues, spermaries, ovaries, brain, muscle, fat body and body wall), with most abundance in Malpighian tubules. By searching in database, the *A. pernyi* LysRS protein homologues were found in various kinds of life organisms, including bacteria, fungi, plants, invertebrates and vertebrates, with 37–84% amino acid sequence identity, suggesting that they are highly conserved during the evolution of life organisms. Phylogenetic analysis based on the LysRS protein homologue sequences clearly separated the known bacteria, fungi, plants, invertebrates and vertebrates, consistent with the topology tree on the classical systematics, suggesting the potential value of LysRS protein in phylogenetic inference of life organisms.

5. Acknowledgment

This work was supported by grants from the National Natural Science Foundation of China (No. 31072082), the National Modern Agriculture Industry Technology System Construction Project (Silkworm and Mulberry), the Support Project for Tip-top Young and Middle-aged Talent of Shenyang Agricultural University, the Scientific Research Project for High School of the Educational Department of Liaoning Province (No. 2008643).

6. References

Chan, V. L. & Bingham, H.L. (1992). Lysyl-tRNA synthetase gene of *Campylobacter jejuni*. *Journal of Bacteriology*, Vol. 174, No. 3, pp. 695-701, ISSN 0021-9193

Desogus, G.; Todone, F.; Brick, P. & Onesti, S. (2000). Active site of lysyl-tRNA synthetase: structural studies of the adenylation reaction. *Biochemistry*, Vol.39, No.29, pp. 8418-8425, ISSN 0006-2960

Duan, J.; Li, R.Q.; Cheng, D.J.; Fan, W.; Zha, X.F.; Cheng, T.C.; Wu, Y.Q.; Wang, J.; Mita, K.; Xiang, Z.H. & Xia, Q.Y. (2010). SilkDB v2.0: a platform for silkworm (*Bombyx mori*) genome biology. *Nucleic Acids Research*, Vol. 38, Suppl 1, pp. D453-456, ISSN 0305-1048

Emmerich, R.V. & Hirshfield, I.N. (1987). Mapping of the constitutive lysyl-tRNA synthetase gene of *Escherichia coli* K-12. *Journal of Bacteriology*, Vol. 169, No. 11, pp. 5311-5313, ISSN 0021-9193

Guo, M.; Ignatov, M.; Musier-Forsyth, K.; Schimmel, P. & Yang, X.L. (2008). Crystal structure of tetrameric form of human lysyl-tRNA synthetase: Implications for multisynthetase complex formation. *Proceedings of the National Academy of Sciences*, Vol. 105, No. 7, pp. 2331-2336, ISSN 1091-6490Hirshfield, I.N., Bloch, P.L. Van Bogelen, R.A. & Neidhardt, F.C. (1981). Multiple forms of lysyl-transfer ribonucleic acid synthetase in *Escherichia coli. Journal of Bacteriology*, Vol. 146, No. 1, pp. 345-351, ISSN 0021-9193

Kepp, O.; Gdoura, A.; Martins, I.; Panaretakis, T.; Schlemmer, F.; Tesniere, A.; Fimia, G.M.; Ciccosanti, F.; Burgevin, A.; Piacentini, M.; Eggleton, P.; Young, P.J.; Zitvogel, L.; van Endert, P. & Kroemer, G. (2010). Lysyl tRNA synthetase is required for the translocation of calreticulin to the cell surface in immunogenic death. *Cell Cycle*, Vol. 9, No. 15, pp. 3072-3077, ISSN 1538-4101

Li, Y.P.; Xia, R.X.; Wang, H.; Li, X.S.; Liu, Y.Q.; Wei, Z.J.; Lu, C. & Xiang, Z.H. (2009). Construction of a full-length cDNA library from Chinese oak silkworm pupa and identification of a KK-42 binding protein gene in relation to pupal-diapause termination. *International Journal of Biological Sciences*, Vol.5, No.4, pp. 451-457, ISSN 1449-2288

Li, Y.P.; Wang, H.; Xia, R.X.; Wu, S.; Shi, S.L.; Su, J.F.; Liu, Y.Q.; Qin, L. & Wang, Z,D. (2010). Molecular cloning, expression pattern and phylogenetic analysis of the *will die slowly* gene from the Chinese oak silkworm, *Antheraea pernyi. Molecular Biology*

Reports, DOI: 10.1007/s11033-010-0495-2, ISSN 0301-4851, Available online: http://www.springerlink.com/index/Y0112G15354345L7.pdf

Liu, Y.Q.; Li, Y.; Li, X.S. & Qin, L. (2010a). The origin and dispersal of the domesticated Chinese oak silkworm, *Antheraea pernyi*, in China: A reconstruction based on ancient texts. *Journal of Insect Science*, 10.180, ISSN 1536-2442, Available online: insectscience.org/10.180

Liu, Y.Q.; Li, Y.P.; Wang, H.; Xia, R.X.; Li, X.S.; Wan, H.L.; Qin, L.; Jiang, D.F.; Lu, C. & Xiang, Z.H. (2010b). cDNA cloning and expression pattern of two enolase genes from the Chinese oak silkworm, *Antheraea pernyi*. *Acta Biochimica et Biophysica Sinica*, Vol.42, No.11, pp. 816-826, ISSN 1672-9145

Liu, Y.Q.; Li, Y.P.; Wu, S.; Xia, R.X.; Shi, S.L.; Qin, L.; Lu, C. & Xiang, Z.H. (2010c). Molecular cloning and expression pattern of a lysophospholipase gene from *Antheraea pernyi*. *Annals of the Entomological Society of America*, Vol.103, No.4, pp. 647-653, ISSN 0013-8746

Park, S.G.; Kim, H.J.; Min, Y.H.; Choi, E.C.; Shin, Y.K.; Park, B.J.; Lee, S.W. & Kim, S. (2005). Human lysyl-tRNA synthetase is secreted to trigger proinflammatory response. *Proceedings of the National Academy of Sciences*, Vol. 102, No. 18, pp. 6356-6361, ISSN 1091-6490

Saitou, N. & Nei, M. (1987). The neighbor-joining method: a new method for reconstructing phylogenetic trees. *Molecular Biology & Evolution*, Vol. 4, pp. 406-425, ISSN 0737-4038

Tamura, K.; Dudley, J.; Nei, M. & Kumar, S. (2007). MEGA4: Molecular Evolutionary Genetics Analysis (MEGA) software version 4.0. *Molecular Biology & Evolution*, Vol. 24, pp. 1596-1599, ISSN 0737-4038

Thompson, J.D.; Gibson, T.J.; Plewniak, F.; Jeanmougin, F. & Higgins, D.G. (1997). The CLUSTAL_X windows interface: flexible strategies for multiple sequence alignment aided by quality analysis tools. *Nucleic Acids Research*, Vol. 25, pp. 4876-4882, ISSN 0305-1048

VanBogelen, R.A.; Vaughn, V. & Neidhardt, F.C. (1983). Gene for heat-inducible lysyl-tRNA synthetase (*lysU*) maps near *cadA* in *Escherichia coli*. *Journal of Bacteriology*, Vol. 153, No. 2, pp. 1066-1068, ISSN 0021-9193

Wolf, Y.I.; Aravind, L.; Grishin, N.V. & Koonin, E.V. (1999). Evolution of aminoacyl-tRNA synthetases--analysis of unique domain architectures and phylogenetic trees reveals a complex history of horizontal gene transfer events. *Genome Research*, Vol. 9, No. 8, pp. 689-710, ISSN 1474-760X

Wu, S.; Xuan, Z.X.; Li, Y.P.; Li, Q.; Xia, R.X.; Shi, S.L.; Qin, L.; Wang, Z.D. & Liu, Y.Q. (2010). Cloning and characterization of the first actin gene in Chinese oak silkworm, *Antheraea pernyi*. *African Journal of Agricultural Research*, Vol. 5, Nol. 10, pp. 1095-1100, ISSN 1991-637X

Part 7

Plant Biology

17

Positional Cloning in *Brassica napus*: Strategies for Circumventing Genome Complexity in a Polyploid Plant

Gregory G. Brown and Lydiane Gaborieau
McGill University, Montreal QC,
Canada

1. Introduction

Positional, or map-based, cloning provides a strategy for isolating genes of interest when there is little or no information available on the molecular characteristics of the gene or its products, i.e. the RNA or protein that it specifies. Instead, the strategy relies on the identification of the chromosomal location of the gene through genetic mapping with molecular markers. Molecular markers, in general, allow for the detection of polymorphisms between the DNA sequences of different individuals of a species. By following the segregation of such polymorphisms through genetic crosses it is possible to construct genetic maps of a species' chromosomes. The first such markers used for this purpose were restriction fragment length polymorphisms (RFLPs, Botstein et al., 1980), in which DNA sequence differences are detected through their capacity to influence the size of fragments generated by restriction endonucleases. Subsequently, a wide range of marker types have been developed, most of which rely on the polymerase chain reaction. These include simple sequence repeat (SSR), or microsatellite, polymorphisms (Hearne et al., 1992) amplified fragment length polymorphisms (AFLPs, Vos et al., 1995) and single nucleotide polymorphisms (SNPs, Brookes, 1999).

Once chromosomal markers positioned on each side of the gene of interest are identified, DNA clones spanning the interval between these are recovered from genomic libraries. The size of the chromosomal interval between flanking markers that must be cloned is dependent on the resolution of the genetic map in the region of interest. This, in turn, depends both on the density of the molecular markers available for the region and the size of, and extent of genetic polymorphism within, the genetic populations employed for mapping.

Until the early 1990s, chromosome walking was usually required to clone specific genomic intervals. This process involved successive rounds of library screening and assembling the recovered clones into progressively larger sets of overlapping fragments (contigs) until the entire interval between the flanking markers was spanned. The development of new methods for cloning larger fragments of DNA, however, especially the development of bacterial artificial chromosomes (BACs, Shizuya et al., 1992) has greatly simplified and to some extent eliminated that need for chromosome walking. For most plant species, a population of several thousand individuals is sufficient to provide the mapping resolution needed to reduce the size of the interval separating the nearest flanking markers to one that can be contained within one or few BAC clones.

To characterize a cloned chromosomal segment, the region is first sequenced and the sequence analyzed with program such as GENSCAN (Burge and Karlin, 1997) to identify potential protein coding sequences. For plants, introduction of subclones containing the various potential genes into an appropriate recipient plant via genetic transformation (e.g. Moloney et al., 1989; Brown et al., 2003) provides the most effective means of gene identification Other methods for gene identification can involve higher resolution mapping and/or sequencing corresponding intervals from genetic strains lacking the gene of interest. The overall strategy of map-based cloning was devised on the premise that the target genome was a conventional diploid, with the target individual or line being homozygous for the gene of interest. Many plant genomes, however, including those several economically important crops, are polyploids in which most chromosomal segments are present in several paralogous copies (Adams and Wendel, 2005). This adds a layer of complexity to map-based cloning projects, since it necessitates the development of an additional strategy for identifying which of the several paralogous regions or clones that may be isolated from a genomic library represents the one that contains the gene of interest.

Cytoplasmic male sterility (CMS) is a widespread trait in flowering plants that is specified by novel, often chimeric genes in the maternally inherited mitochondrial genome (Chase, 2007). The trait can be suppressed by nuclear restorer of fertility (Rf) genes that act to specifically down-regulate the expression of the corresponding novel, CMS-specifying, mitochondrial gene. The phenomenon of CMS and nuclear fertility restoration is of commercial interest because it can be used for the production of higher yielding hybrid crop varieties (Bonen and Brown, 1993), and of fundamental biological interest because it represents a novel evolutionary process termed an "intragenomic arms race" that has apparently been occurring throughout much of angiosperm evolutionary history (Budar et al., 2003; Fuji et al., 2011).

Out of an interest in characterizing nuclear restorer genes for CMS in the oilseed rape, or canola plant *Brassica napus*, we developed strategies for employing map based cloning approaches for the complex, polyploid genome of this plant. In the sections below we briefly discuss the architecture of genomes for the genus *Brassica* in general, and *Brassica napus* specifically, and our approaches for circumventing problems posed by the genomic complexity this presents.

2. *Brassica* crops

Plants of the genus *Brassica* comprise an exceptionally diverse group of crops and include varieties that are grown as oilseeds, vegetables, condiment mustards and forages. The cytogenetic and evolutionary relationships among the major oilseed and vegetable species are commonly depicted as U's triangle, named after the Korean scientist who first formulated it (U, 1935). U speculated that *B. carinata*, *B. juncea* and *B. napus* are each allotetraploids formed by interspecific hybridization events between the parental diploid species *B. nigra*, *B. rapa* and *B. oleracea*. Thus, hybridization between *B. nigra* and *B. rapa* resulted in the formation of *B. carinata*, and between *B. nigra* and *B. oleracea* in the formation of *B. juncea*. *B. napus* is a hybrid between *B. oleracea* and *B. rapa*. The relationships among these species first postulated by U have since been confirmed by a large variety of genetic and molecular analyses. The haploid genomes of *B. rapa*, *B. nigra* and *B. oleracea* are designated A, B and C respectively. Thus diploid *B. rapa* has two copies of the A genome within 20 chromosomes (AA, n=10, 2n=20) and diploid *B. napus* has two copies of both the A and C genomes within 38 chromosomes (AACC, n=19, 2n=38).

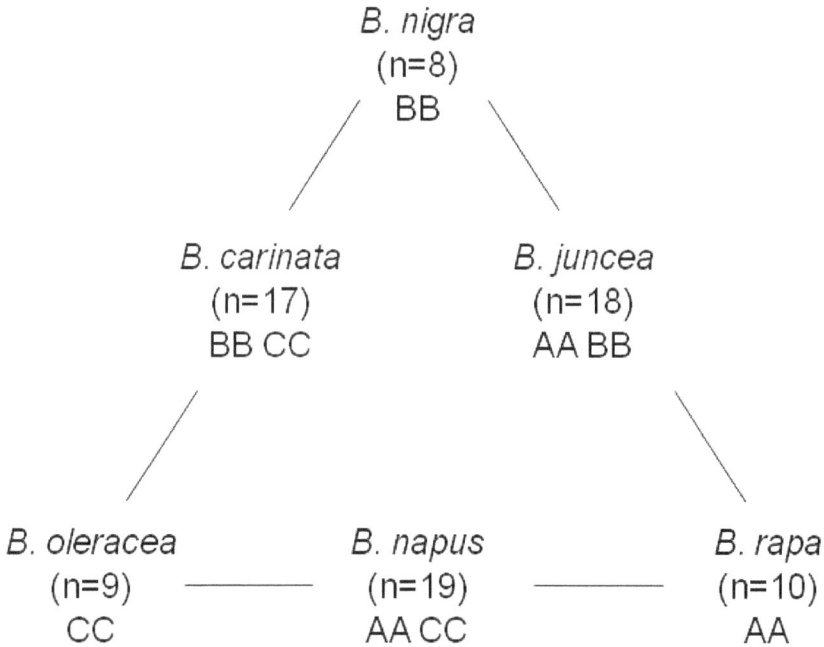

B. nigra
(n=8)
BB

B. carinata
(n=17)
BB CC

B. juncea
(n=18)
AA BB

B. oleracea
(n=9)
CC

B. napus
(n=19)
AA CC

B. rapa
(n=10)
AA

Fig. 1. U's triangle representation of the relationships between the diploid Brassica species *B. nigra*, *B. rapa* and *B. oleracea* and the allotetraploid species *B. carinata*, *B. juncea* and *B. napus*. The haploid genomes of the diploid species of *B. rapa*, *B. nigra* and *B. oleracea*, are referred to as A, B and C respectively. Thus diploid cells of *B. rapa* contain two copies of the A genome, and diploid cells of *B. napus* contain two copies of the A and C genome.

2.1 *Brassica* genomes

The model plant *Arabidopsis thaliana* and the *Brassica* species belong to the same plant family, the *Brassica*ceae. Initial efforts at determining the relationships between the *Brassica* and *Arabidopsis* genomes involved using molecular probes for co-linear sets derived from the developing *Arabidopsis* genomic resources to map RFLP polymorphisms in *Brassica* species (Kowalski et al., 1994). These studies indicated that there is extensive co-linearity between the *Brassica* and *Arabidopsis* genomes, but that most single copy *Arabidopsis* regions exist as multiple, on average 3 copies in modern *Brassica* genomes (Lagercrantz et al., 1995; Sadowski et al., 1996; Cavell et al., 1998). This in turn, gave rise to the hypothesis that the modern diploid *Brassica* species are derived from a hexaploid ancestor whose genome was generated from a diploid, *Arabidopsis*-like genome through polyploidization events. This view has largely been confirmed through subsequent comparative mapping studies involving much larger numbers of markers whose position was known on the sequenced *Arabidopsis* genome (Parkin et al., 2005). The latter study allowed for the identification of a large number of segments of the *B. napus* genome that are co-linear with corresponding regions of *Arabidopsis*. The average length of these co-linear segments was 14.3 cM in *Brassica*, corresponding to 4.3 Mb in *Arabidopsis thaliana*, suggesting that, on average, 1 cM genetic distance in *B. napus* corresponded to 300 kb in physical distance in *Arabidopsis thaliana*. Similar high definition mapping experiments in *B. rapa* indicate that 1 cM in this species corresponds to 341 kb in *Arabidopsis* and thus a similar relationship between physical and genetic distances in the two species.

Recent years have witnessed large scale genome sequencing efforts targeting many crop species, including those of the *Brassica* species *B. oleracea*, *B. napus* and especially *B. rapa* (Town, et al., 2006; Yang et al., 2006; Cheung et al., 2008). On the basis of synonomous base substitution rates in orthologous protein codein sequences, such studies indicate that the *Arabidopsis* and *Brassica* lineages diverged roughly 17 Mya and that the complex structure of the *Brassica* genomes resulted from replication and divergence of three subgenomes derived from polyploidization events occurring roughly 14 Mya. In addition, these studies confirmed the extensive co-linearity between the *Brassica* and *Arabidopsis* genomes of species evident from comparative genetic mapping, but further indicated the occurrence of segmental duplications, interspersed deletions and the occasional insertion of noncolinear genes or gene fragments. Comparison of orthologous A and C genome segments of *B. olearacea*, *B. rapa* and *B. napus* has indicated that the A and C genomes diverged about 3.7 Mya. The timing of the hybridization event between the A and C genomes that gave rise to *B. napus* has proven more difficult to determine, since the rates of silent base substitution between the parental and *B. napus* orthologs varies among different regions, suggesting that modern *B. napus* was derived from multiple progenitor varieties bearing varying degrees of similarity to the sequenced *B. rapa* and *B. oleracea* regions used in these investigations (Cheung et al., 2008). Physical distances between corresponding orthologous sequences in *Arabidopsis* and *Brassica* chromosomes appear, in general, to be similar.

Our current view of the stucture and evolution of the *Brassica napus* genome is illustrated in Figure 2. Following the divergence of the *Brassica* lineage from that leading to modern *Arabidopsis*, its genome underwent triplication of all or most chromosomal segments. For a given region in such an ancestral *Brassica* plant, each *Arabidopsis* gene (a, b, c) was present in 3 copies (a, a', a'', b, b', b'' etc.). Subsequent chromosomal rearrangements reduced the length of the conserved sequence blocks. In addition, as a result of the genetic redundancy of the triplicated genome, chromosomal deletions led to the loss of specific gene copies, illustrated in the Figure 2 as the loss of gene b in triplicated region 1 of both the A and C genomes. Subsequent to the divergence of the A and C genomes, additional gene loss (and duplications) has taken place in both genomes (illustrated by the loss of gene c_A' in region 2 of the A genome).

It is evident from this discussion, that a given "single copy" region of the *Arabidopsis thaliana* genome is present, on average, six times in the *B. napus* genome. Underlying this complexity, it is now widely accepted that more ancient genome duplication events occurred during the evolution of angiosperms (Blanc & Wolfe, 2004; Adams & Wendel, 2005), further increasing the complexity of *Brassica* genomes. Thus, to perform map-based cloning in *B. napus*, it is necessary to be able to distinguish which of the multiple copies of a given genome segment correspond to that linked to the gene of interest. In the sections below, we illustrate the strategy our group has successfully developed to achieve this.

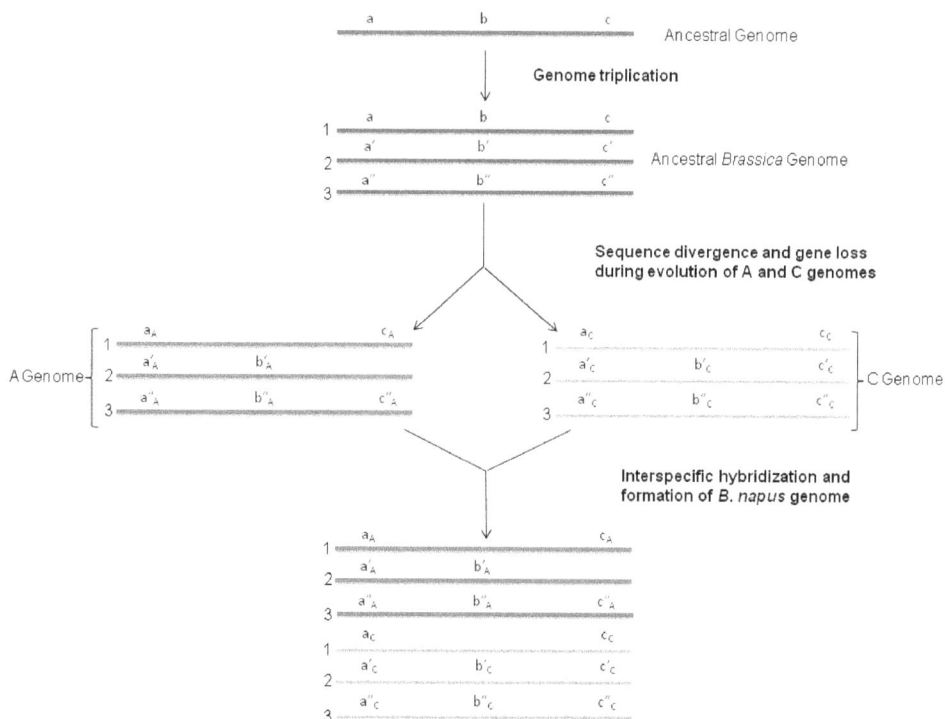

Fig. 2. Structure and evolution of co-linear regions of the *B. napus* genome. Polyploidization events occurring in an Arabidopsis like ancestor ~15 Mya resulted in a genome triplication. These three genomic regions underwent extensive independent rearrangement, sequence divergence (leading to the a, a' and a'' etc. gene sequences for regions 1, 2 and 3 respectively) and sequence loss (symbolized by gene b from region 1) and duplication. Following the divergence of the A and C lineages, the different sub-genomes underwent additional evolutionary diversification, leading to additional rearrangements, sequence divergence and gene loss (symbolized by the loss of gene c' from region 2 of the A genome). Finally, the formation of B. napus by hybridization of the A and C genomes resulted in a genome in which many regions are present in six copies.

3. Map-based cloning in *Brassica napus*

The strategy we have employed to identify genes in *Brassica napus*, knowing only the phenotype they specify, is largely that developed by Tanksley and others for map-based cloning in other plant species, but with several specific modifications necessary for dealing with the highly duplicated nature and *Brassica* genomes and for taking advantage of the extensive co-linearity to between the *Arabidopsis* and *Brassica* genomes. The overall strategy is outlined in Figure 3. The gene in question is first mapped to a resolution of approximately 10 cM using molecular markers. Then a fine mapping stage is initiated that involves developing a large population segregating for the gene, and at the same time, developing flanking markers suitable for high-throughput, PCR based screening and developing a set of ordered genetic markers based on the co-linear region of the *Arabidopsis* genome. The gene is then positioned with respect to the ordered markers at a resolution of < 0.5 cM. The most closely positioned markers flanking the gene are then used to select one or more BAC clones that flank the gene. This is the most critical aspect of the process, since it is essential that the region selected is actually that spanning the target gene and not a homeologous chromosomal segment. Finally the BAC is sequenced, annotated and candidate genes are identified. Several approaches can be used to determine which of the candidate genes correspond to the target, the most rigorous of which is genetic transformation.

3.1 Rough genetic mapping

Our initial rough genetic mapping experiments (Jean et al., 1997; Li et al., 1998) were conducted using RFLP markers developed by a research team led by Benoit Landry, initially at Agriculture and Agri-Food Canada then at DNA LandMarks, a plant genetics research company that is now an operating unit of BASF Plant Science. These markers were based on probes selected from a cDNA library (Landry et al., 1991). Now a wide variety of mapped markers, especially SSR and SNP markers are available for this purpose, and necessary information relevant to the development of such markers is available through public resources (http://*Brassica*.bbsrc.ac.uk/; http://*Brassica*db.org/brad/). In some cases, the *Arabidopsis* genome sequence coordinates corresponding to these markers are also known, which simplifies use of this information for the fine mapping analysis described below. Thus, rough mapping a single gene in *Brassica napus* is now a relatively simple and straightforward process.

Our efforts to map the nuclear restorer gene *Rfp*, for the *B. napus* 'Polima' CMS system, one of two CMS systems native to *B. napus*, were initiated in the early 1990s, when RFLPs were the only available marker system. We used Westar-*Rfp*, a derivative of the standard cultivar "Westar" into which the 'Polima' or *pol* male sterility conferring mitochondrial genome, as well as the corresponding nuclear restorer gene, *Rfp*, had been introgressed through a series of repeated backcrosses, as the male parent for our mapping population. The female parent for the population was Karat-*pol*, a derivative of the cultivar Karat in which the *pol* mitochondrial genome (but not the restorer) had been introgressed. We generated fertile F1 hybrid plants that were then crossed back to the female parental plant to generate a backcross population of approximately180 individuals, roughly half of which, as expected, were male sterile due to the absence of the dominant *Rfp* allele, and the remaining half were fertile due to the presence of *Rfp*. We identified one RFLP marker, termed cRF1, that showed complete linkage to the *Rfp* gene (i.e. all fertile progeny and none of the sterile progeny displayed the RFLP specific to the fertile parent; Jean et al., 1997). The cRF1 probe was then used to initiate a short chromosome walk with a cosmid library constructed from a doubled haploid B. *rapa* line into which the *pol* mt genome and *Rfp* gene had been introduced (Formanova et al., 2006). From these clones we were able to identify additional markers mapping close to *Rfp*, all of which were located on the same side of the gene. The next closest flanking marker on the opposite side of the gene was situated 10.8 cM away from *Rfp*.

A critical strategy for dealing with the complexity of the *Brassica* genome was developed at the stage in which we selected our initial clones for this chromosome walk using the cRF1 probe. First, the complexity of the genome we analyzed was reduced by transferring *Rfp* and the *pol* mitochondrial DNA, from the allotetraploid B. *napus* into the diploid B. *rapa* by genetic crosses and microspore culture. Even so, we found that clones selected with the cRF1 probe could fall into anyone of four contigs, Thus, the sequences homologous to the cRF1 probe were present in at least four genomic locations. We were able to determine which of the four locations corresponded to that linked to *Rfp*, however, because only for this contig did the size of the fragments hybridizing to the cRF1 probe match those of linked the *Rfp*-linked polymorphic fragments detected in genomic DNA with the same probe. This allowed us to "anchor" this cosmid, and the cosmids recovered by extending the walk, into the *Rfp* region. Subsequent analysis of the *Rfp* region, as outlined below, confirmed the linkage between this genomic region and the *Rfp* locus.

The *nap* CMS system, like the pol system, is also indigenous to B. *napus*. Interestingly, most B. *napus* cultivars possess the CMS-conferring *nap* mt genome, but are male sterile due to the presence of the corresponding *Rfn* nuclear restorer gene. A few natural cultivars, such as 'Bronowski' have been found that lack the *nap* mtDNA and *Rfn*; such cultivars possess the *cam* mt genome of B. *rapa* (formerly B. *campestris*), one of the diploid progenitors of B. *napus*. When such cultivars are crossed as males to other B. *napus* varieties and the resulting F1 plants are then backcrossed as females to 'Bronowski' plants, male sterility segregates 1:1 in the first backcross generation.

When we performed mapping experiments on populations derived from crosses such as that described immediately above, we discovered that the *Rfn* restorer locus maps to a position indistinguishable from that of *rfp*. From this, and related experiments on the different effects the two genes have on mt transcript profiles, we postulated that they represent different alleles or haplotypes of a single nuclear locus (Li et al. 1998).

Our strategy for fine mapping genes, as outlined below is based, in part, on exploiting the co-linearity between the *Arabidopsis* and *Brassica* genomes. This can be done most effectively if rough mapping allows the genomic region that could potentially encode the gene to be narrowed to a single co-linear *Arabidopsis* region. As mentioned above, the average length of B. *napus* genome segments that are co-linear with the *Arabidopsis* genome is 14.3 cM. Normally, this degree of resolution can be achieved with mapping populations that allow ~200 meioses to be surveyed, but this is highly dependent on the density of available markers, the frequency of crossovers in the target region, and of course, the extent of co-linearity with *Arabidopsis*, which varies significantly from target region to target region. Given the sequence of linked polymorphic markers (e.g the cDNA sequence of an RFLP probe or the sequence of a linked SSR amplicon) the corresponding genomic location in *Arabidopsis* can be determined using readily available software such as BLAST (Altschul et al., 1990; www.blast.ncbi,nlm.nih.gov/).

1. Rough Mapping
(population size ≈ 100)

2a. Development of PCR
based flanking markers

2b. Development of markers based on
colinear *Arabidopsis* chromosome sequence

3a. Fine mapping stage 1: Screening large population (>1000 individuals with
PCR based flanking markers for informative recombination events

3b. Fine mapping stage 2: Precise positioning of gene using
informative recombinants and *Arabidopsis* derived markers

4. Screening large insert library,
assembly of contigs spanning defined, gene containing interval

5. Sequencing of defined chromosomal interval,
Identification of candidate genes

6. Gene identification via genetic transformation

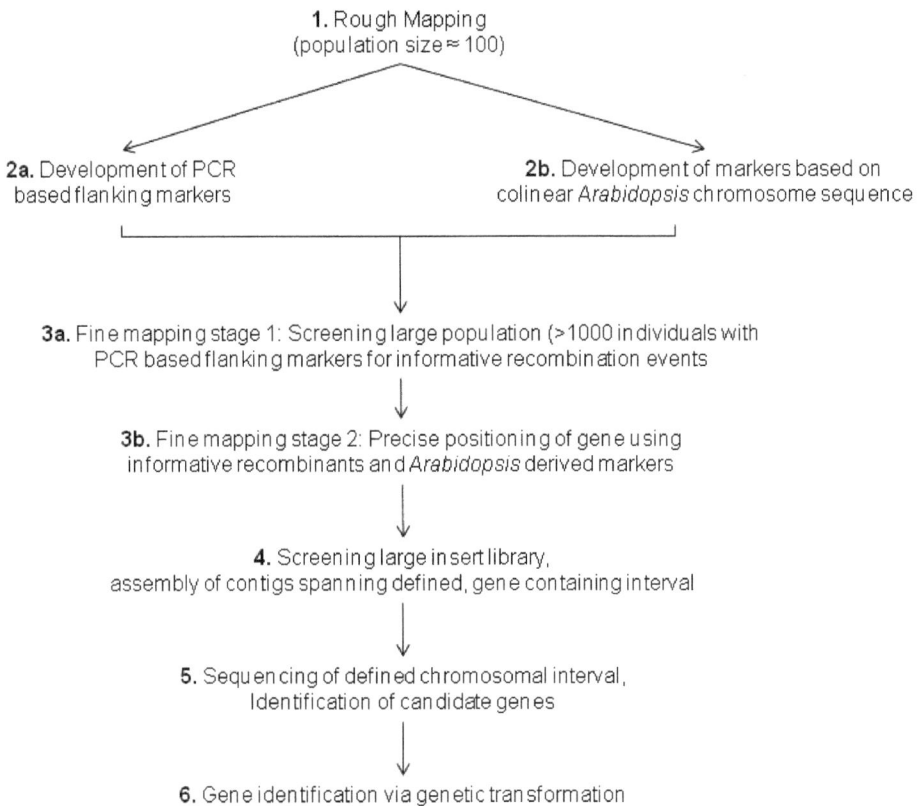

Fig. 3. Strategy for map-based cloning in *Brassica napus*. Following rough mapping with
molecular markers (1), markers most closely flanking the target gene are converted, if
necessary, into markers that can be used for high-throughput screening of small DNA
preparations from a large number of plants (2a) and the sequences of the flanking markers
are used to select corresponding allelic sequences in order to "land" in the target and
identify the corresponding region or regions of the *Arabidopsis thaliana* genome. Additional
markers are then derived from the *Arabidopsis* interval (2b) that will allow more precise
positioning of the target gene. A large population of young plants is then screened with the
flanking markers (3a) and plants with a different allele configuration between the two
markers are raised to maturity, scored for phenotype and scored for genotype using
markers derived from the corresponding interval of the *Arabidopsis* genome (3b). Markers
mapping closest to the target gene are used to select clones from a large insert DNA library
that contain the expected allele (4). The selected clones are then sequenced, the sequence is
annotated and candidate genes are identified using criteria described in the text. Finally, the
identity of the target gene is determined by genetic transformation and phenotype
assessment.

3.2 Fine genetic mapping

Fine genetic mapping is usually accomplished in two stages, the first of which involves raising a large (>1000 individuals) genetic population segregating for the gene in question, then screening that population with PCR-based markers for individuals containing a chromosome that has undergone recombination in the region close to the target gene and which are therefore informative for more precise localization of that gene. If, as is normally now the case, the original mapping was performed using PCR-based markers such as SSRs or SNPs, no additional marker conversion is necessary. If on the other hand, the initial mapping was performed with RFLPs or AFLPs, it is necessary to convert the most closely flanking markers into a form that can be easily interrogated before attempting to screen the population.

3.2.1 Development of PCR based flanking markers

For fine mapping *Rfp*, we already had genomic sequence corresponding to the RFLP probe regions on one side of the gene, and used this to design primers that amplified corresponding regions from plants homozygous for either the *Rfp* or *rfp* alleles. One such pair of primers amplified different sized products from these plants. We used bulked segregant analysis (BSA) to determine if the polymorphism detected between the parent plants of the population was indeed linked to the *Rfp* gene (Michelmore, et al., 1991). In this process DNAs from fertile and sterile plants of the mapping population are pooled prior to analysis. The principle of the process is that linked markers will differentiate between the separately pooled DNA samples, whereas unlinked markers will not. Because the pooled samples from the fertile plants amplified a product that was not observed in the amplification products obtained from sterile plants, we determined the maker was indeed linked to the *Rfp* gene. To identify an PCR-based marker that mapped to the opposite side of *Rfp*, we identified an AFLP that mapped appropriately, then sequenced the differentially amplified product from the fertile and sterile parents to convert the AFLP into a sequenced characterized amplified region (SCAR) polymorphism in which the same primer pair amplified different sized products from pools of DNA from fertile and sterile plants.

In summary, to design PCR based markers to facilitate analysis of the large genetic populations necessary for fine mapping, our overall approach has been to use RFLP probes to select corresponding sequences from a genomic cosmid library, ensure that the selected genomic clones correspond to the region linked to the gene of interest, then sequence these clones and design primers that detect polymorphisms (length polymorphisms, SNPs or cleaved amplified polymorphic sequences [CAPS, Konieczny & Ausubel, 1993]) linked to the gene. Fortunately, the numbers and density of publicly available marker sequences may now obviate the need for the development of such tools for many map based cloning projects in *Brassica napus* and its diploid progenitors. It is critical, however, to ensure that the flanking makers detect a least one recombination event between the marker and the target gene.

3.2.2 Development of markers based on co-linear *Arabidopsis* chromosome sequence

The second important aspect of our fine mapping strategy involves exploiting co-linearity between the *Arabidopsis* and *Brassica* genomes. In general, the strategy for rough mapping outlined above provides mapping resolution such that the closest markers flanking either side of the target gene will match sequences in the same *Arabidopsis* genomic region.

Assuming this is the case, one then simply can amplify coding sequences from the annotated *Arabidopsis* region   using either *Arabidopsis* genomic DNA, or preferably the corresponding BAC DNA. Information on the annotation and BACs is available from the TAIR online resource (www.*Arabidopsis*.org) and the amplicons can be used as RFLP probes to analyze digests of DNA from the ~100 plants whose genomes have experienced informative recombination events. Normally, analysis of digests using three to four restriction endonucleases is sufficient that 10 to 20% of the tested probes will detect a linked polymorphism, although this again, will vary, depending on the extent of genetic difference between the parental lines and the degree of polymorphism within the target region. We have found that in general, the order of sequences in the *Arabidopsis* genome conforms with the genetic localizations of the detected polymorphisms in *Brassica* plants, although we have noted an exception to this rule (Formanova, 2010).

Due to the degree of genetic redundancy in the *B. napus* genome, a given RFLP probe may detect polymorphism between parental lines at multiple genomic sites, only one of which, and possibly none of which, are linked to the target gene. For this reason, it is essential that bulked segregant analysis (BSA) be performed for each probe to confirm that a detected polymorphism is linked to the target region before using it in mapping experiments with the subpopulation of informative recombinant genomes. An example of BSA for an RFLP linked to the *Rfp* gene is shown in Fig. 4, and other examples can be found in Formanova et al. (2010).

In addition to using the *Arabidopsis* sequence as a source of ordered RFLP probles, the sequence may also be used to identify other marker types, particularly SNPs, mapping to the target region. For SNP identification we have selected *B. napus* cDNA and EST sequences by interrogating databases with coding sequences derived from the targeted region of the annotated *Arabidopsis* genome, then designed primers from the selected sequences that amplified corresponding regions from the parental line and bulked segregant DNA samples. For linked polymorphisms, sequence differences are evident in the products amplified from the bulked segregant DNA samples. In this manner we were able to identify one SNP that has provided the closest of flanking markers to the *Rfp* gene (Formanova et al., 2010) and that served as the key probe for recovering a BAC spanning the entire target region.

3.2.3 Screening a large mapping population with PCR based markers

A key advantage of using a two step approach to fine mapping the target gene is that the entire population need be grown only to the point where sufficient tissue can obtained to permit a limited number of PCR based marker assays to be performed while maintaining the viability of each individual plant. For the vast majority of plants in such populations, the marker genotype detected by each of the closely linked markers will be the same. Other than the very small number of plants that may have experienced double crossovers in the region of interst, these plants will not be informative for more precise localization of the gene in question and can be discarded. The remaining plants can be grown to maturity and analyzed for both phenotype and genotype, as described below. Methodology for DNA preparation and analysis by PCR based probes can be found in Formanova et al. (2010).

3.2.4 Precise gene positioning using informative individual and *Arabidopsis* derived markers

Once the markers and informative recombinants have been identified, the actual mapping of the gene is very straightforward. For example, a backross population segregating for the dominant nuclear restorer of fertility gene *Rfp*, will consist of sterile plants, lacking *Rfp* and therefore homozygous for the recessive allele *rfp*, and fertile plants that are heterozygous, i.e. contain a copy of each allele *Rfp/rfp*. The further the marker is from the *Rfp* locus, the larger the number of recombination events observed between the marker and the gene. For example, male-sterile plants in this population will be homozygous for the *rfp* allele. If a recombination event has occurred within an individual's chromosome at a site between a marker and the *Rfp/rfp* locus, then that individual will score as heterozygous for the marker alleles derived from the two parents of the initial cross. A more complete example of marker scoring and fine mapping can be found in Formanova et al. (2010).

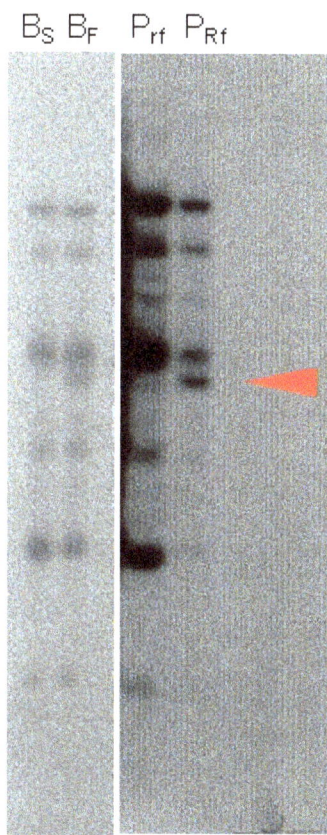

Fig. 4. An example of the use of bulked segregant analysis to demonstrate linkage between a restriction fragment length polymorphism detected between the parents of population designed for mapping population the nuclear restorer gene Rfp (P_{rf} and P_{Rf}) and DNA from 12 pooled male-sterile (sterile bulk or B_s) and 12 pooled fertile segregants (fertile bulk or B_f)of the population. Only linked polymorphisms (red arrow) will appear as differences between the samples pooled on the basis of phenotype. Clones selected from a genomic library that carry the linked polymorphic fragment represent the target gene region.

3.3 Screening a large insert library
In the case of the *Rfp* gene, the closest flanking markers we identified using a BC1 population of roughly 10^3 individuals corresponded to sequences on the *Arabidopsis* genome roughly 100 kb apart, which made it likely that the corresponding region of the *B. napus* genome might be captured within one or at most two overlapping BAC clones from a library with a mean insert size of 120kb (Parkin, I.A. unpublished data). It should be noted, however, that the sequence of the marker separated from cRF1 by 12.7 cM in our rough mapping analysis was located only 2 Mb from the position corresponding to cRF1 in the *Arabidopsis* genome, suggesting that the correspondence between genetic distance in *B. napus* and physical distance in *Arabidopsis* for this region of the genome might be considerably less than the 300kb estimated overall. It is suggested therefore, that larger fine mapping populations for future gene identification endeavors be constructed, if possible.

Our initial efforts at cloning the *Rfp* region were directed at assembling a contig of overlapping cosmid clones constructed from the library generated using the doubled haploid *B. rapa* line homozygous for *Rfp* described above (Formanova et al., 2006). The clone insert size in this library was quite small (~25 kb) and given that most probes recovered clones from at least 4 different genomic regions, the task of assembling contigs that could cover the target region proved more difficult than we initially envisioned. Instead, we adopted a different strategy and used the sequence of one cosmid clone, which contained the SNP polymorphism most closely linked to *Rfp*, to design primers to screen an ordered array of a *B. napus* BAC library generated from a *B. napus* homozygous for the *Rfn* allele. We were able to recover several clones from this library that amplified sequences from the library that matched the sequence of the cosmid over 1kb or more and, importantly, contained the expected SNP allele characteristic of the *Rfn* genotype. Thus the clone was anchored in the targeted *Rfn/Rfp* region.

A more generalized version of this strategy would be to generate a fosmid library from the targeted genotype, as we did for the cloning of the radish *Rfo* restorer gene (Brown et al, 2003). This library could then be used to recover clones corresponding to marker sequences by hybridization to colony lifts (Sambrook et al, 1989), which could then be sequenced; sequences of clones carrying linked polymorphisms could then be used to design primers useful for screening available BAC libraries.

3.4 Sequencing the defined chromosomal interval, identification of candidate genes
With the advent of very high throughput sequencing technologies, the sequencing of one or two BAC clones is a relatively straightforward process now best left to service providers. From the standpoint of the cloning effort it is recommended that very highly purified BAC DNA be provided. For *Brassica* genomic regions that have not yet been sequenced, it is recommended that dideoxy sequencing be used to construct a scaffold sequence. After shearing the BAC DNA this to an appropriate size, linkers are attached to individual fragments, and these are amplified and sequenced. Normally 5 to 10 fold coverage is sufficient. Alignment of the sequence contigs with the corresponding *Arabidopsis* sequence can be useful for determining the relationships among the contigs, which can then be used for designing primers that will allow amplification and sequencing of gaps between the contigs. At the time this chapter is being written the *B. rapa* genome sequencing project is well underway, and for many A genome regions it is possible that a reference sequence will

be available that will allow higher throughput, shorter read sequencing methodologies such as pyroseqeuncing to be employed.

Once a complete sequence for the genetically defined interval containing the target gene has been assembled, it is annotated with a bioinformatics tool such as GENSCAN (Burge and Karlin, 1997), which allows for the identification of open reading frames and likely sites of introns, polyadenylation signals and promoters. If some knowledge of the sequence characteristics of the target gene product are known, it is usually possible to quickly identify likely candidates using only the annotation output. For example, most nuclear restorer genes have been found to encode a subclass of P-type PPR proteins (Fuji et al., 2011), and these can be easily identified from an annotated BAC sequence. An additional strategy involve amplification of various gene sequences from genotypes that differ with respect to the allele of the candidate gene using a high fidelity thermostable DNA polymerase. Sequences showing allelic differences would obviously represent strong candidates for further investigation, although, in the case of *B. napus*, distinguishing alleles from closely related paralogs can proved to pose a significant challenge.

3.5 Gene identification via genetic transformation

Certain cultivars of *Brassica napus*, e.g. the spring variety "Westar", are highly amenable to genetic transformation with Agrobacterium tumefaciens. Although explants of different plant tissues can be employed in such experiments, perhaps the most widely used protocol employs petiolar tissue of young cotyledons (Moloney et al., 1989). With this protocol, and using as a vector pRD400, which carries a wild-type kanamycin resistance gene, we have routinely been able to achieve transformation/regeneration efficiencies of >10% of tissue inoculants. Normally, some degree of selection of candidate genes should be attempted. However, this is not essential; in the case of the *Rfo* restorer gene, since we had no information as to the nature of the gene or its product, we successfully identified the gene after testing each predicted gene from a ~100 kb interval by evaluating the floral phenotype of over 100 transformed regenerants generated from 19 different genetic constructs. In our experience, transformation with genomic DNA has proven sufficient for generating the altered phenotype expected of the target gene, although in some cases increased expression using a cDNA and high-level expression vector may be necessary.

4. Conclusion

Positional cloning has been successful at cloning only a few genes thus far from plants of the genus *Brassica* and allied genera. However, the rapid development of high throughput markers and the accumulation of large amounts of genomic sequence, particularly in *Brassica rapa*, should greatly facilitate the application of the approaches described in this chapter. Nevertheless, because of the complexity of these genomes, researchers must be exceedingly careful to ensure that the regions they are analyzing are genetically linked to the target gene before further extending their analyses.

5. Acknowledgments

G.G.B. would like to thank the many collaborators who have helped develop approaches explained in this chapter. Particular thanks are owed to Benoit Landry, Natasa Formanova,

Wing Cheung, Martin Laforest, Prahant Patil, Xiu-Qing Li and Isobel Parkin. Financial support from the Natural Sciences and Engineering Council of Canada and BASF Plant Sciences are gratefully acknowledged.

6. References

Adams, K.L & Wendel, J.F. (2005). Novel patterns in gene expression in polyploid plants. *Trends Genet.* 21: 539-543.

Altschul, S.F.; Gih, W.; Miller, W.; Myers, E.W. & Lipman, D.J. (1990) Basic local alignment search tool. *J. Mol. Biol.* 215:403-410

Blanc, G. & Wolfe, K.H. (2004). Widespread paleopolyploidy in model plant species inferred from age distributions of duplicate genes. *Plant Cell* 16:1667-1678.

Bonen, L. & Brown, G.G. (1993). Genetic plasticity and its consequences: perspectives on gene organization and expression in plant mitochondria. *Can. J. Bot.* 71, 645-660.

Botstein, D.; White, R.L.; Skolnick, M.; Davis, R.W. (1980). Construction of a genetic linkage map in man using restriction fragment length polymorphisms. *Am. J. Hum. Genet.* 32: 314-331.

Brown, G.G.; Formanova, N.; Jin, H.; Wargachuk, R.; Dendy, C.; Patil, P.; Laforest, M.; Zhang, J.; Cheung, W.Y. & Landry, B.S. (2003). The radish *Rfo* restorer gene of Ogura cytoplasmic male sterility encodes a protein with multiple pentatricopeptide repeats. *Plant J.* 35: 262-72.

Budar, F.; Touzet, P. & DePaepe R. (2003). The nucleo-mitochondrial conflict in cytoplasmic male sterilities revisited. *Genetica* 117: 3-16.

Burge, C. & Karlin, S. (1997). Prediction of complete gene structures in human genomic DNA. *J. Mol. Biol.* 2668: 78-94.

Cavell, A.C.; Lydiate D.J.; Parkin I.A.; Dean, C. & Trick M. (1998). Collinearity between a 30-centimorgan segment of *Arabidopsis thaliana* chromosome 4 and duplicated regions within the *Brassica napus* genome. *Genome* 41: 62-69.

Chase, C.D. (2007). Cytoplasmic male sterility: a window to the world of plant mitochondrial-nuclear interactions. *Trends Genet.* 23: 81-90.

Cheung, F.; Trick, M.; Drou, N.; Lim, Y.P.; Kwib, S.-J.; Kim, J.-A.; Scott, R.; Pires, J.C.; Paterson, A.H.; Town, C. & Bancroft, I. (2008). Comparative analysis between homeologous genome segments of *Brassica napus* and its progenitor species reveals extensive, sequence level divergence. *Plant Cell* 21: 1912-1928.

Formanova, N.; Li, X.-Q.; Ferrie, A.M.R.; DePauw, M.; Keller, W.A.; Landry, B. & Brown, G.G. (2006). Towards positional cloning in *Brassica napus*: generation and analysis of doubled haploid *B. rapa* possessing the *B. napus pol* CMS and *Rfp* nuclear restorer gene. *Plant Mol. Biol.* 61: 269-281.

Formanova, N.; Stollar, R.; Geddy, R.; Mahe, L.; Laforest, M.; Landry, B.S. & Brown, G.G. (2010). High resolution mapping of the *Brassica napus Rfp* restorer locus using *Arabidopsis*-derived molecular markers. *Theor. Appl. Genet.* 120: 843-851.

Fuji S.; Bond, C.S. & Small, I.D. (2011) Selection patterns on restorer-like genes reveal a conflict between nuclear and mitochondrial genomes throughout angiosperm evolution. *Proc. Natl. Acad. Sci. USA* 108:1723-1728.

Jean, M.; Brown, G.G. & Landry, B.S. (1997). Genetic mapping of fertility restorer genes for the Polima cytoplasmic male sterility in canola (*Brassica napus* L.) using DNA markers. *Theor. Appl. Genet.* 95, 321-328.

Konieczny, N. & Ausubel, F.M. (1993). A procedure for mapping *Arabidopsis* mutations using co-dominant ecotype-specific PCR-based markers. Plant J. 4:403-410.

Kowalski, S.D.; Lan, T.-H.,; Feldman, K.A. & Paterson, A.H. (1994). Comparative mapping of *Arabidopsis thaliana* and *Brassica oleracea* chromosomes reveals islands of conserved gene order. *Genetics* 138: 499-510.

Lagercranz, U.; Putterill, J.; Coupland G. & Lydiate, D. (1996). Comparative mapping in *Arabidopsis* and *Brassica*, fine scale genome collinearity and congruence of genes controlling flowering time. *Plant J.* 9:13-20.

Landry, B.S.; Etoh, T.; Harada, J.J. & Lincoln, S.E. (1991). A genetic map for *Brassica napus* based on restriction fragment length polymorphisms detected with expressed DNA sequences. *Genome* 34:543-552.

Li, X-Q.; Jean, M.; Landry, B.S. & Brown, G.G. (1998) Restorer genes for different forms of *Brassica* cytoplasmic male sterility map to a single nuclear locus that modifies transcripts of several mitochondrial genes. *Proc. Natl. Acad. Sci. USA* 95, 10032-10037.

Michelmore R.W.; Paran, I. & Keseli, R.V. (1991). Identification of markers linked to disease resistance genes by bulked segregant analyis: a rapid method to detect makers in specific genomic regions using segregating populations. *Proc. Natl. Acad. Sci. USA* Vol. 88, pp. 9828-9832.

Moloney, M.M.; Walker, J.M. & Sharma, K.K. (1989) High efficiency transformation of *Brassica napus* using *Agrobacterium* vectors. *Plant Cell Rep.* Vol. 8, pp. 238-242.

Parkin, I.A.; Gulden, S.M.; Sharpe, A.G.; Lukens, L.; Trick, M.; Osborn, T.C. & Lydiate, D.J. (2005). Segmental structure of the *Brassica napus* genome based on comparative analysis with *Arabidopsis thaliana*. Segmental structure of the B. napus genome based on comparative analysis with Arabidopsis thaliana. Genetics Vol. 171, pp. 765-781.

Sadowski, J.; Gaubier, P.; Delseny, M. & Quiros, C.F. (1996) Genetic and physical mapping in *Brassica* diploid species of a gene cluster defined in *Arabidopsis thaliana*. *Mol. Gen. Genet.* Vol. 12, pp. 298-306

Sambrook, J.; Frisch, E.F. & Maniatis, T. (1989). *Molecular Cloning: a Laboratory Manual, 2nd edn.* Cold Spring Harbor, N.Y; Cold Spring Harbor Laboratory Press.

Shizuya, H.; Birren, B.; Kim, U.J.; Mancino, V.; Slepak, T.; Tachiiri, Y. & Simon, N. (1992). Cloning and stable maintenance of 300-kilobase-pair fragments of human DNA in Escherichia coli using an F-factor-based vector. *Proc. Natl. Acad. Sci. USA* Vol. 89, pp. 8794-8797.

Town, C.D. et al., (2006). Comparative genomics of *Brassica oleracea* and *Arabidopsis thaliana* reveals gene loss, fragmentation and dispersal following polyploidy. *Plant Cell* Vol 18, pp. 1348-1359.

Yang, T.J. et al. (2006). Sequence level analysis of the diploidization process in the triplicated FLC region of *Brassica rapa*. *Plant Cell* Vol. 18, pp. 1339-1347.

U, N. (1935). "Genome analysis in *Brassica* with special reference to the experimental formation of *B. napus* and peculiar mode of fertilization". *Japan. J. Bot* Vol. 7, pp. 389–452.

Vos, P.; Hogers, R.; Bleeker, M.; Reijans, M.; Lee, T.V.D.; Hornes M *et al.* (1995). AFLP: a new technique for DNA fingerprinting. *Nucleic Acids Res* Vol 23, pp. 4407–4414.

18

Cloning and Characterization of a Candidate Gene from the Medicinal Plant *Catharanthus roseus* Through Transient Expression in Mesophyll Protoplasts

Patrícia Duarte et al.[*]

IBMC – Instituto de Biologia Molecular Celular, Universidade do Porto, Portugal

1. Introduction

Catharanthus roseus (L.) G. Don. accumulates in the leaves the dimeric terpenoid indole alkaloids (TIAs) vinblastine and vincristine, which were the first natural anticancer products to be clinically used, and are still among the most valuable agents used in cancer chemotherapy. The great pharmacological importance of the TIAs, associated with the low abundance of the anticancer alkaloids in the plant, stimulated intense research on the TIA pathway, and *C. roseus* has now become one of the most extensively studied medicinal plants. About 130 TIAs have already been isolated from *C. roseus*, and it has been shown that the biosynthesis of vinblastine is highly complex, involving at least 30 steps from the amino-acid tryptophan and the monoterpenoid geraniol (Loyola-Vargas et al., 2007; van der Heijden et al., 2004). All the TIAs of *C. roseus* derive from the common precursor strictosidine, after which the TIA pathway splits into several branches including a short one leading to ajmalicine and serpentine (used as an antihypertensive and as a sedative respectively), and two long branches leading to vindoline and catharanthine - the leaf abundant monomeric precursors of vinblastine and vincristine (Loyola-Vargas et al., 2007; van der Heijden et al., 2004). In our lab, we have performed the characterization of a key biosynthetic step of the anticancer TIAs - the biosynthesis of the first dimeric TIA, α-3′,4′-anhydrovinblastine (AVLB), from vindoline and catharanthine. We identified a leaf class III peroxidase (Prx) with AVLB synthase activity and we purified and characterized this enzyme, which was named *Catharanthus roseus* peroxidase 1, CroPrx1 (Bakalovic et al., 2006; Peroxibase, http://peroxibase.toulouse.inra.fr index.php). We have further shown the localization of *CroPrx1* in the same subcellular compartment where alkaloids accumulate, the vacuole, through biochemical and molecular methodologies,

[*]Diana Ribeiro[1,2], Gisela Henriques[1], Frédérique Hilliou[3], Ana Sofia Rocha[1,4],
Francisco Lima[1,4], Isabel Amorim[4,5] and Mariana Sottomayor[1,4]
[1]*IBMC – Instituto de Biologia Molecular Celular, Universidade do Porto, Portugal*
[2]*Departamento de Biologia, Universidade do Minho, Portugal*
[3]*INRA, Université de Nice-Sophia Antipolis, France*
[4]*Departamento de Biologia, Faculdade de Ciências, Universidade do Porto, Portugal*
[5]*Environment and Society, Centro de Geologia da Universidade do Porto, Portugal*

and we performed the cloning and molecular characterization of *CroPrx1* (Costa et al., 2008; Duarte et al., 2010; Sottomayor et al., 1996; Sottomayor et al., 1998). In the course of this work, we have discovered, cloned and characterized another Prx gene from *C. roseus*, *CroPrx3*, which will be addressed in this chapter.

Class III peroxidases (Prxs; EC 1.11.1.7) are typical of plants and form large multigenic families. They are multifunctional enzymes that catalyze the oxidation of small molecules at the expense of H_2O_2, and are capable of recognizing a broad range of substrates. Prxs show an extraordinary diversity, with the presence of a high number of isoenzymes in a single plant, and have mainly been implicated in key processes determining the architecture and defense properties of the cell wall. They are also thought to play a role in biotic and abiotic stress resistance, in secondary metabolism, and in hydrogen peroxide scavenging and production (Ferreres et al., 2011; Fry, 2004; Passardi et al., 2005; Ros Barceló et al., 2004; Sottomayor et al., 2004). Regarding subcellular localization, Prxs are either vacuolar or extracellular. They are targeted to the secretory pathway by an N-terminal signal peptide (SP) that directs the protein to the endoplasmic reticulum (ER), from where they follow the default pathway to the cell wall, or they are sorted to the vacuole if an additional sorting signal exists (Costa et al., 2008). When compared to any characterized mature plant Prx, all vacuolar Prxs show the presence of a C-terminal extension (CTE) in their deduced amino-acid sequence (Welinder et al., 2002). Previous results obtained in our laboratory revealed that the CTE of CroPrx1 is both necessary and sufficient to target the sGFP reporter to the central vacuole, both in *C. roseus* bombarded cells and in Arabidopsis protoplasts, and must therefore constitute the Prx vacuolar sorting signal (Costa et al., 2008; Duarte et al., 2010). In contrast with the high number of studies on cell wall Prxs, much less is known about their vacuolar counterparts (Costa et al., 2008; Welinder et al., 2002).

On the other hand, although much is known about the *C. roseus* TIA pathway and its regulation, so far, enzyme/gene characterization is still lacking for many biosynthetic steps, no effective master switch of the pathway has been identified, and the membrane transport mechanisms of TIAs are basically uncharacterized (Sottomayor et al., 2004; van der Heijden et al., 2004; Verpoorte et al., 2007). Recently, omic approaches and *in silico* data-mining of EST libraries are being intensively used to enable the identification of candidate genes implicated in TIA metabolism (Carqueijeiro et al., 2010; Liscombe et al., 2010; Murata et al., 2008; Rischer et al., 2006). This task has now been made easier due to the release of the first assemblies of *C. roseus* transcriptomes by the consortium Medicinal Plant Genomics Resource, together with data for 10 more medicinal plants (http://medicinalplant genomics.msu.edu/index.shtml). Therefore, the use of molecular cloning strategies to retrieve the cDNAs of *C. roseus* candidate genes and the development of tools to characterize these genes is of great importance.

1.1 Scientific problem

We first became interested in *C. roseus* due to the high importance credited to the coupling reaction leading to the first dimeric TIA, AVLB, due to its regulatory importance and potential application for the semisynthetic production of the anticancer alkaloids. We have succeeded to identify, characterize and clone a leaf vacuolar Prx with AVLB synthase activity, CroPrx1 (Costa et al., 2008; Duarte et al., 2010; Sottomayor et al., 1996; Sottomayor et al., 1998), and while seeking to characterize the full genomic sequence of

CroPrx1, we have found a different highly homologous Prx gene, *CroPrx3*. This gene codes for two alternative splicing transcripts/proteins, and given the high similarity of CroPrx3-a and CroPrx3-b with CroPrx1, it was hypothesized that they could be involved in TIA metabolism, since for instance the conversion of ajmalicine into serpentine has also been suggested to be mediated by an uncharacterized Prx (Blom et al., 1991; Sottomayor et al., 2004). Important for the implication of a Prx in TIA metabolism, is its subcellular localization in the vacuole, where the TIA substrates and products accumulate. Considering all this, we decided to clone the *CroPrx3* cDNAs and to investigate the subcellular sorting of the CroPrx3 proteins using *GFP* fusions and their transient expression in protoplasts. As such, we decided to also develop a methodology for the isolation and transformation of *C. roseus* mesophyll protoplasts, since the use of an homologous system is far more reliable.

2. Materials and methods

2.1 Outline of the experimental approach

The experimental approach used to clone and characterize *CroPrx3* involved three main steps: i) cloning of *CroPrx3*, ii) construction of *GFP-CroPrx3* fusions, and iii) transient expression in *C. roseus* mesophyll protoplasts, as outlined below.

i.   Cloning of *CroPrx3* involved the characterization of *CroPrx1* positive clones from a *C. roseus* genomic library by a strategy combining subcloning of restriction fragments, PCR, sequencing, and assembly of sequence data. This enabled the *in silico* generation of a new Prx gene highly similar to *CroPrx1, CroPrx3,* which coded for two alternative splicing transcripts, *CroPrx3-a* and *CroPrx3-b*. Since RT-PCR revealed that both *CroPrx3* transcripts were highly expressed in the roots, RT-PCR using root RNA and specific primers designed to cover the full coding sequences, enabled to retrieve *CroPrx3s* full cDNAs.

ii. In order to study the subcellular sorting of CroPrx3s, which has functional relevance, a number of *GFP-CroPrx3-b* fusions were designed and generated, including putative sorting signals correctly cloned at the N- and C-terminus of GFP. Concerning the choice of FP and the rules for design of FP fusions to investigate subcellular localization of candidate gene products, please see Duarte et al. (2010).

iii. For transient expression of the *GFP-CroPrx3* fusions in *C. roseus* cells, a fast and efficient technique for isolation of mesophyll protoplasts was developed and a procedure for their PEG-mediated transformation was optimized resulting in high transformation yields. The localization of fluorescence was monitored using a confocal microscope.

2.2 Plant material and growth conditions

Plants of *Catharanthus roseus* (L.) G. Don cv. Little Bright Eye were grown at 25°C, in a growth chamber, under a 16 h photoperiod, using white fluorescent light with a maximum intensity of 70 μmol m^{-2} s^{-1}. Seeds were acquired from AustraHort (Australia) and voucher specimens are deposited at the Herbarium of the Department of Biology of the Faculty of Sciences of the University of Porto (PO 61912). Callus cultures of the *C. roseus* line MP183L were kindly provided by Johan Memelink from the University of Leiden, The Netherlands, and were grown as described by Pasquali et al. (1992).

2.3 Cloning of *CroPrx3*
2.3.1 Isolation and sequencing of *CroPrx3*
The *CroPrx3* gene was found while screening a genomic library for *CroPrx1* (Genbank accession number AM236087). Screening was carried out using a *C. roseus* genomic library kindly provided by Johan Memelink from the University of Leiden, The Netherlands, which was produced using *C. roseus* genomic DNA partially digested with Sau3AI and cloned in λGEM11 (Promega) BamHI arms (Goddijn et al., 1994). Around 600,000 primary phage plaques grown on *Escherichia coli* host strain KW251 were screened by hybridization with a mixture of two radiolabelled probes specific for *CroPrx1*, using standard plaque lift methods (Sambrook et al., 1989). After three rounds of screening, four distinct positive clones were encountered corresponding to genomic inserts of 15 to 20 kbp, and two clones, λ8 and λ17, were characterized further. These clones were not accessible to direct sequencing due to their big size, therefore they were digested with EcoRI and the restriction fragments were analyzed by Southern blotting using the same probes as above (Sambrook et al., 1989). The positive and/or common restriction fragments of the two clones were subcloned either in the plasmid pGEM4 (Promega) or in pBluescript SK+ (Stratagene), and were sequenced using first the vectors universal primers, followed by subsequently designed internal primers, until coverage of the entire sequences was obtained. From the first data acquired, it became clear that the gene present in the clones was not *CroPrx1* but a highly similar Prx new gene, which was named *CroPrx3* (Bakalovic et al., 2006; Peroxibase, http://peroxibase.toulouse.inra.fr index.php). Sequence data of overlapping regions obtained from both strands was assembled and analyzed using PDRAW 32 version3, to produce the complete gene sequence of *CroPrx3* (Genbank accession number AM937226).

2.3.2 RT-PCR analysis of *C. roseus* organs
RNA was isolated using the hot phenol-LiCl procedure previously described by van Slogteren et al. (1983) and Menke et al. (1996) from various organs of five month old *C. roseus* plants, from 7-day old *C. roseus* seedlings, and from callus cultures. The different tissues were ground under liquid nitrogen and total RNA was isolated by a 5 min extraction with two volumes of hot phenol buffer (80 °C, 1:1 mixture of phenol containing 0.1 % hydroxyquinoline with 100 mM LiCl, 10 mM EDTA, 1 % SDS, 100 mM Tris pH 9.0) and one volume of chloroform. The mixture was microcentrifuged 30 min at 5000 rpm at room temperature (RT), and the aqueous phase was mixed with one volume of chloroform, shaken for 5 min and centrifuged as before. The aqueous phase volume was measured, mixed with one-third of its volume of 8 M LiCl (2 M final concentration) and the RNA was precipitated overnight at 4 °C. The RNA was then collected by microcentrifugation at 10000 rpm for 30 minutes at 4 °C. The pellet was washed twice with 70 % ethanol and dried under vacuum. Alternatively, RNA can be extracted with the RNeasy Plant Mini Kit (QIAGEN). The dried RNA was dissolved in water, microcentrifuged at 10000 rpm for 10 minutes, at 4 °C, to precipitate impurities (polysaccharides), the RNase inhibitor RNAsin (Promega) was added, and the RNA was stored at –20 °C. Integrity and equal loading of RNA was verified by ethidium bromide staining of ribosomal RNA bands after agarose electrophoresis. Quantification of the RNA samples was also performed spectrophotometrically.

RNA was treated with DNase I (Promega) and RT-PCR was performed as follows. Reverse transcription polymerase chain reactions were performed using the Promega Reverse Transcription System (Promega), using a poly(dT)12-18 as a first strand primer. PCR reactions to detect the presence of transcripts from *CroPrx1*, *CroPrx3-a* and *CroPrx3-b* were performed with DFS-Taq DNA Polymerase (Bioron) for 35 cycles, using the primer pairs described in Table 1. These were cross-tested for specificity using cloned DNA from each gene as template for PCR reactions. The amount of cDNA included in the PCR reactions of each set of experiments was estimated after standardization using the reference gene *Rps9* encoding the 40 S ribosomal protein S9 (Genbank accession number AJ749993; Menke et al., 1999). The RT-PCR products were subjected to electrophoresis in 1.5% agarose gels and amplification products were visualized by staining with ethidium bromide. Band intensity was assessed both visually and with the help of the image analysis software Kodak DC120 Gel Electrophoresis Analysis System.

Amplification product	Primer name	Primer sequence 5'-3'
CroPrx3-a fragment	forward	CTCATCTGCTCTCTTCTATTGG
	reverse	CAGCACAAAGACATGGCTGA
CroPrx3-b fragment	forward	GCTGGCTGGAAGGAACGAG
	reverse	CAGCACAAAGACATGGCTGA
Rps9 fragment	forward	GAGCGTTTGGATGCTGAGTT
	reverse	TCATCTCCATCACCACCAGA

Table 1. Sequence of the primers used in the RT-PCR analysis of *C. roseus* organs for the presence of *CroPrx3-a* and *CroPrx3-b*.

2.3.3 Isolation of the complete cDNAs of *CroPrx3-a* and *CroPrx3-b*

Total RNA was extracted from *C. roseus* roots as described in the previous section. Reverse transcription was performed using the Transcriptor Reverse Transcriptase (Roche) using a poly(dT)12-18 as a first strand primer. The generated cDNA was then PCR amplified using 5'UTR specific primers for the two alternative splicing transcripts and a common 3'UTR reverse primer (Table 2, Fig. 3). PCRs were carried out using 0.5 µg of cDNA as template and Pfu DNA Polymerase (Promega). The generated PCR products were cloned into the vector pGEM-Teasy (Promega) and the clones obtained were sequenced to verify the inclusion of the complete coding regions of *CroPrx3-a* and *CroPrx3-b*.

Amplification product	Primer	Primer sequence 5'-3'
CroPrx3-a	forward	GGAACTATGGCTTTGATTGC
	reverse	CAGCACAAAGACATGGCTGA
CroPrx3-b	forward	GCTGGCTGGAAGGAACGAG
	reverse	CAGCACAAAGACATGGCTGA

Table 2. Sequence of the primers used for the isolation of the *CroPrx3-a* and *CroPrx3-b* complete cDNAs.

2.4 Generation of *GFP-CroPrx3* fusions

In order to study the subcellular localization of CroPrx3s, a series of fusion constructs of *CroPrx3-b* regions with *GFP* were designed and generated, as shown in Fig. 1. The *CroPrx3-b* sequences used in the fusions were amplified by PCR using a proofreading Pfu DNA polymerase (Fermentas) and primers including specific *CroPrx3-b* sequences plus the restriction sites required for directional cloning, including additional nucleotides to guarantee efficient cleavage close to fragments termini, according to manufacturer indications. The primers are shown in Table 3 and their localization relative to Prxs domains are shown in Fig. 1A. Primers were designed to be in frame with *GFP*, and the fusion sequences were analyzed with the ExPASy Translate tool (http://expasy.org/tools/dna.html) to confirm the generation of a correct open reading frame.

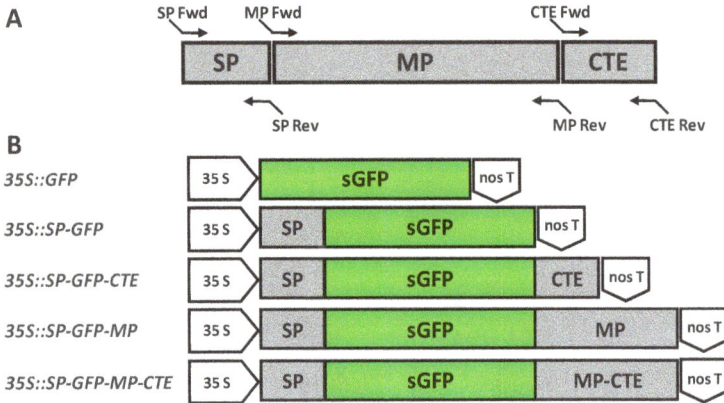

Fig. 1. Schematic representation of the cloning strategy followed to obtain the *GFP-CroPrx3-b* constructs. A) Representation of the different regions of *CroPrx3-b* and the primers used for their amplification. B) Schematic representation of the constructs generated to study the subcellular localization of CroPrx3-b. 35S - cauliflower mosaic virus strong promoter, SP - signal peptide, CTE - C-terminal extension, MP - mature protein, nos T - *Agrobacterium tumefaciens* nopaline synthase terminator. The two first constructs were generated in the plasmid pTH-2, while the remaining were generated in pTH-2BN.

CroPrx3-b amplification product	Primer name	Primer sequence 5'-3'
SP	SP Fwd	AGCCGTCGACAAAATGGTTTTTATGAGTTCCTTTTC
	SP Rev	GATGCCATGGTTGTTTGAGCTTCGATATGG
CTE	CTE Fwd	CAAGATCTTTCGGAATGCCGCCAGCGGACGTTCTT
	CTE Rev	CGCCTCGAGTTAAAACATGGACAAGCCAACTTCTGC
MP	MP Fwd	CAAGATCTTGCCACCTATAGTGAGTGGACTTTCATT
	MP Rev	CGCCTCGAGTTAGGCATTCCGAACTGAACAATT
MP-CTE	MP Fwd	CAAGATCTTGCCACCTATAGTGAGTGGACTTTCATT
	CTE Rev	CGCCTCGAGTTAAAACATGGACAAGCCAACTTCTGC

Table 3. Sequence of the primers used for the generation of the *GFP-CroPrx3-b* fusion constructs. Engineered restriction sites are underlined. SP - signal peptide, MP - mature protein, CTE - C-terminal extension.

The fusion of *CroPrx3-b* sequences at the N-terminus coding sequence of *GFP* were generated using the plasmid pTH-2 corresponding to pUC18 carrying the 35SΩ-sGFP(S65T)-nos construct and an ampicillin/carbenicillin-resistance marker (Niwa et al., 1999). This plasmid may be requested from Yasuo Niwa (niwa@fns1.u-shizuokaken.ac.jp). Whenever fusions at the C-terminus coding sequence were required, the plasmid used was pTH-2BN (Kuijt et al., 2004), which lacks a stop codon in the end of the GFP sequence and may be requested from Johan Memelink (j.memelink@biology.leidenuniv.nl).

The amplified CroPrx3-b signal peptide coding sequence (*SP*) was cloned in frame in pTH-2 and in pTH-2BN, using SalI and NcoI, to generate *35S::SP-GFP* constructs in each of the plasmids. The *35S::SP-GFP* construct in pTH-2 (Fig.1B) was used directly for transformation, while the *35S::SP-GFP* construct in pTH-2BN was used to further insert at the C-terminus coding sequence of *GFP*, using BglII and XhoI, the following sequences:

i.   the CTE coding sequence of *CroPrx3-b* to generate the construct *35S::SP-GFP-CTE* (Fig.1B)
ii.   the mature protein coding sequence of *CroPrx3-b* (excluding the SP and CTE coding sequences) to generate the construct *35S::SP-GFP-MP* (Fig.1B)
iii. the mature protein plus CTE coding sequences of *CroPrx3-b* to generate the construct *35S::SP-GFP-MP-CTE* (Fig.1).

All the generated constructs were sequenced with universal and internal primers in order to certify that all clones were error-free.

2.5 Transient expression in *C. roseus* mesophyll protoplasts
2.5.1 Isolation of *C. roseus* mesophyll protoplasts

C.roseus mesophyll protoplasts were obtained using a protocol adapted from Negrutiu et al. (1987), Sottomayor et al. (1996) and Yoo et al. (2007). Approximately 8-10 leaves (~1.5 - 2 g) of adult plants (usually 2nd and 3rd pairs) were cut with frequently renewed scalpel blades into ~1 mm strips excluding the central vein, and immediately transferred, abaxial face down, to a Petri dish with 10 mL of digestion medium composed of 2 % (w/v) cellulase (Onozuka R-10, Duchefa), 0.3 % (w/v) macerozyme (Onozuka R-10, Serva) and 0.1 % pectinase (Sigma) dissolved in MM buffer (0.4 M mannitol and 20 mM Mes, pH 5.6-5.8). The medium was vacuum infiltrated during 15 min, applying slow disruptions of the vacuum every ½ a min. Leaf strips were incubated in the digestion medium for *ca.* 3 h at 25°C, in the dark, after which the Petri dishes were placed on an orbital shaker (~60 rpm) for 15 min in the dark and at RT, to help release the protoplasts. The suspension was then filtered through a 100 μm nylon mesh and the filtrate was gently transferred into 15 mL falcon tubes, using sawn-off plastic Pasteur pipettes. The protoplast suspension was centrifuged at 65 g for 5 min at 20 °C, the supernatant was removed, and the protoplasts were washed twice in MM buffer and once in cold W5 solution (154 mM NaCl, 125 mM $CaCl_2.2H_2O$, 5 mM KCl and 2 mM Mes, pH 5.7) – the ressuspension of the protoplasts must be performed gently, by flicking the tube after addition of a small volume of medium, and only after that adding the full volume for washing. The last pellets were all ressuspended in a minimum volume of W5 and pooled together. Protoplasts were counted using a haemocytometer and were left to rest on ice for 30 min. After this incubation, the protoplasts were pelleted as above and ressuspended to a protoplast concentration of 5×10^6 cells mL^{-1} using the adequate volume of MMg buffer (0.4 M mannitol, 15 mM $MgCl_2$ and 4 mM Mes, pH 5.7). At this point, the protoplasts were ready for transformation. The integrity of the isolated protoplasts was checked by observation under an optical microscope (Olympus) and images were acquired by a coupled Olympus DP 25 Digital Camera and respective software (Cell B, Olympus).

2.5.2 Transformation of *C. roseus* mesophyll protoplasts

Transformation of *C. roseus* mesophyll protoplasts was adapted from Yoo et al. (2007). In all cases, 10 µL of 2 µg µL^{-1} of ultrapure plasmid DNA were mixed with 100 µL of protoplast suspension with 5×10^5 cells, using 2 mL round bottom eppendorfs (the protoplast suspension was pipetted using a P1000 sawn-off tip). Plasmid DNA used for protoplast transformation was isolated using the QIAGEN Plasmid Midi Kit following the manufacturer's instructions. One volume (110 µL) of 40 % (w/v) PEG, 0.2 M mannitol and 0.1 M CaCl$_2$.2H$_2$O, was slowly added (drop by drop) to the DNA-protoplast mixture, gently flicking the tube after each drop, and the tubes were left to incubate for 15 min at RT. After this incubation time, four volumes of W5 solution were slowly added and the tubes were centrifuged at 600 rpm for 2 min, with acceleration and deceleration set at the minimum (1). The supernatant was removed, the pellet was gently ressuspended in 100 µL of W5 solution by flicking the tube, the protoplasts were transferred to 15 mL falcon tubes containing 900 µL of W5, and were incubated in the dark at 25°C, with the tubes lying in a slight slope, for at least 2 days. Prior to confocal observation, cells were subjected to a 3 h incubation at 35 °C, since this showed to increase GFP expression and/or fluorescence. GFP fluorescence was examined using a Leica SP2 AOBS SE confocal microscope equipped with a scanhead with an argon laser. Visualization of GFP was performed using an excitation wavelength of 488 nm and an emission wavelength window from 506 to 538 nm, and visualization of chloroplast autofluorescence was performed using the same excitation wavelength and an emission wavelength window from 648 to 688 nm.

3. Results and discussion

3.1 Cloning of *CroPrx3*

The new Prx gene *CroPrx3* was found while trying to obtain the full genomic sequence of *CroPrx1*, implicated in the biosynthesis of the *C. roseus* anticancer alkaloids. *CroPrx1* positive clones from a *C. roseus* genomic library were obtained through standard screening methods, and were analysed by restriction digestion using an enzyme cutting only once in the previously known coding sequence. This showed that the three smallest bands from clones λ8 and λ17 were similar (data not shown), which was interpreted as the possible presence of the common *CroPrx1* gene in these fragments. This was indeed confirmed by Southern blotting (data not shown), and they were subcloned in two stages. First the λ8 fragments were cloned in pGEM4 with limited success, and in a second stage the λ17 equivalent fragments were successfully cloned in pBluescriptII SK+. These last clones were sequenced using the plasmid universal primers (T3, T7, Fig. 2) and the sequences obtained were used to design internal primers for further rounds of sequencing (IPs, Fig. 2). From the first data acquired it became clear that the gene present in the clones was not *CroPrx1* but a highly similar Prx new gene, which was named *CroPrx3*, using the nomenclature followed by Peroxibase (Bakalovic et al., 2006; http://peroxibase.toulouse.inra.fr/index.php). The whole sequencing data obtained was assembled and analyzed using PDRAW 32 version3, to produce a gene sequence of *CroPrx3* spanning a 5242 bp region (Genbank accession number AM937226), which revealed the gene structure depicted in Fig. 3. *CroPrx3* revealed a striking structural difference to *CroPrx1*, due to the presence of two exons 1 in tandem, separated by a spacer sequence, suggesting this is a gene with alternative splicing coding for two proteins with distinct N-terminal regions, CroPrx3-a and CroPrx3-b (manuscript in preparation).

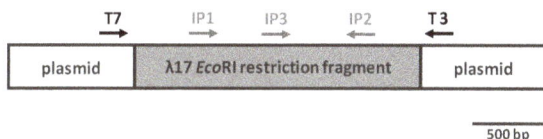

Fig. 2. Schematic representation of the λ17 EcoRI restriction fragments subcloned in the plasmid pBluescriptII SK+ showing the position of the plasmid universal primers (T7,T3) and the newly designed internal primers (IPs) used to obtain the full sequence of the inserts.

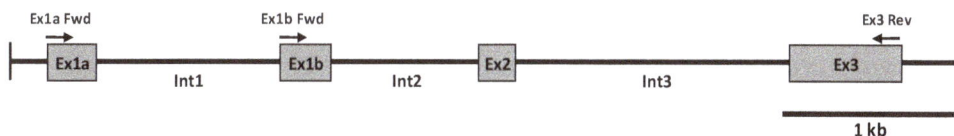

Fig. 3. Schematic representation of the *CroPrx3* gene with the localization of the primers used to obtain the full length cDNAs of *CroPrx3-a* and *CroPrx3-b*. Ex, exon; Int, intron.

In order to investigate the subcellular localization of the proteins codified by *CroPrx3*, the next step was to isolate and clone the cDNAs of the two transcripts putatively resulting from *CroPrx3*. For this, it was necessary to use RNA from a tissue/organ where the transcripts were expressed. Therefore, expression analysis by RT-PCR was performed with primers specific for both forms of *CroPrx3*, revealing that both transcript variants are highly expressed in seedlings, roots and callus cultures, and are generally not expressed in the aerial parts of the plant (Fig. 4).

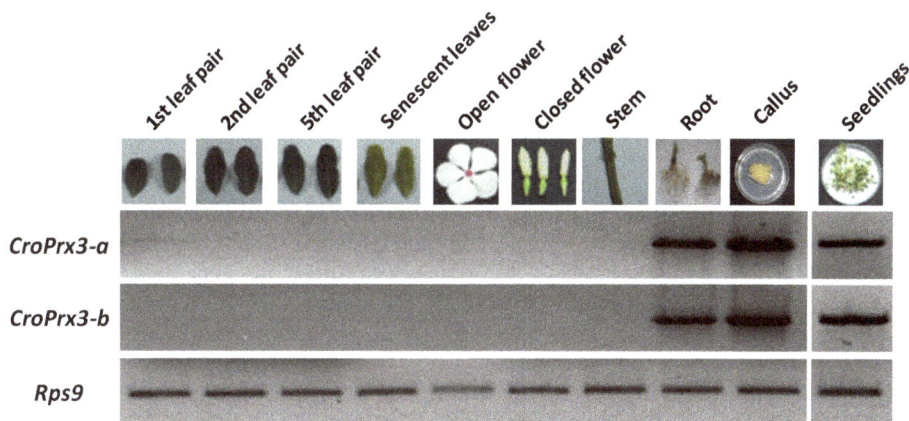

Fig. 4. RT-PCR analysis of *C. roseus* organs to assess the presence of the alternative splicing forms of *CroPrx3* (*CroPrx3-a* and *CroPrx3-b*). *Rps9* is commonly used as a house-keeping gene for this species (Menke et al., 1999).

Given the expression results, mRNA extracted from roots was used to isolate the *CroPrx3* cDNAs. RT-PCR was performed using 5'UTR specific primers for the two alternative splicing transcripts with a common 3'UTR reverse primer (Table 2, Fig. 3), and enabled the isolation of two complete cDNAs with a 100% sequence match with the two predicted alternative splicing forms. The complete cDNAs obtained for *CroPrx3-a* and *CroPrx3-b* were cloned into the vector pGEM-Teasy (Promega).

3.2 Design of *GFP-CroPrx3* fusions for the characterization of CroPrx3 subcellular sorting

The subcellular localization of CroPrx3s is highly relevant to investigate their function, since it will indicate which potential substrates are accessible *in vivo*, namely vacuolar TIAs. Prxs are targeted to the secretory pathway by an N-terminal signal peptide (SP) that directs the protein to the endoplasmic reticulum (ER), from where they follow the default pathway to the cell wall, or they are sorted to the vacuole if an additional sorting signal exists. We have previously shown that CroPrx1 is targeted to the vacuole by a C-terminal extension (CTE) in its amino-acid sequence, which is present in all vacuolar Prxs, but not in cell wall Prxs (Costa et al., 2008; Welinder et al., 2002). CroPrx3-a and CroPrx3-b differ only in the N-terminal regions, with different SPs and N-terminus of the mature proteins (Fig. 5), and share most of the core polypeptide sequence and the C-terminal sequence.

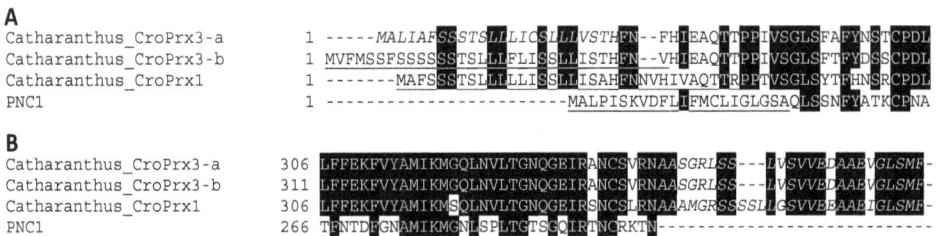

A

```
Catharanthus_CroPrx3-a    1  -----MALIAFSSSTSLLLICSLLLVSTHFN--FHIEAQTTPPIVSGLSFAFYNSTCPDL
Catharanthus_CroPrx3-b    1  MVFMSSFSSSSSSTSLLLFLISSLLISTHFN--VHIEAQTTPPIVSGLSFTFYDSSCPDL
Catharanthus_CroPrx1      1  -------MAFSSSTSLLLLLLISSLLISAHFNNVHIVAQTTRPPTVSGLSYTPHNSRCPDL
PNC1                      1  --------------------MALPISKVDFLTFMCLIGLGSAQLSSNFYATKCPNA
```

B

```
Catharanthus_CroPrx3-a  306  LFFEKFVYAMIKMGQLNVLTGNQGEIRANCSVRNAASGRLSS---LVSVVBDAABVGLSMF
Catharanthus_CroPrx3-b  311  LFFEKFVYAMIKMGQLNVLTGNQGEIRANCSVRNAASGRLSS---LVSVVBDAABVGLSMF
Catharanthus_CroPrx1    306  LFFEKFVYAMIKMSQLNVLTGNQGEIRSNCSLRNAAAMGRSSSSLLGSVVBEAABIGLSMF
PNC1                    266  TENTDFGNAMIKMGNLSPLTGTSGQLRTNGRKTN--------------------------
```

Fig. 5. Multiple alignment of the N-terminal (A) and C-terminal (B) regions of the *C. roseus* Prxs, CroPrx3-a, CroPrx3-b and CroPrx1, and PNC1, a peanut Prx localized at the cell wall. The regions highlighted in black correspond to conserved amino acid residues in at least three of the proteins. SPs determined experimentally are underlined, and predicted SPs and CTEs are in italic letters.

Alignment of the predicted CroPrx3s amino acid sequences with CroPrx1, and the well characterized cell wall peanut Prx PNC1 (Genbank accession number M37636) (Fig. 5), shows that both CroPrx3 proteins include a common CTE, and should therefore be sorted to the vacuole. To confirm this, several *GFP* fusions were designed and generated for *CroPrx3-b* (Fig.1), in which the coding sequences for the SP, the mature CroPrx3-b protein (MP), and the CTE were cloned in the adequate sides of the GFP coding sequence (Duarte et al., 2010). CroPrx3-b was chosen, since we also purified the protein and obtained MS/MS amino acid sequence data (manuscript in preparation) indicating the precise location of the N-terminal sequence of the mature CroPrx3-b protein, after cleavage of the SP (Fig. 5A). Therefore, the SP used to fuse at the GFP N-terminus was the sequence in italic in Fig. 5A, plus the subsequent 6 amino acids, to make sure that the cleavage site is correctly recognized. The SP sequence represented for CroPrx3-b in Fig. 5A was predicted using PSORT (http://psort.hgc.jp/form.html) and SignalP (http://www.cbs.dtu.dk/services/SignalP/). The length of the CroPrx3 CTE was deduced from sequence alignment with a well characterized cell wall Prx lacking a vacuolar sorting signal, peanut PCN1 (Fig. 5B), and bearing in mind that all mature Prx proteins terminate four to six residues after the last conserved cysteine (Welinder et al., 2002). In this case, the CTE used included two more amino acid residues than the predicted CTE sequence (Fig. 5B).

3.3 Subcellular sorting of GFP-CroPrx3 fusions in *C. roseus* mesophyll protoplasts

In order to assess the subcellular localization of CroPrx3, a method for transient expression of the *GFP-CroPrx3* fusions in *C. roseus* mesophyll protoplasts was developed. To obtain *C. roseus* mesophyll protoplasts, we adapted the general method described by Negrutiu et al. (1987) for Arabidopsis, using an enzyme composition adapted from a previous method used in our lab for *C. roseus* mesophyll (Sottomayor et al., 1996), and adapting the last steps of the procedure to the subsequent PEG mediated transformation method from Yoo et al. (2007). This enabled to obtain a high yield of healthy and pure protoplasts that, under the light microscope, showed no apparent membrane damage or disintegration of the internal structure (Fig. 6). This methodology usually enabled the retrieval of a total of 5×10^7 protoplasts from a Petri dish prepared as described above, which is enough for up to 100 independent transformations – we usually prepare this amount, use the protoplasts for around 6 transformations per researcher, and use the rest for vacuole isolation or biochemical determinations, but the protoplast isolation method can be scaled down.

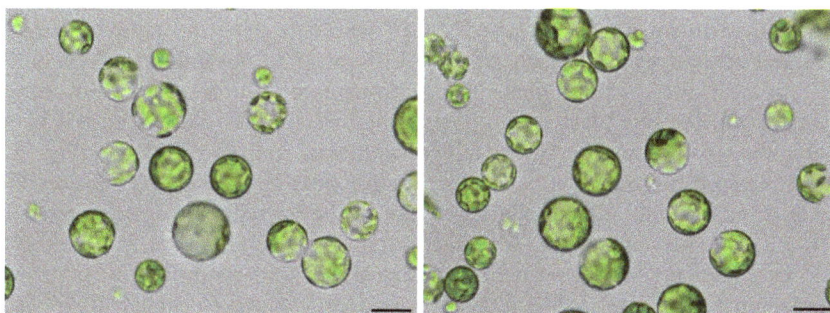

Fig. 6. Bright field images of *C. roseus* mesophyll protoplasts. Bars = 20 μm.

The method used for transient expression of the *GFP-CroPrx3* fusions in *C. roseus* mesophyll protoplasts was adapted from Yoo et al. (2007) with just a few modifications, since the method reported by these authors for Arabidopsis revealed to work quite well when it was applied directly to *C. roseus*, indicating that its transference to further species may also be simple. The main modifications that were introduced were a 25 fold increase in protoplast concentration, enabling a significant increase in transformation yield, and an incubation at 35 °C during the 3 h preceding observation, to boost GFP expression and/or fluorescence. The improved method yielded transformation rates of up to 70 - 80% when transforming with the cytosolic GFP construct. It was evident that both the yield and GFP fluorescence intensity decreased with the size of the GFP fusion construct/protein.

For the investigation of the vacuolar sorting of CroPrx3-b by transient expression, the following constructs were used (Fig. 1B), with the following predicted subcellular localizations:

i.   GFP → cytosol
ii.   SP-GFP → ER, Golgi, extracellular space as final destination
iii. SP-GFP-CTE → ER, Golgi, vacuole as final destination
iv.   SP-GFP-MP → ER, Golgi, extracellular space as final destination
v.   SP-GFP-MP-CTE → ER, Golgi, vacuole as final destination

The localization of fluorescence was monitored from 24 to 72 h after transformation and the results obtained are shown in Fig. 7. In fact, the time-course of the transient expression and GFP peak accumulation/fluorescence depended on the construct, possibly as a consequence of the different protein paths and destination compartments involved. Thus, cytosolic GFP already appeared in a few protoplasts 24 h after transformation with a strong fluorescence signal, with the number of fluorescent protoplasts continuing to increase until at least 72 h after transformation (Fig. 7A). All the remaining secretory GFPs (ii to v, above) were not visible at 24 h, with ER and/or Golgi labeling being well recognized at 48 h (Fig. 7B to E). For the construct SP-GFP-MP, it was observed that fluorescence had completely disappeared 72 h after transformation (data not shown), possibly indicating a more transient expression for this construct than for SP-GFP, which putatively codes for the same final subcellular destination. On the other hand, this disappearance of fluorescence may be interpreted as an indication that the GFP synthesized during that period in fact had the extracellular space as final destination, since in this localization the dilution effect prevents fluorescence detection, while in a contained compartment the fluorescence of GFP would be expected to be observed for a longer time. For the constructs including the CTE, at 48 h, not only the ER/Golgi were labeled, but also the vacuole presented a clear green fluorescence, indicating this compartment was the final destination (Fig. 7C and E). At 72 h, fluorescence was mainly observed in the vacuole, and had practically disappeared from the ER/Golgi (Fig. 7F). The vacuolar fluorescence signal was never very strong, possibly due to the dilution effect in such a big compartment, and to GFP instability in the vacuole, as a consequence of the acid pH and/or vacuolar proteases. In fact, according to Tamura et al. (2003), GFP is degraded by vacuolar proteases if plant cells are under light conditions, a fact prevented if cells are maintained in the dark. Likewise, we can only observe vacuolar fluorescence in *C. roseus* protoplasts if they are kept in the dark.

The SP of CroPrx3-b clearly determined the entrance of GFP in the secretory system, with fluorescence observed in the ER and/or Golgi (Fig. 7B), and therefore corresponds to an ER signal peptide, as generally assumed for Prxs. The addition of the CTE of CroPrx3-b to the secretory SP-GFP was enough to sort the fluorescent fusion to the central vacuole, which could also be observed for the fusion including the MP-CTE on the GFP C-terminus (Fig. 7C, E and F). On the other hand, deletion of the CTE from this latter construct prevented the fusion protein to be sorted to the vacuole, since fluorescence was observed only in ER/Golgi-like patterns (Fig. 7D) and disappeared at 72 h (data not shown). These results show that CroPrx3-b is localized in the vacuole and that the vacuolar sorting information for this protein is indeed localized in its CTE. The vacuolar localization of CroPrx3 means that the vacuolar accumulated TIAs are potential substrates for this enzyme.

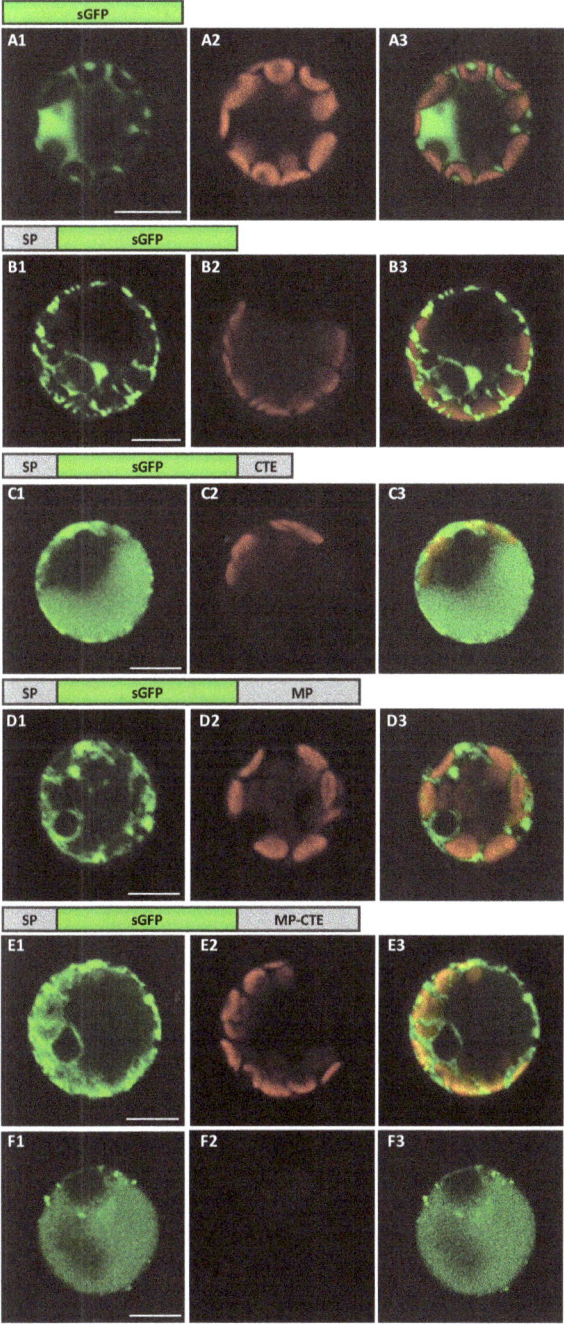

Fig. 7. Transient transformation of *C. roseus* mesophyll protoplasts with *GFP-CroPrx3*
fusions. The schematic representation of each construct used for transformation is depicted

right above the respective set of confocal images. A) GFP fluorescence pattern observed for the transformation with the control construct *35S::sGFP*. GFP accumulates in the cytosol and the nucleus. B) GFP fluorescence pattern observed for the transformation with the construct *35S::SP-sGFP*. GFP fluorescence is observed in the ER and possibly Golgi indicating sorting to the secretory pathway. C) GFP fluorescence pattern observed for the transformation with the construct *35S::SP-sGFP-CTE*. The presence of the CTE of CrPrx3-b targets secretory GFP to the central vacuole. D) GFP fluorescence pattern observed for the transformation with the construct *35S::SP-sGFP-MP*. GFP fluorescence is observed in the ER and possibly Golgi indicating sorting to the secretory pathway but no fluorescence is observed in the vacuole indicating that the absence of the CrPrx3-b CTE impairs the vacuolar sorting of GFP. E and F) GFP fluorescence pattern observed for the transformation with the construct *35S::SP-sGFP-MP-CTE*. Addition of the CTE to the CrPrx3-b MP successfully targets secretory GFP to the central vacuole. E1-E3, 48 h after transformation. F1-F3, 72 h after transformation. Images on the left column – GFP channel. Images on the middle column – red channel showing chloroplast autofluorescence. Images on the right column – merged images of the two channels. Bars = 10 μm.

4. Conclusions

Here, we have cloned a new Prx gene from the medicinal plant *C. roseus*, *CroPrx3*, and we have characterized its subcellular sorting through transient expression of GFP fusions in *C. roseus* mesophyll protoplasts, using a newly developed method for this important medicinal plant.

Initial cloning of *CroPrx3* was achieved through a classical approach involving screening of a genomic library. Currently, for the isolation of new genomic sequences we have been using, and would suggest, PCR based strategies such as inverse PCR (IPCR) (Costa et al., 2008) and genome walking (Gutiérrez et al., 2009).

The transient expression of *GFP-CroPrx3* fusions indicated a vacuolar localization for this enzyme, and a vacuolar sorting function for the CroPrx3 CTE, since the presence of this region was sufficient and necessary for the accumulation of green fluorescence in the central vacuole of *C. roseus* mesophyll protoplasts. The investigation of the subcellular localization of the proteins codified by candidate genes can thus be efficiently investigated by a strategy involving a careful design of fusions with fluorescent proteins (FPs), followed by transient expression (Duarte et al., 2010).

In fact, a highly useful tool for the characterization of candidate genes, now constantly arising from omic and *in silico* approaches, is the use of transient expression methodologies to investigate subcellular localization, protein-protein interactions, and to perform diverse functional assays like enzymatic activities, transport, etc. These methodologies are particularly relevant if they have been developed for the homologous organism as done in this work. Here, we have developed an easy and fast method for transient expression in the important medicinal plant *C. roseus*, contributing with a significant breakthrough for its future research. Moreover, the transference and adaptation of the developed methodology to other species of interest should be possible and encouraged.

5. Acknowledgment

The authors would like to thank the following institutions for their financial support: i) Programa Operacional Ciência e Inovação (POCI 2010) financed by the European Union and

Fundação para a Ciência e Tecnologia (FCT) from the Portuguese Government; ii) the FCT grants PRAXIS/P/BIA/10267/1998 and POCTI/BIO/38369/2001; iii) an FCT grant to Patrícia Duarte (SFRH/BPD/20669/2004), iv) a Scientific Mecenate award from Grupo Jerónimo Martins SA. The authors would also like to thank Yasuo Niwa (University of Shizuoka, Japan) for the pTH-2 plasmid, and Johan Memelink (University of Leiden, The Netherlands) for the *C. roseus* callus culture line MP183L, for the *C. roseus* genomic library, and for the pTH-2BN plasmid.

6. References

Bakalovic, N., Passardi, F., Ioannidis, V., Cosio, C., Penel, C., Falquet, L. & Dunand, C. (2006). PeroxiBase: A class III plant peroxidase database. *Phytochemistry*, Vol. 67, No. 6, (March 2006), pp. 534-539, ISSN 0031-9422

Blom, T.J.M., Sierra, M., van Vliet, T.B., Franke-van Dijk, M.E.I., de Koning, P., van Iren, F., Verpoorte, R. & Libbenga, K.R. (1991). Uptake and accumulation of ajmalicine into isolated vacuoles of cultured cells of *Catharanthus roseus* (L.) G. Don. and its conversion into serpentine. *Planta*, Vol. 183, No. 2, (January 1991), pp. 170-177, ISSN 0032-0935

Carqueijeiro, I., Gardner, R., Lopes, T., Duarte, P., Goossens, A. & Sottomayor, M. (2010). An omic approach to unravel the metabolism of the highly valuable medicinal alkaloids from *Catharanthus roseus, Book of abstracts. BioTrends 2010: New trends in green chemistry,* Dortmund, Germany, December 1-2, 2010

Costa, M.M.R., Hilliou, F., Duarte, P., Pereira, L.G., Almeida, I., Leech, M., Memelink, J., Barcelo, A.R. & Sottomayor, M. (2008). Molecular cloning and characterization of a vacuolar class III peroxidase involved in the metabolism of anticancer alkaloids in *Catharanthus roseus. Plant Physiology*, Vol. 146, No. 2, (February 2008), pp. 403-417, ISSN 0032-0889

Duarte, P., Memelink, J. & Sottomayor, M. (2010). Fusions with fluorescent proteins for subcellular localization of enzymes involved in plant alkaloid biosynthesis, In: *Methods in Molecular Biology, Metabolic Engineering of Plant Secondary Pathways*, Fett-Netto, A., ed, pp. 275-290, ISSN 1940-6029, Humana Press Inc, Totowa NJ, USA

Ferreres, F., Figueiredo, R., Bettencourt, S., Carqueijeiro, I., Oliveira, J., Gil-Izquierdo, A., Pereira, D., Valentão, P., Andrade, P., Duarte, P., Ros Barceló, A. & Sottomayor, M. (2011). Identification of phenolic compounds in isolated vacuoles of the medicinal plant *Catharanthus roseus* and their interaction with vacuolar class III peroxidase - an H_2O_2 affair? *Journal of Experimental Botany*, DOI 10.1093/jxb/erq458, ISSN 0022-0957

Fry, S.C. (2004). Oxidative coupling of tyrosine and ferulic acid residues: Intra- and extra-protoplasmic occurrence, predominance of trimers and larger products, and possible role in inter-polymeric cross-linking. *Phytochemistry Reviews*, Vol. 3, No. 1-2, (January 2004), pp. 97-111, ISSN 1568-7767

Goddijn, O.J.M., Lohman, F.P., De Kam, R.J., Schilperoort, R.A. & Hoge, J.H.C. (1994). Nucleotide sequence of the tryptophan decarboxylase gene of *Catharanthus roseus* and expression of tdc-gusA gene fusions in *Nicotiana tabacum.* *Molecular Gene Genetics,* Vol. 242, No. 2, (January 1994), pp. 217-225, ISSN 0026-8925

Gutiérrez, J., Nunez-Flores, M.J.L., Gomez-Ros, L.V., Uzal, E.N., Carrasco, A.E., Diaz, J., Sottomayor, M., Cuello, J. & Barcelo, A.R. (2009). Hormonal regulation of the basic peroxidase isoenzyme from *Zinnia elegans. Planta,* Vol. 230, No. 4, (September 2009), pp. 767-778, ISSN 0032-0935

Kuijt, S.J.H., Lamers, G.E.M., Rueb, S., Scarpella, E., Ouwerkerk, P.B.F., Spaink, H.P. & Meijer, A.H. (2004). Different subcellular localization and trafficking properties of KNOX class 1 homeodomain proteins from rice. *Plant Molecular Biology,* Vol. 55, No. 6, (January 2005), pp. 781-796, ISSN 0167-4412

Liscombe, D.K., Usera, A.R. & O'Connor, S.E. (2010). Homolog of tocopherol C methyltransferases catalyzes N methylation in anticancer alkaloid biosynthesis. *Proceedings of the National Academy of Sciences of the United States of America,* Vol. 107, No. 44, (November 2010), pp. 18793-18798, ISSN 0027-8424

Loyola-Vargas, V.M., Galaz-Ávalos, R.M. & Kú-Cauich, R. (2007). Catharanthus biosynthetic enzymes: The road ahead. *Phytochemistry Reviews,* Vol. 6, No. 2-3, (July 2007), pp. 307-339, ISSN 1568-7767

Menke, F.L., Champion, A., Kijne, J.W. & Memelink, J. (1999). A novel jasmonate- and elicitor-responsive element in the periwinkle secondary metabolite biosynthetic gene Str interacts with a jasmonate- and elicitor-inducible AP2-domain transcription factor, ORCA2. *EMBO Journal,* Vol. 18, No. 16, (August 1999), pp. 4455-4463, ISSN 0261-4189

Menke, F.L.H., Kijne, J.W. & Memelink, J. (1996). Digging for gene expression levels in *Catharanthus roseus*: Nonradioactive detection of plant mRNA levels, In: *Biochemica 2,* Leous, M., Matter, K., Schröder, C. & Ziebolz, B. (Eds.), pp. 16-18, Boehringer-Mannheim, Mannheim, Germany, ISSN 0946-1310

Murata, J., Roepke, J., Gordon, H. & De Luca, V. (2008). The leaf epidermome of *Catharanthus roseus* reveals its biochemical specialization. *The Plant Cell,* Vol. 20, No. 3, (March 2008), pp. 524-542, ISSN 1040-4651

Negrutiu, I., Shillito, R., Potrykus, I., Biasini, G. & Sala, F. (1987). Hybrid genes in the analysis of transformation conditions. 1. Setting up a simple method for direct gene-transfer in plant protoplasts. *Plant Molecular Biology,* Vol. 8, No. 5, (September 1987), pp. 363-373, ISSN 0167-4412

Niwa, Y., Hirano, T., Yoshimoto, K., Shimizu, M. & Kobayashi, H. (1999). Non-invasive quantitative detection and applications of non-toxic, S65T-type green fluorescent protein in living plants. *The Plant Journal,* Vol. 18, No. 4, (May 1999), pp. 455-463, ISSN 0960-7412

Pasquali, G., Goddijn, O.J., de Waal, A., Verpoorte, R., Schilperoort, R.A., Hoge, J.H. & Memelink, J. (1992). Coordinated regulation of two indole alkaloid biosynthetic genes from *Catharanthus roseus* by auxin and elicitors. *Plant Molecular Biology,* Vol. 18, No. 6, (April 1992), pp. 1121-1131, ISSN 0167-4412

Passardi, F., Cosio, C., Penel, C. & Dunand, C. (2005). Peroxidases have more functions than a Swiss army knife. *Plant Cell Reports,* Vol. 24, No. 5, (July 2005), pp. 255-265, ISSN 0721-7714

Rischer, H., Oresic, M., Seppanen-Laakso, T., Katajamaa, M., Lammertyn, F., Ardiles-Diaz, W., Van Montagu, M.C.E., Inze, D., Oksman-Caldentey, K.M. & Goossens, A. (2006). Gene-to-metabolite networks for terpenoid indole alkaloid biosynthesis in *Catharanthus roseus* cells. *Proceedings of the National Academy of Sciences of the United States of America,* Vol. 103, No. 14, (April 2006), pp. 5614-5619, ISSN 0027-8424

Ros Barceló, A., Gómez Ros, L.V., Gabaldón, C., López-Serrano, M., Pomar, F., Carrión, J.S. & Pedreño, M.A. (2004). Basic peroxidases: The gateway for lignin evolution? *Phytochemistry Reviews,* Vol. 3, No. 1-2, (January 2004), pp. 61-78, ISSN 1568-7767

Sambrook, J., Fritsch, E. & Maniatis, T. (1989). *Molecular Cloning: A Laboratory Manual* (3rd edition), Cold Spring Harbor Laboratory Press, ISBN 978-087969577-4, New York

Sottomayor, M., dePinto, M., Salema, R., DiCosmo, F., Pedreno, M. & AR, B. (1996). The vacuolar localization of a basic peroxidase isoenzyme responsible for the synthesis of alpha-3',4'-anhydrovinblastine in *Catharanthus roseus* (L) G. Don leaves. *Plant Cell and Environment,* Vol. 19, No. 6, (June 1996), pp. 761-767, ISSN 0140-7791

Sottomayor, M., Lopez-Serrano, M., DiCosmo, F. & Ros Barcelo, A. (1998). Purification and characterization of alpha-3',4'-anhydrovinblastine synthase (peroxidase-like) from *Catharanthus roseus* (L.) G. Don. *FEBS Letters,* Vol. 428, No. 3, (May 1998), pp. 299-303, ISSN 0014-5793

Sottomayor, M., Lopes Cardoso, I., Pereira, L. & Ros Barceló, A. (2004). Peroxidase and the biosynthesis of terpenoid indole alkaloids in the medicinal plant *Catharanthus roseus* (L.) G. Don. *Phytochemistry Reviews,* Vol. 3, No. 1-2, (January 2004), pp. 159-171, ISSN 1568-7767

Tamura, K., Shimada, T., Ono, E., Tanaka, Y., Nagatani, A., Higashi, S., Watanabe, M., Nishimura, M. & Hara-Nishimura, I. (2003). Why green fluorescent fusion proteins have not been observed in the vacuoles of higher plants. *The Plant Journal,* Vol. 35, No. 4, (August 2003), pp. 545-555, ISSN 0960-7412

van der Heijden, R., Jacobs, D.I., Snoeijer, W., Hallared, D. & Verpoorte, R. (2004). The Catharanthus alkaloids: Pharmacognosy and biotechnology. *Current Medicinal Chemistry,* Vol. 11, No. 5, (March 2004), pp. 607-628, ISSN 0929-8673

van Slogteren, G.M.S., Hoge, J.H.C., Hooykaas, P.J.J. & Schilperoort, R.A. (1983). Clonal analysis of heterogeneous crown gall tumor tissues induced by wild-type and shooter mutant strains of Agrobacterium tumefaciens-expression of T-DNA genes. *Plant Molecular Biology,* Vol. 2, No. 6, (November 1983), pp. 321-333, ISSN 0167-4412

Verpoorte, R., Lata, B. & Sadowska, A., (Eds.). (2007). *Biology and Biochemistry of Catharanthus roseus (L.) G. Don,* Phytochemistry Reviews, Vol. 6, Springer, ISSN 1568-7767, The Netherlands

Welinder, K., Justesen, A., Kjaersgard, I., Jensen, R., Rasmussen, S., Jespersen, H. & Duroux, L. (2002). Structural diversity and transcription of class III peroxidases from *Arabidopsis thaliana*. *European Journal of Biochemistry*, Vol. 269, No. 24, (December 2002), pp. 6063-6081, ISSN 0014-2956

Yoo, S.D., Cho, Y.H. & Sheen, J. (2007). Arabidopsis mesophyll protoplasts: a versatile cell system for transient gene expression analysis. *Nature Protocols*, Vol. 2, No. 7, (July 2007), pp. 1565-1572, ISSN 1754-2189

Permissions

The contributors of this book come from diverse backgrounds, making this book a truly international effort. This book will bring forth new frontiers with its revolutionizing research information and detailed analysis of the nascent developments around the world.

We would like to thank Gregory G. Brown, for lending his expertise to make the book truly unique. He has played a crucial role in the development of this book. Without his invaluable contribution this book wouldn't have been possible. He has made vital efforts to compile up to date information on the varied aspects of this subject to make this book a valuable addition to the collection of many professionals and students.

This book was conceptualized with the vision of imparting up-to-date information and advanced data in this field. To ensure the same, a matchless editorial board was set up. Every individual on the board went through rigorous rounds of assessment to prove their worth. After which they invested a large part of their time researching and compiling the most relevant data for our readers. Conferences and sessions were held from time to time between the editorial board and the contributing authors to present the data in the most comprehensible form. The editorial team has worked tirelessly to provide valuable and valid information to help people across the globe.

Every chapter published in this book has been scrutinized by our experts. Their significance has been extensively debated. The topics covered herein carry significant findings which will fuel the growth of the discipline. They may even be implemented as practical applications or may be referred to as a beginning point for another development. Chapters in this book were first published by InTech; hereby published with permission under the Creative Commons Attribution License or equivalent.

The editorial board has been involved in producing this book since its inception. They have spent rigorous hours researching and exploring the diverse topics which have resulted in the successful publishing of this book. They have passed on their knowledge of decades through this book. To expedite this challenging task, the publisher supported the team at every step. A small team of assistant editors was also appointed to further simplify the editing procedure and attain best results for the readers.

Our editorial team has been hand-picked from every corner of the world. Their multi-ethnicity adds dynamic inputs to the discussions which result in innovative outcomes. These outcomes are then further discussed with the researchers and contributors who give their valuable feedback and opinion regarding the same. The feedback is then collaborated with the researches and they are edited in a comprehensive manner to aid the understanding of the subject.

Apart from the editorial board, the designing team has also invested a significant amount of their time in understanding the subject and creating the most relevant covers. They scrutinized every image to scout for the most suitable representation of the subject and create an appropriate cover for the book.

The publishing team has been involved in this book since its early stages. They were actively engaged in every process, be it collecting the data, connecting with the contributors or procuring relevant information. The team has been an ardent support to the editorial, designing and production team. Their endless efforts to recruit the best for this project, has resulted in the accomplishment of this book. They are a veteran in the field of academics and their pool of knowledge is as vast as their experience in printing. Their expertise and guidance has proved useful at every step. Their uncompromising quality standards have made this book an exceptional effort. Their encouragement from time to time has been an inspiration for everyone.

The publisher and the editorial board hope that this book will prove to be a valuable piece of knowledge for researchers, students, practitioners and scholars across the globe.

List of Contributors

Abbas Padeganeh, Babak Bakhshinejad and Majid Sadeghizadeh
Department of Genetics, Faculty of Biological Sciences, Tarbiat Modares University, Tehran

Mohammad Khalaj-Kondori
Department of Biology, Faculty of Natural Sciences, University of Tabriz, Tabriz, Iran

Sriram Padmanabhan, Sampali Banerjee and Naganath Mandi
Lupin Limited, Biotechnology, R & D, Ghotawade Village, Mulshi Taluka, India

Srinivas Jayanthi, Beatrice Kachel, Jacqueline Morris and Thallapuranam K. Suresh Kumar
Department of Chemistry & Biochemistry, University of Arkansas, Fayetteville, USA

Igor Prudovsky
Maine Medical Centre Research Institue, Scarborough, USA

Lisa Wen
Department of Chemistry, Western Illinois University, USA

Rui-Li ZHAO
Department of Animal Science, Tianjin Agricultural University, Tianjin. China
College of Animal Science and Veterinary Medicine, Jilin University, Changchun, China

Jun-You HAN
College of Plant Science, Jilin University, Changchun, China

Wen-Yu HAN
College of Animal Science and Veterinary Medicine, Jilin University, Changchun, China

Hong-Xuan HE
Research Center for Wildlife Borne Diseases, Institute of Zoology, Chinese Academy of Sciences, China

Ji-Fei MA
Department of Animal Science, Tianjin Agricultural University, Tianjin, China

Takashi Sonoki
Hematology/Oncology, Wakayama Medical University, Japan

Gloria Esteso, Ángeles Jiménez-Marín, Gema Sanz, Juan José Garrido and Manuel Barbancho
Unidad de Genómica y Mejora Animal, Departamento de Genética, Universidad de Córdoba, Spain

Elena Crespo, and José M. Pérez de la Lastra
Instituto de Investigación en Recursos Cinegéticos (UCLM-CSIC-JCCLM), Ronda Toledo, Spain

José de la Fuente
Instituto de Investigación en Recursos Cinegéticos (UCLM-CSIC-JCCLM), Ronda Toledo, Spain
Department of Veterinary Pathobiology, Center for Veterinary Health Sciences, Oklahoma State University, Stillwater, USA

Denise V. Tambourgi
Immunochemistry Laboratory, Butantan Institute, São Paulo, Brazil

Gandhi Rádis-Baptista
Institute of Marine Sciences, Federal University of Ceará, Brazil

Yasser Shahein
Department of Molecular Biology, National Research Centre, Egypt

Amira Abouelella
Department of Biochemistry, National Centre for Radiation Research and Technology Egypt

Patrick Ogbunude and Joy Ikekpeazu
University of Nigeria, Enugu-Campus, Enugu, Nigeria

Joseph Ugonabo
Department of Microbiology, University of Nigeria, Nsukka, Nigeria

Michael Barrett
University of Glasgow, Glasgow, Scotland

Patrick Udeogaranya
Department of Pharmacy, University of Nigeria, Nsukka, Nigeria

Blanca R. Jarilla
Department of Immunology, Research Institute for Tropical Medicine (RITM), Philippines

Takeshi Agatsuma
Department of Environmental Health Sciences, Kochi University, Kochi, Japan

Chen Chen and Quanxi Wang
Shanghai Normal University, Shanghai, China

Yan-Qun Liu and Li Qin
Department of Sericulture, College of Bioscience and Biotechnology, Shenyang Agricultural University, China

Gregory G. Brown and Lydiane Gaborieau
McGill University, Montreal QC, Canada

Patrícia Duarte
IBMC – Instituto de Biologia Molecular Celular, Universidade do Porto, Portugal

www.ingramcontent.com/pod-product-compliance
Lightning Source LLC
Chambersburg PA
CBHW070727190326
41458CB00004B/1073